# Quantum Physics Without Quantum Philosophy

Detlef Dürr • Sheldon Goldstein • Nino Zanghì

# Quantum Physics Without Quantum Philosophy

Springer

*Authors*
Dr. Detlef Dürr
Mathematisches Institut
Universität München
München, Germany

Nino Zanghì
Istituto Nazionale Fisica Nucleare,
Sezione di Genova (INFN)
Università di Genova,
Genova, Italy

Sheldon Goldstein
Department of Mathematics, Rutgers
State University of New Jersey
Piscataway
New Jersey, USA

ISBN 978-3-642-30689-1     ISBN 978-3-642-30690-7 (eBook)
DOI 10.1007/978-3-642-30690-7 1
Springer Heidelberg Dordrecht London New York

Library of Congress Control Number: 2012952411

© Springer-Verlag Berlin Heidelberg 2013
This work is subject to copyright. All rights are reserved, whether the whole or part of the material is concerned, specifically the rights of translation, reprinting, reuse of illustrations, recitation, broadcasting, reproduction on microfilm or in any other way, and storage in data banks. Duplication of this publication or parts thereof is permitted only under the provisions of the German Copyright Law of September 9, 1965, in its current version, and permission for use must always be obtained from Springer. Violations are liable to prosecution under the German Copyright Law.
The use of general descriptive names, registered names, trademarks, etc. in this publication does not imply, even in the absence of a specific statement, that such names are exempt from the relevant protective laws and regulations and therefore free for general use.

Printed on acid-free paper

Springer is part of Springer Science+Business Media (www.springer.com)

# Foreword

In an ideal world, this book would not occasion any controversy. It provides the articulation and analysis of a physical theory, presented with more clarity and precision than is usual in a work of physics. A reader might start out predisposed towards the theory, or skeptical, or neutral, but should in any case be impressed by the pellucid explication. The theory, in various incarnations, postulates exact physical hypotheses about what exists in the world, and precise, universal, mathematically defined laws that determine how those physical entities behave. Large, visible objects (such as planets, or rocks, or macroscopic laboratory equipment) are postulated to be collections of small objects (particles). Since the theory specifies how the small objects behave, it automatically implies how the large, visible objects behave. It is then just a matter of analysis to determine what the theory predicts about the outcomes of experiments and other sorts of observable phenomena, and to compare these predictions with empirical data. So long as those predictions prove accurate (they do), the theory must be regarded as a candidate for the true theory of the physical universe. It has to face competition from other empirically accurate theories, and there might be disputes over which of the various contenders is the most promising. But a fair competition requires that all the contestants be judged on their merits, which demands that each be clearly and sympathetically presented. This book supplies such a presentation.

Unfortunately, we do not live in an ideal world. On certain topics, cool rational judgment is hard to find, and quantum mechanics is one of those topics. For reasons rooted in the tortuous history of the theory[1], clear and straightforward physical theories that can account for the phenomena treated by quantum mechanics are viewed with suspicion, if not downright hostility. These phenomena, it is said, admit of no clear or "classical" explanation, and anyone who thinks that they do has not appreciated the revolutionary character of the quantum world. In order to "understand" quantum phenomena, it is said, we must renounce classical logic, or amend classical probability theory, or admit a plethora of invisible universes, or recognize the central role that conscious observers play in production of the physical world. Lest the

---

[1] A useful account of that history may be found in the book of James Cushing [1].

reader think I am exaggerating, there are many clear examples of each of these. The Many Worlds interpretation posits that whenever quantum theory seems to present a probability, there is in fact a multiplicity: Schrödinger's cat splits into a myriad of cats in each experiment, some of which are alive and some dead. Defenders of the "consistent histories" approach insist that classical logic must be abandoned: in some cases, the claim P can be true and the claim Q can be true but the conjunction "P and Q" be not only not true, but meaningless.[2] David Mermin, in a famous article on Bell's theorem [2], asserts that "[w]e now know that the moon is demonstrably not there when nobody looks."

These sorts of extraordinary claims should not be dismissed out of hand. Perhaps the world is so strange that "classical" modes of thought are incapable of comprehending it. But extraordinary claims require extraordinary proof. And one would hope that such extreme positions would only be advocated if one were certain that nothing less radical could be correct. Surely, one imagines, these sorts of claims would not be made if some clear, precise theory that uses classical probability theory and classical logic, a theory that postulates only one, commonplace world in which observers are just complicated physical systems interacting by the same physical laws as govern everything else, actually existed. Surely, one imagines, respected physicists would not be driven to these extreme measures unless no alternative were available. But such an alternative is available, and has been for almost as long as the quantum theory itself has existed. It was first discovered by Louis de Broglie, and later rediscovered by David Bohm. It goes by the names "pilot wave theory" and "causal interpretation" and "ontological interpretation" and "Bohmian mechanics." It is the main subject of this book.

How could the most prominent physicists of the last century have failed to recognize the significance of Bohmian mechanics? This is a fascinating question, but subsidiary to our main task. The important thing is to become convinced that they did fail to recognize its significance. Consider one example.

In his classic *Lectures on Physics*, Richard Feynman introduces his students to quantum theory by means of an experiment:

> In this chapter, we shall tackle immediately the mysterious behavior in its most strange form. We choose to examine a phenomenon which is impossible, absolutely impossible, to explain in any classical way, and which has in it the heart of quantum mechanics. In reality, it contains the only mystery. We cannot explain the mystery in the sense of "explaining" how it works. We will tell you how it works. In telling you how it works we will have told you about the basic peculiarities of all quantum mechanics. [3, p. 37-2]

The experiment Feynman describes is the two-slit interference experiment for electrons. An electron beam, shot through a barrier with two holes in it, forms interference bands on a distant screen. The bands are formed by individual marks on the screen, one for each electron. They form even when the intensity of the beam is so low that only one electron at a time passes through the device.

What does Feynman find so mysterious about this phenomenon? He considers the most straightforward, obvious attempt to understand it: "The first thing we would say

---

[2] Cf. Omnes [4] or Griffiths [5].

is that since they come in lumps, each lump, which we may as well call an electron, has come either through hole 1 or hole 2. [3, p. 37–6 ]" This leads him to what he calls *Proposition A*:

> Each electron either goes through hole 1 or it goes through hole 2. [3, p. 37-6]

The burden of Feynman's argument is then to show that Proposition A is false: it is not the case that each electron goes through exactly one of the two holes. In order to prove this, he suggests that we "check this idea by experiment." First, block up hole 2 and count the number of electrons that arrive at each part of the screen, yielding a distribution P1. Then, block up hole 1 and count the arrival rates to get another distribution P2. Finally, note that the distribution of electrons on the screen when both holes are open, P12, is not the sum of P1 and P2. This is an empirical result that cannot be denied. But what, exactly, does it imply?

Feynman asserts that these phenomena simply cannot be explained if we accept Proposition A:

> It is all quite mysterious. And the more you look at it, the more mysterious it seems. Many ideas have been concocted to try to explain the curve for P12 in terms of individual electrons going around in complicated ways through the holes. None of them has succeeded. None of them can get the right curve for P12 in terms of P1 and P2. [3. p. 37–6]

The reader should now turn to p. 13 of the introduction and study the diagram to be found on that page. The diagram depicts the trajectories of individual electrons in the two-slit experiment according to Bohmian mechanics. Note that each electron does go through exactly one slit, validating Proposition A. Note also that the trajectories do not look particularly "complicated." And the visual simplicity of the diagram fails to convey the mathematical simplicity of the exact equations: the trajectories of the electrons are guided by the wavefunction in the simplest, most straightforward possible way. Feynman's claim about the impossibility of understanding this experiment consistent with electrons following continuous trajectories, and hence consistent with Proposition A, is flatly false. It had been known to be false since 1927, when de Broglie first presented the pilot wave theory, and its falsity was reinforced in 1952 when Bohm published the theory again.

And yet, according to Feynman, this experiment presents the fundamental mystery of quantum mechanics, a phenomenon that cannot possibly be understood in any "classical" way. How did he manage to get so confused?

The argument Feynman gives above turns on a plausible idea that cannot be correct, namely that if an electron goes through hole 1, then its later trajectory cannot be depend on whether hole 2 was open or closed. This is refuted directly by the experimental data: there are places on the screen where electrons arrive when only hole 1 is open and electrons arrive when only hole 2 is open, but where no electrons arrive when both holes are open. Hence the behavior of every single electron depends upon whether only one hole is open or both are open. But once we accept this surprising fact, forced on us by experiment, we see that there is no reason at all to expect P12 to be any sort of function (much less a simple sum) of P1 and P2. Experiments done with only one hole open give us no guidance about what will happen with both holes open, because something in the physical world is sensitive to the fact that both holes are open. It is this thing that gives rise to the interference phenomena in the first place.

If we want to accept the obvious explanation of the fact that electrons "arrive in lumps," namely that electrons are particles that follow continuous trajectories, then we need to postulate something else that is somehow sensitive to the fact that both holes are open. Every version of quantum mechanics postulates such a thing: the wave function. And every version of quantum mechanics postulates a linear dynamics for the wave function, such as the Schrödinger equation. All that is needed is a "guidance equation" for the particles, which specifies how the particles move given the wave function. Bohmian mechanics provides just such a guidance equation, which completes the theory.

Feynman passes on to other mysteries. What happens if we try to experimentally determine which hole each electron goes through by constructing some sort of detector at the holes? If the detector works, then each electron is detected at one hole or the other, but the interference pattern also disappears. Apparently, the behavior of the electrons is changed if we "look for" them. Doesn't this somehow show that the act of observation, and the existence of observers, plays a central role in quantum theory that is absent entirely from classical physics? Can Mermin's remark about the moon be far behind?

Once again, Bohmian mechanics provides a rather prosaic answer. In any physical theory, the outcome of an experiment depends on how the experiment is constructed. What one wants is a physics that treats experiments, and observers, and observations, in just the same way as it treats all physical systems and interactions. Placing detectors in the experimental set-up changes its physical construction. Any physical theory—classical or otherwise—must take account of those changes when predicting the outcome. In Bohmian mechanics, the presence of detectors necessarily results in the entanglement of the wave function of the electron with the wave function of the detector: otherwise the detector can't detect anything about the electron. This same entanglement suppresses the formation of interference bands at the screen by means of the same physical laws that create the bands in the original experiment. No magic is involved, and no special status is ascribed to "observation:" rather, it is all simple physical analysis of the situation.

In sum, none of the "inexplicable mysteries" that Feynman cites present any problems at all for Bohmian mechanics. The phenomena follow from the application of precise dynamical laws to the appropriate physical systems. No changes in logic or probability theory are required, no special status ascribed to observers or "measurements," worlds are not multiplied. The only mystery is how Feynman could have been so mystified, and why he insisted that his students should be mystified as well.

I have not used Feynman as an example in order to particularly criticize him. His remarks are typical of even the most distinguished physicists. Since the inception of quantum theory, claims have been made that the phenomena themselves admit of no possible "classical" explanation, and that the microscopic details of the physical world either don't exist at all or are somehow beyond our grasp. Bohmian mechanics serves as a counterexample to such claims. Mermin's assertion that the moon is demonstrably not there when no one looks is itself demonstrably false: Bohmian mechanics is the proof. But the theory serves an even more important function than this. It reminds us just what a perfectly comprehensible physical theory is, a theory in which all of the physics is contained in precise hypotheses about what exists and exact mathematical descriptions of how it behaves. It is a theory in which every sharply stated physical question receives an answer, and every macroscopic phenomenon is understood in terms of the behavior of its microscopic parts. It is a theory that appeals to mathematical analysis rather than philosophical pontification. In short, it provides a standard of clarity that every physical theory should aspire to.

For me, this realization was fortuitous. I had the sheer, blind good fortune to have Shelly Goldstein as a colleague, and to get to know Nino Zanghì and Detlef Dürr through Shelly. When I arrived at Rutgers University some quarter century ago, I brought with me the same prejudices that any student of physics would acquire. I thought that quantum mechanics had established that the world is fundamentally indeterministic, that microscopic systems typically have no "classical" physical properties at all, and that observers and measurements (and possibly even conscious experience!) had permanently assumed a central place in physics. These views were intertwined: if the microscopic world is somehow intrinsically "fuzzy" or indistinct, then something remarkable must happen to yield sharp macroscopic outcomes, such as the survival of Schrödinger's cat.

Shelly's probing criticism awoke me from my dogmatic slumbers. Over the years, Detlef, Nino and Shelly (with other colleagues as well) have produced a series of papers, many collected in this volume, showing how the mathematical formalism of quantum theory can be understood in terms of precise physical hypotheses. They have also demonstrated how the basic ideas of de Broglie and Bohm can be extended to cover the phenomena associated with field theory (particle creation and annihilation) and even gravity.

You, dear reader, now share my good fortune. You hold in your hands the hard-won result of decades of creativity and careful thought. If you harbor a deep desire to understand how the phenomena treated by quantum theory could possibly be produced by exact physical law, you can do no better than to study these papers. But you must come with an open mind, prepared to discover that the most common pronouncements about quantum theory are simply, flatly false.

John Stewart Bell, in his own presentation of Bohm's theory, describes how, as a student, he had difficulty understanding quantum theory. He wondered whether a more complete account of the physics of a system than that provided by standard quantum mechanics might be possible, but was informed that von Neumann had proven the impossibility of such a completion.

> Having read this, I relegated the question to the back of my mind and got on with more practical things.
>
> But in 1952, I saw the impossible done. It was in papers by David Bohm. Bohm showed explicitly how parameters could be introduced, into nonrelativistic wave mechanics, with the help of which the indeterministic description could be transformed into a deterministic one. More importantly, in my opinion, the subjectivity of the orthodox version, the necessary reference to the 'observer', could be eliminated.
>
> Moreover, the essential idea was one that had been advanced already by de Broglie in 1927, in his 'pilot wave' picture.

But why then had Born not told me of this 'pilot wave'? If only to point out what was wrong with it? Why did von Neumann not consider it? More extraordinarily, why did people go on producing 'impossibility proofs', after 1952 and as recently as 1978? When even Pauli, Rosenfeld, and Heisenberg, could produce no more devastating criticism of Bohm's theory than to brand it as 'metaphysical' and 'ideological'? Why is the pilot wave picture ignored in text books? Should it not be taught, not as the only way, but as an antidote to the prevailing complacency? To show that vagueness, subjectivity and indeterminism are not forced on us by experimental facts but by deliberate theoretical choice?

Bell, as usual, says concisely just what needs to be said. But here, at the end of this preface, even more appropriate are the words that once came to Augustine: *tolle, lege*.

# References

1. J. T. Cushing. Quantum Mechanics: Historical Contingency and the Copenhagen Hegemony. University of Chicago Press, Chicago, 1994.
2. N. D. Mermin. Quantum Mysteries for Anyone. The Journal of Philosophy, 78:397–408, 1981.
3. R. Feynman, R. B. Leighton, M. Sands. The Feynman Lectures on Physics, I. Addison-Wesley, New York, 1963.
4. R. Omnes. Logical Reformulation of Quantum Mechanics. Journal of Statistical Physics, 53:893–932, 1988.
5. R. B. Griffiths. Consistent Histories and the Interpretation of Quantum Mechanics. Journal of Statistical Physics, 36:219–272, 1984.

Princeton                                                                                      Tim Maudlin

# Authors' Preface

In this book we have collected, as chapters following the introduction, our basic articles on Bohmian mechanics, which have appeared in the course of the last twenty years in various journals and books as given below. Our hope is that the result is a self-contained presentation of the theory, accessible to mathematicians, physicists, and philosopher of physics with some basic knowledge of quantum mechanics. We have made some very minor revisions of many of the articles in order to reduce redundancy, enhance clarity, and eliminate some errors.

**Articles:**

- Quantum Equilibrium and the Origin of Absolute Uncertainty. *J. Statist. Phys.* **67**, 843–907 (1992).
- Quantum Equilibrium and the Role of Operators as Observables in quantum theory. *J. Statist. Phys.* **116**, 959–1055 (2004).
- Quantum Philosophy: The Flight from Reason in Science (without D.D. and N.Z). Contribution to *The Flight from Science and Reason*, edited by P. Gross, N. Levitt, and M.W.Lewis, Annals of the New York Academy of Sciences 775, 119-125 (1996). Reproduced with permission of Blackwell Publishing Ltd.
- Seven Steps Towards the Classical World (with V. Allori). *Journal of Optics B* **4**, 482-488 (2002). Reproduced with kind permission of IOP Publishing Ltd.
- On the Quantum Probability Flux Through Surfaces (with M. Daumer). *J. Statist. Phys.* **88**, 967-977 (1997).
- On the Weak Measurement of Velocity in Bohmian Mechanics. *J. Statist. Phys.* **134**, 1023-1032 (2009).
- Topological Factors Derived From Bohmian Mechanics (with J. Taylor and R. Tumulka). *Ann. H. Poincaré* **7**, 791-807 (2006).
- Hypersurface Bohm-Dirac Models (with K. Münch-Berndl). *Physical Review A* **60**, 2729-2736 (1999). Reprinted with permission. Copyright 1999 by the American Physical Society.

- Bohmian Mechanics and Quantum Field Theory (with R. Tumulka). *Physical Review Letters* **93**, 090402 (2004). Reprinted with permission. Copyright 2004 by the American Physical Society.
- Quantum Spacetime without Observers: Ontological Clarity and the Conceptual Foundations of Quantum Gravity (with S. Teufel, without D.D. and N.Z.). In *Physics meets Philosophy at the Planck Scale*, edited by C. Callender and N. Huggett, 275-289. Copyright by Cambridge University Press, 2001. Reproduced with permission.
- Reality and the Role of the Wavefunction (without D.D.).

We thank our coauthors Martin Daumer, Karin Münch-Berndl, Stefan Teufel, James Taylor, Valia Allori, and Roderich Tumulka for their kind permission to reprint here the articles on which they were involved, as well as for their significant contributions to these articles and the pleasure of working with them.

We are especially grateful to Tim Maudlin, GianCarlo Ghirardi and Angela Lahee, without whom this book would not exist.

We also thank for their help, suggestions, and support David Albert, Sergio Albeverio, Enrico Beltrametti, Guido Bacciagaluppi, Jean Bricmont, Kai-Uwe Bux, Bruno Caprile, Maurice de Gosson, Gianfausto Dell'Antonio, Mauro Dorato, Fay Dowker, Gregory Eyink, William Faris, Rodolfo Figari, Jürg Fröhlich, Giovanni Gallavotti, Pedro Garrido, Rebecca Goldstein, Ned Hall, Doug Hemmick, Michael Kiessling, Martin Kruskal, Antti Kupiainen, Reinhard Lang, Federico Laudisa, Dustin Lazarovici, Joel Lebowitz, Barry Loewer, Frank Loose, Stephen Lyle, Christian Maes, Vishnya Maudlin, Tilo Moser, Travis Norsen, Giuseppe Olivieri, Folker Schamel, Penny Smith, Avy Soffer, Eugene Speer, Herbert Spohn, Hector Sussmann, Ward Struyve, Daniel Tausk, Alessandro Teta.

We are grateful for the hospitality that some of us have enjoyed, on more than one occasion, at the following institutions: Mathematisches Institut of Ludwig-Maximilians-Universität München (Germany), Dipartimento di Fisica dell'Università degli Studi di Genova (Italy), Institut des Hautes Études Scientifiques in Bures-sur-Yvette (France), Mathematics Department of Rutgers University (USA), and "the garden."

Arquata Scrivia

Detlef Dürr
Sheldon Goldstein
Nino Zanghì

# Contents

**1 Introduction** .................................................. 1
   1.1 The Title ................................................ 1
   1.2 What is Wrong with Quantum Mechanics? .................... 3
   1.3 History .................................................. 5
   1.4 Impossibility ............................................ 7
   1.5 The Dismissal ............................................ 9
   1.6 Roads to Bohmian Mechanics ............................... 9
   1.7 Questions ................................................ 11
   1.8 The Book ................................................. 14
   References .................................................. 19

## Part I Quantum Equilibrium

**2 Quantum Equilibrium and the Origin of Absolute Uncertainty** ...... 23
   2.1 Introduction ............................................. 23
   2.2 Reality and the Role of the Wave Function ................ 26
   2.3 Bohmian Mechanics ........................................ 29
   2.4 The Problem of Quantum Equilibrium ....................... 34
   2.5 The Effective Wave Function .............................. 37
   2.6 The Fundamental Conditional Probability Formula .......... 44
   2.7 Empirical Distributions .................................. 46
   2.8 Multitime Experiments: the Problem ....................... 50
   2.9 Random Systems ........................................... 51
   2.10 Multitime Distributions ................................. 54
   2.11 Absolute Uncertainty .................................... 57
   2.12 Knowledge and Nonequilibrium ............................ 59
   2.13 Quantum Equilibrium and Thermodynamic (non)Equilibrium .. 60
   2.14 Global Equilibrium Beneath Nonequilibrium ............... 63
   2.15 Appendix: Random Points ................................. 65
   References .................................................. 74

## 3 Quantum Equilibrium and the Role of Operators as Observables in Quantum Theory .......... 79
- 3.1 Introduction .......... 79
  - 3.1.1 Notations and Conventions .......... 82
- 3.2 Bohmian Experiments .......... 83
  - 3.2.1 Equivariance and Quantum Equilibrium .......... 83
  - 3.2.2 Conditional and Effective Wave Functions .......... 85
  - 3.2.3 Decoherence .......... 86
  - 3.2.4 Wave Function and State .......... 88
  - 3.2.5 The Stern-Gerlach Experiment .......... 89
  - 3.2.6 A Remark on the Reality of Spin in Bohmian Mechanics ... 92
  - 3.2.7 The Framework of Discrete Experiments .......... 92
  - 3.2.8 Reproducibility and its Consequences .......... 94
  - 3.2.9 Operators as Observables .......... 96
  - 3.2.10 The General Framework of Bohmian Experiments .......... 97
- 3.3 The Quantum Formalism .......... 99
  - 3.3.1 Weak Formal Measurements .......... 100
  - 3.3.2 Strong Formal Measurements .......... 101
  - 3.3.3 From Formal Measurements to Experiments .......... 104
  - 3.3.4 Von Neumann Measurements .......... 105
  - 3.3.5 Preparation Procedures .......... 106
  - 3.3.6 Measurements of Commuting Families of Operators .......... 106
  - 3.3.7 Functions of Measurements .......... 108
  - 3.3.8 Measurements of Operators with Continuous Spectrum .... 110
  - 3.3.9 Sequential Measurements .......... 110
  - 3.3.10 Some Summarizing Remarks .......... 112
- 3.4 The Extended Quantum Formalism .......... 114
  - 3.4.1 POVMs and Bohmian Experiments .......... 115
  - 3.4.2 Formal Experiments .......... 116
  - 3.4.3 From Formal Experiments to Experiments .......... 119
  - 3.4.4 Measure-Valued Quadratic Maps .......... 120
- 3.5 The General Emergence of Operators .......... 122
  - 3.5.1 "No Interaction" Experiments .......... 123
  - 3.5.2 "No $X$" Experiments .......... 124
  - 3.5.3 "No $Y$" Experiments .......... 126
  - 3.5.4 "No $Y$ no $\Phi$" Experiments .......... 127
  - 3.5.5 The Basic Operators of Quantum Mechanics .......... 127
  - 3.5.6 From Positive-Operator-Valued Measures to Experiments .. 130
  - 3.5.7 Invariance Under Trivial Extension .......... 130
  - 3.5.8 POVMs and the Positions of Photons and Dirac Electrons .. 131
- 3.6 Density Matrices .......... 132
  - 3.6.1 Density Matrices and Bohmian Experiments .......... 133
  - 3.6.2 Strong Experiments and Density Matrices .......... 134
  - 3.6.3 The Notion of Instrument .......... 135
  - 3.6.4 On the State Description Provided by Density Matrices .... 136

3.7 Genuine Measurements ................................................. 137
   3.7.1 A Necessary Condition for Measurability ............... 138
   3.7.2 The Nonmeasurability of Velocity, Wave Function and Deterministic Quantities ............................. 139
   3.7.3 Initial Values and Final Values ....................... 140
   3.7.4 Nonlinear Measurements and the Role of Prior Information . 141
   3.7.5 A Position Measurement that Does not Measure Position ... 142
   3.7.6 Theory Dependence of Measurement .................... 143
3.8 Hidden Variables ..................................................... 144
   3.8.1 Experiments and Random Variables .................... 145
   3.8.2 Random Variables, Operators, and the Impossibility Theorems ........................................... 146
   3.8.3 Contextuality ........................................ 148
   3.8.4 Against "Contextuality" .............................. 151
   3.8.5 Nonlocality, Contextuality and Hidden Variables ......... 154
3.9 Against Naive Realism About Operators .............................. 157
References .............................................................. 159

## 4 Quantum Philosophy: The Flight from Reason in Science ........... 163
References .............................................................. 167

## Part II Quantum Motion

## 5 Seven Steps Towards the Classical World ......................... 171
5.1 Introduction ......................................................... 171
5.2 Motion in an External Potential ..................................... 172
5.3 The Classical Limit in Bohmian Mechanics ........................... 173
5.4 Wave Packets ......................................................... 175
5.5 Local Plane Waves .................................................... 176
5.6 General Wave Functions .............................................. 177
5.7 Limitations of the Model: Interference and the Role of the Environment ......................................................... 179
5.8 Towards a Mathematical Conjecture ................................... 180
5.9 The Classical Limit in a Nutshell ................................... 182
References .............................................................. 182

## 6 On the Quantum Probability Flux through Surfaces ................ 183
6.1 Introduction ......................................................... 183
6.2 Standard Scattering Theory .......................................... 184
6.3 Near Field Scattering ............................................... 187
6.4 Bohmian Mechanics ................................................... 189
References .............................................................. 191

## 7 On the Weak Measurement of Velocity in Bohmian Mechanics ...... 193
7.1 Introduction ............................................. 193
7.2 Bohmian Analysis of Weak Measurement of Velocity............. 196
7.3 A More Careful Analysis .................................. 199
7.4 Bohmian Mechanics and the Crucial Condition................. 200
7.5 The Impossibility of Measuring the Velocity in Bohmian Mechanics ............................................... 201
7.6 Conclusion .............................................. 202
References .................................................. 203

## 8 Topological Factors Derived From Bohmian Mechanics ............ 205
8.1 Introduction ............................................. 205
8.2 Bohmian Mechanics in Riemannian Manifolds ................. 206
8.3 Scalar Wave Functions on the Covering Space ................. 209
8.4 Vector-Valued Wave Functions on the Covering Space ........... 213
8.5 Conclusions ............................................. 218
References .................................................. 219

## Part III Quantum Relativity

## 9 Hypersurface Bohm-Dirac Models ............................... 223
9.1 Introduction ............................................. 223
9.2 The Hypersurface Bohm-Dirac Model ......................... 227
9.3 Statistical Analysis of the HBD model ........................ 231
  9.3.1 Quantum Equilibrium ................................ 231
  9.3.2 Comparison with Quantum Mechanics .................. 235
9.4 Perspective .............................................. 237
References .................................................. 238

## 10 Bohmian Mechanics and Quantum Field Theory ................... 239
10.1 Introduction ............................................. 239
10.2 Configuration Space ...................................... 240
10.3 The Laws of Motion ...................................... 242
10.4 Field Operators .......................................... 244
10.5 Conclusions ............................................. 245
References .................................................. 246

## 11 Quantum Spacetime without Observers: Ontological Clarity and the Conceptual Foundations of Quantum Gravity ............... 247
11.1 Introduction ............................................. 247
11.2 The Conceptual Problems of Quantum Gravity ................. 248
11.3 The Basic Problem of Orthodox Quantum Theory: the Lack of a Coherent Ontology ....................................... 252
11.4 Bohmian Quantum Gravity ................................. 253
11.5 A Universal Bohmian Theory................................ 256
References .................................................. 260

## Contents

**12 Reality and the Role of the Wave Function in Quantum Theory** ...... 263
    12.1 Questions About the Wave Function ........................... 263
    12.2 The Wave Function of a Subsystem............................ 264
    12.3 The Wave Function as Nomological ........................... 266
        12.3.1 Comparison of $\psi$ with the Classical Hamiltonian $H$ ....... 267
        12.3.2 $\psi$ versus $\Psi$ ........................................ 268
        12.3.3 The Universal Level ................................. 268
        12.3.4 Schrödinger's Equation as Phenomenological (Emergent) .. 270
        12.3.5 Two Transitions ..................................... 271
        12.3.6 Nomological versus Nonnomological ................... 272
        12.3.7 Relativistic Bohmian Theory........................... 272
        12.3.8 Wave Function as Nomological and Symmetry............ 274
        12.3.9 Possible Resolutions ................................. 275
        12.3.10 $\psi$ as Quasi-Nomological ............................... 275
    12.4 The Status of the Wave Function in Quantum Theory ............. 276
    References ................................................... 277

**Index** .......................................................... 279

# Chapter 1
# Introduction

## 1.1 The Title

When physics students learn about quantum mechanics, they may be intrigued by the notion that the possibility of a classical understanding of nature ended with the quantum revolution. Such a classical understanding presupposes the existence of an external real world *out there*, as well as the belief that it is the task of physics to find the basic constituents of this exterior real world and the laws that govern them. In quantum mechanics the situation seems to be quite different. One of the founders of quantum mechanics, Werner Heisenberg, writes that

> the idea of an objective real world whose smallest parts exist objectively in the same sense as stones or trees exist, independently of whether or not we observe them ... is impossible .... [1, p. 129]

In quantum mechanics one does, in fact, talk about small particles, for example electrons in a diffraction experiment; nonetheless, one is cautioned against thinking of those particles as particles in the ordinary sense. In the textbook of Landau and Lifshitz one reads that

> It is clear that this result can in no way be reconciled with the idea that electrons move in paths.... In quantum mechanics there is no such concept as the path of a particle. [2, p. 2]

Statements of this kind abound, raising the question: What then is it that quantum physics is about? To students it must seem first of all to be about a whole lot of mathematics, for example, the abstract mathematics of Hilbert spaces and self-adjoint operators. As much as some students might hope that with the advanced mathematics insight into the genuine physical content of quantum mechanics advances as well, they soon learn that that is not so. At best they learn of the ongoing debate about the *interpretations* of quantum mechanics, about Schrödinger's cat and other paradoxes. Some students will learn that the debate is somehow related to a genuine problem in the formulation of quantum mechanics, namely the measurement problem. At the end of the day, the student will likely feel that the understanding of nature in any ordinary sense is impossible: the hallmarks of orthodox quantum theory are the denial of determinism and, more importantly, the denial of an objective reality. Such denials, even though they are often not well thought out, became the accepted wisdom concerning quantum mechanics and quantum reality, though there were some prominent exceptions such as Albert Einstein and Erwin Schrödinger.

Quantum philosophy provides a philosophical foundation for the claim that a fundamental physics focused on an objective reality is impossible. At the same time, quantum philosophy is hard to grasp. One of its most distinguished proponents was Niels Bohr, who insisted that a rational description of nature is impossible. Instead, he proposed the new concept of "complementarity," according to which

> A complete elucidation of one and the same object may require diverse points of view which defy a unique description. Indeed, strictly speaking, the conscious analysis of any concept stands in a relation of exclusion to its immediate application. [3, p. 102]

Schrödinger, who had intense discussions with Bohr on the meaning of quantum mechanics, in a letter to Wilhelm Wien says that

> Bohr's ... approach to atomic problems ... is really remarkable. He is completely convinced that any understanding in the usual sense of the word is impossible. Therefore the conversation is almost immediately driven into philosophical questions, and soon you no longer know whether you really take the position he is attacking, or whether you really must attack the position he is defending. [4, p. 228]

When all is said and done, the core of quantum philosophy is that physics is about measurement and observation, and not about an objective reality, about what seems and not about what is.

Along with Einstein and Schrödinger, John Stewart Bell was one of the few physicists who felt compelled to reject quantum philosophy. Bell preferred instead a "theory of local beables," writing:

> This is a pretentious name for a theory which hardly exists otherwise, but which ought to exist. The name is deliberately modelled on 'the algebra of local observables.' The terminology, beable as against observable, is not designed to frighten with metaphysic those dedicated to realphysic. It is chosen rather to help in making explicit some notions already implicit in, and basic to, ordinary quantum theory. For, in the words of Bohr, 'it is decisive to recognize that, however far the phenomena transcend the scope of classical physical explanation, the account of all evidence must be expressed in classical terms'. It is the ambition of the theory of local beables to bring these 'classical terms' into the equations, and not relegate them entirely to the surrounding talk.
>
> The concept of 'observable' lends itself to very precise mathematics when identified with 'self-adjoint operator'. But physically, it is a rather wooly concept. It is not easy to identify precisely which physical processes are to be given status of 'observations' and which are to be relegated to the limbo between one observation and another. So it could be hoped that some increase in precision might be possible by concentration on the beables, which can be described in 'classical terms,' because they are there. The beables must include the settings of switches and knobs on experimental equipment, the currents in coils, and the readings of instruments. 'Observables' must be made, somehow, out of beables. The theory of local beables should contain, and give precise physical meaning to, the algebra of local observables. [5, p. 52]

This book is about a precise quantum theory in which the 'classical terms'—the "local beables"—are brought into the equations. It is a book about an objective

## 1.2 What is Wrong with Quantum Mechanics?

quantum description of nature. It is a book about quantum physics without quantum philosophy.

As soon as one frees oneself from the quantum philosophical notion that physics must not be about an objective reality and recognizes instead that a quantum theory must describe such a reality, the choice for its local beables, of particles moving in physical space, their motion being guided by waves, requires little imagination. This book is about such a theory, called Bohmian mechanics, the de Broglie-Bohm theory, or the pilot-wave theory.

In this theory particles move; how they move is determined by the wave function obeying Schrödinger's equation. In Bohmian mechanics both the wave function and the particles are real physical objects "out there." The structure of the theory is rather simple: The complete description of an $N$-particle system is provided by its configuration $Q$, defined by the positions $\mathbf{Q}_1, \ldots, \mathbf{Q}_N$ of its particles, together with its wave function $\psi = \psi(\mathbf{q}_1, \ldots, \mathbf{q}_N)$. The equations of motion are of the form

$$\begin{cases} \dfrac{dQ}{dt} = v^\psi(Q) & \text{(guiding equation)} \\ i\hbar \dfrac{d\psi}{dt} = H\psi & \text{(Schrödinger's equation)}, \end{cases}$$

where $H$ is the usual nonrelativistic Schrödinger Hamiltonian and $v^\psi$ is a velocity field on configuration space determined by the wave function—its explicit form will be given later in this chapter and, in more detail, in the subsequent chapters of this book.

Bohmian mechanics is a counterexample to all claims that a rational account of quantum phenomena is impossible. It is also a counterexample to the claim that quantum mechanics proves that nature is intrinsically random—that there is no way that determinism can ever be reinstated in the fundamental description of nature. This book is about Bohmian mechanics, its applications, its prospects for relativistic extensions, and how it gives rise to the quantum mechanical rules one learns about in classes and textbooks.

## 1.2 What is Wrong with Quantum Mechanics?

It is often suggested that the fundamental problem with quantum mechanics is the *measurement problem*, or, more or less equivalently, Schrödinger's cat paradox. However, these are but a dramatic symptom of a more fundamental problem: that the usual description of the state of a system in a quantum mechanical universe is of a rather unusual sort. The state of a quantum system is said to be given by a rather abstract mathematical object, namely the wave function or the quantum state vector (or maybe the density matrix) of the system, an object whose physical meaning is rather obscure in traditional presentations of quantum theory.

The measurement problem is this: if one accepts that the usual quantum mechanical description of the state of a quantum system is indeed the complete description of that system, it seems hard to avoid the conclusion that quantum measurements typically fail to have results. Pointers on measurement devices typically fail to point, computer printouts typically fail to have anything definite written on them, and so on. More generally, macroscopic states of affairs tend to be grotesquely indefinite, with cats seemingly both dead and alive at the same time, and the like. This is not good!

These difficulties can be largely avoided by invoking the measurement axioms of quantum theory, in particular the collapse postulate. According to this postulate, the usual quantum mechanical dynamics of the state vector of a system—given by Schrödinger's equation, the fundamental dynamical equation of quantum theory—is abrogated whenever measurements are performed. The deterministic Schrödinger evolution of the state vector is then replaced by a random collapse to a state vector that can be regarded as corresponding to a definite macroscopic state of affairs: to a pointer pointing in a definite direction, to a cat that is definitely dead or definitely alive, etc.

But doing so comes at a price. One then has to accept that quantum theory involves special rules for what happens during a measurement, rules that are in addition to, and not derivable from, the quantum rules governing all other situations. One has to accept that the notions of measurement and observation play a fundamental role in the very formulation of quantum theory, in sharp conflict with the much more plausible view that what happens during measurement and observation in a quantum universe, like everything else that happens in such a universe, is a consequence of the laws governing the behavior of the constituents of that universe—say the elementary particles and fields. These laws apply directly to the microscopic level of description, and say nothing directly about measurement and observation, notions that arise and make sense on an entirely different level of description, the macroscopic level.

We believe, however, that the measurement problem, as important as it is, is nonetheless but a manifestation of a more basic difficulty with standard quantum mechanics: it is not at all clear what quantum theory is about. Indeed, it is not at all clear what quantum theory actually says. Is quantum mechanics fundamentally about measurement and observation? Is it about the behavior of macroscopic variables? Or is it about our mental states? Is it about the behavior of wave functions? Or is it about the behavior of suitable fundamental microscopic entities, elementary particles and/or fields? Quantum mechanics provides us with formulas for lots of probabilities. What are these the probabilities of? Of results of measurements? Or are they the probabilities for certain unknown details about the state of a system, details that exist and are meaningful prior to measurement?

It is often said that such questions are the concern of the foundations of quantum mechanics, or of the interpretation of quantum mechanics—but not, somehow, of quantum mechanics itself, of quantum mechanics simpliciter. We think this is wrong. We think these and similar questions are a reflection of the fact that quantum mechanics, in the words of Bell, is "unprofessionally vague and ambiguous.

Professional theoretical physicists ought to be able to do better. Bohm has shown us a way" [5, p. 160].

What is usually regarded as a fundamental problem in the *foundations* of quantum mechanics, a problem often described as that of *interpreting* quantum mechanics, is, we believe, better described as the problem of finding a sufficiently precise *formulation* of quantum mechanics, of finding a *version* of quantum mechanics that, while expressed in precise mathematical terms, is also clear as physics. Bohmian mechanics provides such a precise formulation.

## 1.3 History

Einstein never accepted quantum mechanics as the last word of physics on nature. He adhered to determinism and an objective description of the world, writing that

> I am, in fact, firmly convinced that the essentially statistical character of contemporary quantum theory is solely to be ascribed to the fact that this [theory] operates with an incomplete description of physical systems. [6, p. 666]

The random outcomes of quantum measurements had to arise, according to Einstein, from our ignorance about the values of certain variables that had not yet been introduced into quantum theory, variables describing the real state of affairs.

In the 1970s Bell introduced for such variables the notion of beables, and in particular of *local beables*, those beables describing the configuration of matter in space-time. A variety of choices for the fundamental local beables may be possible, and each choice may admit a variety of laws to govern the behavior of the fundamental local beables. Each such choice, of local beables and laws governing them, corresponds to a different theory, even when these different theories yield the same predictions for the results of experiments—even, that is, when they are *empirically equivalent*. When these predictions are those of orthodox quantum theory, the different theories are different *versions of quantum theory*.

The way such theories yield experimental predictions is via the fundamental local beables, in terms of which macroscopic variables are defined. Some of these macroscopic variables describe the results of experiments, so that the laws governing the behavior of the fundamental local beables have empirical implications. The orientation of a pointer, for example, is determined by the configuration of the fundamental local beables associated with the pointer, the configuration, say, of its particles, and the behavior of the pointer is determined by that of its particles.

This may seem rather obvious. We think it is. What is not so obvious, perhaps, and what, given the history of quantum mechanics and the surrounding controversy, is perhaps surprising, is that a choice of fundamental local beables and law for them yielding a version of quantum mechanics—yielding a theory empirically equivalent to quantum mechanics—should be possible at all.

But, as we've indicated, such a choice is possible and, insofar as nonrelativistic quantum mechanics is concerned, rather obvious. The local beables are the positions of the particles, and these move according to an equation of motion, the *guiding equation*, that involves the wave function of standard quantum theory. The resulting theory is the pilot-wave theory or Bohmian mechanics. It was discovered by Louis de Broglie not long after Schrödinger's creation, in 1926, of wave mechanics (Schrödinger's equation) [7]. In 1952, it was independently rediscovered and analyzed in measurement situations in two papers by David Bohm [8, 9], who showed that the theory is empirically equivalent to quantum mechanics.

Beginning in the 1960s, the theory was popularized by John Bell. Noting that the theory involved a manifestly nonlocal description of nature, Bell asked the question: Can one improve the theory so that it would involve only a local description? In an ingenious argument, in which Bell joined the famous EPR paradox with a simple probabilistic estimate, the famous Bell's inequality, he concluded in fact that any description of nature which predicts for experiments the quantum probabilities for their results must be nonlocal. The measured statistics for such nonlocality experiments do reproduce the quantum probabilities and therefore establish the fact that nature is indeed nonlocal. In other words, Bohmian mechanics is, arguably, just what the doctor ordered.

The pilot-wave approach to quantum theory was in fact initiated even before the discovery of quantum mechanics itself, by Einstein, who hoped that interference phenomena involving particle like photons could be explained if the motion of the photons were somehow guided by the electromagnetic field—which would thus play the role of what he called a "Führungsfeld" or guiding field [10]. While the notion of the electromagnetic field as guiding field turned out to be rather problematical, the possibility that for a system of electrons the wave function might play this role, of guiding field or pilot wave, was explored by Max Born in his early papers founding quantum scattering theory [11, 12]—a suggestion to which Heisenberg was profoundly unsympathetic. Born presented in those papers the probability interpretation of the wave function, for which he eventually received the Nobel prize. He suggested that, guided by the wave function, particles move around in such a way that the distribution of their positions at a given time is given by the modulus squared of the wave function at that time.

Earlier still, the relation between geometric and wave optics had led to a reformulation of classical mechanics by Hamilton and Jacobi, in which the Hamilton-Jacobi function $S$, defined on the configuration space of a system of particles, defines via its gradient the momenta of the particles and acts like a guiding field. Bohmian mechanics can be regarded as a nonlocal generalization of the Hamilton-Jacobi theory, where $S$ is replaced by the phase of the wave function. It turns out that the phase $S$, in Bohmian mechanics, obeys a modified Hamilton-Jacobi partial differential equation, modified by the addition of an extra potential term, called the *quantum potential* by Bohm, to the classical potential energy term of the usual Hamilton-Jacobi equation. However, while the classical Hamilton-Jacobi theory can be rephrased as Newtonian mechanics, with the Hamilton-Jacobi function $S$ eliminated, this is not possible for Bohmian mechanics, for which a function $S$ on configuration space plays an essential role.

The inclusion of local beables in quantum theory does not imply that the theory is thereby deterministic, though Bohmian mechanics happens to be so. Another version of quantum mechanics, called *stochastic mechanics* and invented by Edward Nelson around 1966 [13, 14], involves the same local beables as Bohmian mechanics—particles, described by their positions in space. But in this theory the particles evolve randomly, according to a diffusion process defined in terms of the wave function.

## 1.4 Impossibility

At the 1927 Solvay Congress, de Broglie presented his pilot-wave theory, which he regarded as an oversimplified version of what he hoped to be able to construct in the future, namely the so-called theory of the double solution, a theory about only waves and not particles. The reception of de Broglie's ideas was not enthusiastic. To begin with, the guiding law for a many-particle system was not formulated on three-dimensional physical space, but on the abstract high-dimensional configuration space: the law seemed to involve, as part of "physical reality," the wave function on that space, the meaning of which had been declared by Born to be a probability amplitude just before the Solvay Congress. While probability densities were acceptable as objects on configuration space, physical fields which determine the motion of particles were not.

Furthermore, de Broglie had not yet analyzed the theory for measurement situations and its relation to quantum mechanics remained unclear. In particular, he responded poorly to an objection of Wolfgang Pauli concerning an application to inelastic scattering, no doubt making a rather bad impression on the illustrious audience gathered for the occasion. After the Solvay Congress, de Broglie did not, for several decades, pursue these ideas further.

However in 1953, after the appearance of Bohm's rediscovery of the guiding equations, de Broglie returned to his idea of the double solution. Commenting on Bohm's revival of the pilot-wave theory, he expressed again his old doubts, which were shared by almost all physicists:

> The wave $\psi$ used in wave mechanics cannot be a physical reality: it has arbitrary normalization, it is supposed to propagate in general in a visibly fictional configuration space, and, in conformity with the ideas of Mr. Born, it is merely a representation of probability depending on our state of knowledge and suddenly modified by information supplied by any new measurement. A causal and objective interpretation of wave mechanics cannot therefore be obtained on the sole basis of the pilot wave theory by assuming that the particle is guided by the wave. For this reason, I have been in full agreement with the purely probabilistic interpretation of Messrs. Born, Bohr, and Heisenberg since 1927. [15, p. 22][1]

---

[1] L'onde $\psi$ utilisé en Mécanique ondulatoire ne peut pas être une réalité physique: Sa normalisation est arbitraire, sa propagation est censée s'effectuer en général dans un espace de configuration visiblement fictif, et, conformément aux idées de M. Born, elle n'est qu'une représentation de probabilité dépendant de l'état de nos connaissances et brusquement modifiée par les informations que nous apporte toute nouvelle mesure. One ne peut donc obtenir à l'aide de la seule théorie de l'onde-pilote une interprétation causale et objective de la mécanique ondulatoire en supposant que le corpuscule est guidé par l'onde $\psi$. Pour cette raison, je m'étais entièrement rallié depuis 1927 à l'interprétation purement probabiliste de MM. Born, Bohr et Heisenberg (English translation by S. Lyle).

Also Bohm, who had shown how the measurement rules of quantum mechanics emerge from his theory, in a discussion with Maurice Pryce in the 1960s described his theory as profoundly incomplete and preliminary [16].

Earlier on, it became widely accepted that quantum randomness could not be accounted for by averaging over additional variables—the so-called *hidden variables*. The notion of hidden variables originated in 1932, in the book of the great mathematician John von Neumann, *On the Mathematical Foundations of Quantum Mechanics*, where he claimed that any attempt to introduce hidden variables into quantum mechanics and thereby restore determinism would inevitably lead to contradictions with quantum mechanical predictions and thus was hopeless. On the authority of von Neumann, this became common wisdom. In particular, the almost universal opinion arose that Bohmian mechanics must be flawed. After 1932 various other "no-go theorems," among them the celebrated paradox of Kochen and Specker [17], were obtained, all seeming to preclude the possibility of introducing hidden variables into quantum mechanics without contradictions.

The true status of the no-go theorems was clarified by Bell [18], who found that all such theorems involved unreasonable assumptions. About his personal discovery of Bohm's theory, Bell wrote:

> But in 1952 I saw the impossible done. It was in papers by David Bohm. Bohm showed explicitly how parameters could indeed be introduced, into nonrelativistic wave mechanics, with the help of which the indeterministic description could be transformed into a deterministic one. . . .
>
> . . . Moreover the essential idea was one that had been advanced already by de Broglie in 1927, in his "pilot wave" picture.
>
> But why then had Born not told me of this "pilot-wave"? If only to point out what was wrong with it? Why did von Neumann not consider it? More extraordinarily, why did people go on producing "impossibility'" proofs, after 1952, and as recently as 1978? [5, p. 160]

Bohmian mechanics is of course as much a counterexample to the Kochen-Specker argument for the impossibility of hidden variables as it is to the one of von Neumann. It is obviously a counterexample to any such argument. The assumptions of any such argument, however compelling they may seem to be, must fail for Bohmian mechanics.

Bell's clarification of the status of the no-go theorems had some impact on the accepted wisdom about the achievements of von Neumann and the others. It became somewhat less common for those no-go theorems to be used to justify the dismissal of Bohmian mechanics. Instead Bell's own theorem, his nonlocality proof, was more often used for that purpose. Bell's analysis became widely regarded as demonstrating that a realistic reformulation of quantum theory was impossible, or at least unacceptable.

But, as we have already indicated, Bell did not establish the impossibility of a realistic reformulation of quantum theory, nor did he ever claim to have done so. On the contrary, over the course of several decades, until his untimely death in 1990, Bell was the prime proponent, for a good part of this period almost the sole proponent, of the very theory, Bohmian mechanics, that he is supposed to have demolished.

It is worth noting that in Bohmian mechanics the positions of the particles are not really hidden. Again Bell:

> Absurdly, such theories are known as "hidden variable" theories. Absurdly, for there it is not in the wavefunction that one finds an image of the visible world, and the results of experiments, but in the complementary "hidden" (!) variables. Of course the extra variables are not confined to the visible "macroscopic" scale. For no sharp definition of such a scale could be made. The "microscopic" aspect of the complementary variables is indeed hidden from us. But to admit things not visible to the gross creatures that we are is, in my opinion, to show a decent humility, and not just a lamentable addiction to metaphysics. In any case, the most hidden of all variables, in the pilot wave picture, is the wavefunction, which manifests itself to us only by its influence on the complementary variables. [5, p. 201]

## 1.5 The Dismissal

Bohmian mechanics is a version of quantum mechanics that, while expressed in precise mathematical terms, is clear as physics. Why isn't it taught? Why isn't it part of a quantum physics education, if only to point out that there is a version of quantum mechanics which is mathematically precise and physically clear? A possible answer could be that Bohmian mechanics is too contrived or that it is too complicated to be taken seriously as physics. This can be judged, of course, only by examining the formulation of the theory.

Many other reasons for dismissing Bohmian mechanics have been given, for example that because it is deterministic it is incompatible with free will, or that the notion of particles has been discredited or that it is a regression to classical modes of thought. If a clever person looks for reasons to dismiss something he does not like, he will find them.

At the end of the day, one of the fundamental reasons for the dismissal seems to be this: *Bohmian mechanics is against the spirit of quantum mechanics*. That is, an objective physical description, a return to determinism, a return to physical clarity clash with the tenets of quantum philosophy.

To students such an attitude must seem rather puzzling. But academia and scholarly behavior are not always rational. In the matter of Bohm, Robert Oppenheimer Bohm's Ph.D. advisor, reportedly has said that "if we cannot disprove Bohm, then we must agree to ignore him" [19, p. 133].

## 1.6 Roads to Bohmian Mechanics

To the wave function and Schrödinger's equation of orthodox quantum theory, Bohmian mechanics adds the actual positions of particles and a first-order equation of motion for the positions, given by a velocity function $v = v^\psi$ depending on the wave function $\psi$ and on the positions. By roads to Bohmian mechanics we mean basically ways of guessing a formula for $v^\psi$. There are many ways to do so. Here are some of them.

1. The simplest route, for particles without spin, is the following: Begin with the de Broglie relation $\mathbf{p} = \hbar \mathbf{k}$, a remarkable and mysterious distillation of the experimental facts associated with the beginnings of quantum theory—and itself a relativistic reflection of the first quantum equation, namely the Planck relation $E = h\nu$. The de Broglie relation connects a particle property, the momentum $\mathbf{p} = m\mathbf{v}$, with a wave property, the wave vector $\mathbf{k}$. Understood most simply, it says that the velocity of a particle should be the ratio of $\hbar \mathbf{k}$ to the mass of the particle. But the wave vector $\mathbf{k}$ is defined only for a plane wave. For a general wave $\psi$, the obvious generalization of $\mathbf{k}$ is the local wave vector $\nabla S(\mathbf{q})/\hbar$, where $S$ is the phase of the wave function (defined by its polar representation, see below). With this choice the de Broglie relation becomes $v = \nabla S/m$, the right hand side of which is our first guess for $v^\psi$.

    We note that the de Broglie relation also immediately yields Schrödinger's equation, giving the time evolution for $\psi$, as the simplest wave equation that reflects this relationship. This is completely standard. In this simple way, the defining equations of Bohmian mechanics can be regarded as flowing in a natural manner from the first quantum equation $E = h\nu$.

2. Another route, for particles without spin, starts with writing Schrödinger's wave function in polar form $\psi = R e^{\frac{i}{\hbar} S}$, with $R$ and $S$ real. When one introduces this polar form into Schrödinger's equation for $\psi$, one obtains two equations: the modified Hamilton-Jacobi equation with the extra quantum potential for $S$ and a continuity equation for $\rho = R^2 = |\psi|^2$. Since in classical Hamilton-Jacobi theory the momentum $m v = p = \nabla S$, it is natural to set $v = \nabla S/m$ in the quantum case as well.[2]

3. The quantum continuity equation is the key for a route which is meaningful also for particles with spin. This equation, an immediate consequence of Schrödinger's equation, involves a quantum probability density $\rho$ and a quantum probability current $J$. Since densities and currents are classically related by $J = \rho v$, it requires little imagination to set $v^\psi = J/\rho$.[3]

4. Another way to arrive at a formula for $v^\psi$ is to invoke symmetry. Since the spacetime symmetry of the non-relativistic Schrödinger equation is that of rotations, translations, time-reversal, and invariance under Galilean boosts, it is natural to demand that this Galilean symmetry be retained when Schrödinger's equation is combined with the guiding equation. As described in Sect. 2.3, this leads to a specific formula for $v^\psi$ as the simplest possibility.

---

[2] This was basically the route followed by Bohm in his 1952 paper, though Bohm went further and recast the first-order theory defined by $v^\psi = \nabla S/m$ into a second-order theory involving accelerations. By differentiating the modified Hamilton-Jacobi equation with respect to time, Bohm obtained a modified Newton equation which involves, in addition to the usual force arising from the classical potential energy, an extra "quantum force" arising from the quantum potential. Bohm formulated his theory in terms of this second-order equation. We believe that the first-order form is preferable.

[3] This route to $v^\psi$ could be called the Pauli-Bell "derivation" because this way of presenting Bohmian mechanics, Bell's favorite, originated with Pauli at the 1927 Solvay Congress ([20], p. 134; English translation [21], p. 365).

5. We mention one last route. With any wave function $\psi(q)$, one can associate a Wigner distribution $W^\psi(q, p)$, a sort of quantum-mechanical joint distribution for position and momentum, which, however, need not be non-negative. Setting

$$mv^\psi(q) = \frac{\int p\, W^\psi(q, p)\, dp}{\int W^\psi(q, p)\, dp}$$

yields our last formula for $v^\psi$.

There are several other natural routes to $v^\psi$, but we shall give no more. All these routes, those we've explicitly mentioned and those to which we've alluded, yield in fact exactly the same formula for $v^\psi$, though the explicit form may appear different in some cases.

## 1.7 Questions

Bohmian mechanics is not a particularly difficult theory. It is defined by two equations: the Schrödinger equation for the wave function $\psi$ and the guiding equation for the positions of the particles. Moreover, the guiding equation is easy to find and is quite simple. There are no axioms about observers, observation, or measurement of observables. There are no axioms about the collapse of the wave function during measurement. This is in sharp contrast with orthodox quantum mechanics.

However, what is not at all easy to understand about Bohmian mechanics is its precise relationship with standard quantum theory. What is at first not at all clear is how the predictions of Bohmian mechanics are related, if at all, to those of that theory. For example:

1. Quantum interference for a particle in a two-slit experiment seems to show that the particle must go through both slits at once. But in Bohmian mechanics a single particles passes through only one of the two slits. Can Bohmian mechanics account for the two-slit experiment? (See below.)
2. Are Bohmian trajectories or velocities observable? (See Chaps. 2, 3, and 7.)
3. How can the intrinsic randomness of quantum theory arise from a deterministic theory such as Bohmian mechanics? (See Chap. 2.)
4. How can the collapse rule for the wave function be compatible with Bohmian mechanics, one of whose axioms is Schrödinger's equation for the evolution of the wave function, which is incompatible with its collapse? (See Chap. 2, where, among other things, it is shown that the collapse rule is a theorem of Bohmian mechanics.)
5. In Bohmian mechanics a particle always has a well-defined position and velocity. How can this be compatible with Heisenberg's uncertainty principle? (See Chap. 2.)
6. Spin, unlike position, has no classical analogue. How can Bohmian mechanics deal with spin? (See Chap. 3.)

7. In the formulation of Bohmian mechanics there are no operators as observables, which play so prominent a role in quantum theory. How, if at all, do operators as observables arise in Bohmian mechanics? (See Chap. 3.)
8. There are many theorems precluding the possibility of hidden variables for quantum mechanics. These theorems seem to demonstrate that a deterministic reformulation of quantum mechanics is impossible, and that the values of quantum observables are normally created only by measurements and do not correspond to any pre-existing values for the observables. How then can Bohmian mechanics be compatible with the predictions of quantum theory? (See Chap. 3.)
9. Isn't the curious indistinguishability of quantum mechanical identical particles incompatible with particles having definite positions, and following definite trajectories? (See Chap. 8.)
10. Similarly, how can quantum entanglement be compatible with Bohmian mechanics? (See Chap. 2.)
11. A widely debated problem in quantum theory is that of time measurements, for example the dwell time of a particle within, and its time of escape from, a certain region of physical space. In quantum mechanics there is in fact no self-adjoint time observable of any suitable sort and there is a large and controversial literature on what to do about this. On the other hand, Bohmian mechanics makes specific predictions for the statistics of such times. How can Bohmian mechanics be then empirically equivalent to standard quantum mechanics? (See Chap. 6.)
12. In Bohmian mechanics the motion of even a single particle requires an infinite-dimensional system for its description—infinite-dimensional because it involves the wave function, whose specification requires an infinite collection of real numbers. How can this be compatible with the emergence of a classical regime, in which the motion of such a particle in physical space is determined by a 6-dimensional system, involving only position and momentum? (See Chap. 5.)
13. Given the prominent role that configuration space—a highly anti-relativistic structure—plays in its formulation, how can a relativistic version of Bohmian mechanics be possible? (See Chap. 9, 11 and 12.)
14. In quantum field theory particles are created and annihilated. Doesn't that preclude a description in terms of genuine particles? (See Chap. 10.)
15. In orthodox quantum mechanics the wave function of a system is usually regarded as a probability amplitude and thus is widely regarded as corresponding only to our knowledge of that system. But in Bohmian mechanics, in which it plays a crucial dynamical role, the wave function should be regarded as entirely objective. Is this a reasonable thing to do? (See Chap. 2, 11 and 12.)[4]

These are some of the questions that will be addressed in this book. Most will be addressed in the ensuing chapters, but the first will be dealt with right now.

According to Richard Feynman, the two-slit experiment for electrons is

---

[4] Very recently, Matthew Pusey, Jonathan Barrett, and Terry Rudolph have developed a nice new argument [25] showing that the Schrödinger wave function can't consistently be interpreted as merely statistical, i.e., subjective. (Rudolph's grandfather, incidentally, was Schrödinger.)

## 1.7 Questions

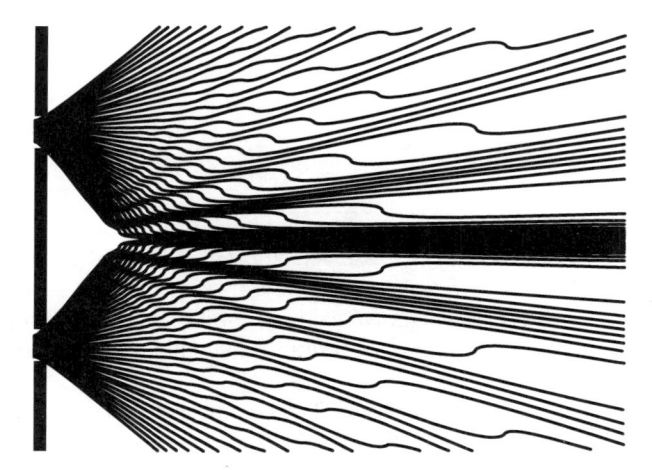

**Fig. 1.1** A family of Bohmian trajectories for the two-slit experiment. (Adapted by Gernot Bauer from Hiley, et al. [26].)

> a phenomenon which is impossible, absolutely impossible, to explain in any classical way, and which has in it the heart of quantum mechanics. In reality it contains the only mystery. [22, p. 37.2]

As to the question:

> How does it really work? What machinery is actually producing this thing? Nobody knows any machinery. Nobody can give you a deeper explanation of this phenomenon than I have given; that is, a description of it. [23, p. 145]

Now consider the Bohmian trajectories for the two-slit experiment (see Fig. 1.1). One sees that Bohmian mechanics resolves the dilemma of the appearance, in one and the same phenomenon, of both particle and wave properties in a rather trivial manner. Notice that while each trajectory passes through but one of the slits, the wave passes through both, and the interference profile that therefore develops in the wave generates a similar pattern in the trajectories guided by this wave.

Finally, compare Feynman's presentation with that of Bell:

> Is it not clear from the smallness of the scintillation on the screen that we have to do with a particle? And is it not clear, from the diffraction and interference patterns, that the motion of the particle is directed by a wave? De Broglie showed in detail how the motion of a particle, passing through just one of two holes in screen, could be influenced by waves propagating through both holes. And so influenced that the particle does not go where the waves cancel out, but is attracted to where they cooperate. This idea seems to me so natural and simple, to resolve the wave-particle dilemma in such a clear and ordinary way, that it is a great mystery to me that it was so generally ignored. [5, p. 191]

With regard to the second question, the following figure, similar to Fig. 1.1, recently appeared in Science [24]:

In the accompanying article by Aephraim Steinberg et al., the trajectories depicted in the figure are reported to be the result of a suitable weak measurement on a quantum particle. According to the authors:

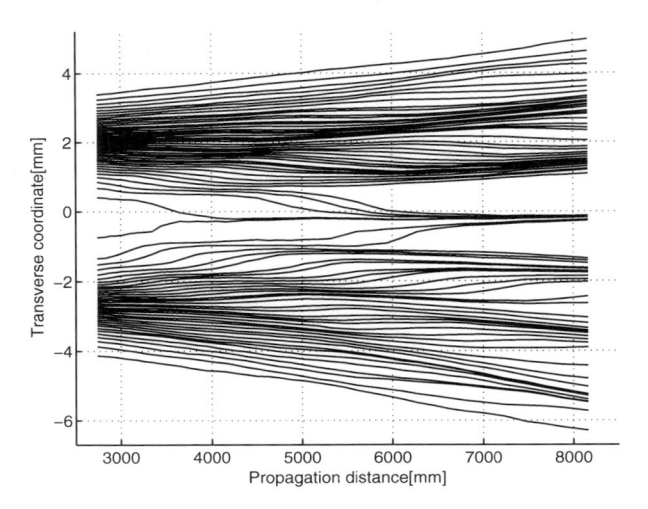

Single-particle trajectories measured in this fashion reproduce those predicted by the Bohm-de Broglie interpretation of quantum mechanics, although the reconstruction is in no way dependent on a choice of interpretation. [24]

Chapter 7 provides a Bohmian analysis of this sort of experiment.

## 1.8 The Book

The book is divided into three parts: Part I, Quantum Equilibrium, Part II, Quantum Motion, and Part III, Quantum Relativity. While what we mean by "quantum motion" should be rather clear by now, "quantum equilibrium" presumably requires a bit of explanation. Quantum equilibrium is a concept analogous to, but quite distinct from, thermodynamic equilibrium. It provides a precise and natural notion of *typical behavior* of a Bohmian system—the behavior which holds true for the overwhelming majority of histories of a universe governed by Bohmian mechanics. A large part of Chap. 2 is devoted to an elaboration of the various facets of this rather subtle notion.

The goal of Part I is to present Bohmian mechanics and to establish in very general terms its relation with standard quantum mechanics. Succinctly, its content can be summarized as follows: the statistical description in quantum theory indeed takes, in Bohmian mechanics, "an approximately analogous position to the statistical mechanics within the framework of classical mechanics" [6, p. 672].

Part II is devoted to more technical aspects of Bohmian systems. Each chapter relates Bohmian mechanics to a basic issue in quantum mechanics: the classical limit (Chap. 5), quantum scattering in the mesoscopic regime and time observables (Chap. 6), the observability and measurability of Bohmian trajectories (Chap. 7), identical particles and quantum statistics (Chap. 8).

## 1.8 The Book

Part III concerns various aspects of the issue of extending Bohmian mechanics to quantum field theory and to include relativity: Bohmian mechanics and Lorentz invariance (Chap. 9 and 12), Bohmian quantum field theory (Chap. 10), and Bohmian quantum gravity (Chap. 11).

Each part ends with a chapter picking up themes discussed in this introduction or providing a broader perspective (Chap. 4, Chap. 8 and 12).

We now describe, chapter by chapter, the detailed contents of this book.

### Chapter 2. Quantum Equilibrium and the Origin of Absolute Uncertainty

We begin with an analysis of the fundamental probability formula of quantum mechanics, $\rho = |\psi(q)|^2$—not just why it should hold for Bohmian mechanics, but what it could possibly mean. We argue that a Bohmian universe, though deterministic, evolves in such a manner that an *appearance* of randomness emerges, precisely as described by the quantum formalism and given by $\rho = |\psi(q)|^2$. A key ingredient of our analysis of the origin of this quantum randomness is the notion of the *wave function of a subsystem*—in addition to that of quantum equilibrium, at whose meaning we briefly hinted above.

The notion of the wave function of a subsystem is rather elusive. Surprisingly, it is not clear a priori what should be meant by the wave function of a subsystem of a Bohmian universe, for which there is fundamentally but one wave function, that of the universe. (In standard quantum mechanics the state of a subsystem is usually described in terms of its reduced density matrix.)

However, it turns out that in Bohmian mechanics (ignoring spin) subsystems do have indeed, at any time, a well-defined wave function $\psi$, what we call the *conditional wave function*, which in general does not evolve according to Schrödinger's equation. Part of the chapter is devoted to clarifying this notion, its status and its role in explaining the emergence of quantum randomness. In particular, we show that such a wave function $\psi$ collapses according to the usual textbook rules with the usual textbook probabilities in the usual measurement situations. (This is done more explicitly in Chap. 3.)

We discuss some consequences of quantum equilibrium, most notably the fact that in a universe governed by Bohmian mechanics there are sharp, precise, and irreducible limitations on the possibility of obtaining knowledge, limitations that can in no way be diminished through technological progress leading to better means of measurement. This *absolute uncertainty* is in precise agreement with Heisenberg's uncertainty principle. Finally we compare and contrast quantum equilibrium with thermodynamic equilibrium and nonequilibrium.

In this comparison, as in this whole chapter, we adopt the common standard viewpoint where an external absolute time is treated as an accessible and physically relevant variable. From a more realistic viewpoint—rarely, if ever, taken in a non-relativistic setting—in which the focus would be on configurations and physical states, not at a given value of external time, but at a given value of a physical time variable (such as, in quantum gravity, might correspond to some feature of the

3-geometry, for example the radius of the universe), a different sort of analysis would be required. In such an analysis, the conditional wave function relative to the value of the physical time variable would presumably replace the wave function of the universe at some external time $\tau$. (Thus the wave function of the universe itself could be stationary and non-normalizable; see the last paragraphs of Sect. 12.3.3 and of Sect. 11.4.) The implementation of such ideas is an open and difficult problem. We shall not deal with it in this book; at most we shall make a few vague gestures at various places in later chapters.

## Chapter 3. Quantum Equilibrium and the Role of Operators as Observables in Quantum Theory

We then turn to the question of how the entire quantum formalism, operators as observables and all the rest, naturally emerges in Bohmian mechanics from an analysis of general experiments. This analysis illuminates the status of operators as observables in the description of quantum phenomena, and facilitates a clear view of the range of applicability of the usual quantum mechanical formulas.

It would appear that because orthodox quantum theory supplies us with probabilities for a huge class of quantum observables and not merely for positions, it is a much richer theory than Bohmian mechanics, which seems exclusively concerned with positions. In this regard, as with so much else in the foundations of quantum mechanics, the crucial remark was made by Bell:

> [I]n physics the only observations we must consider are position observations, if only the positions of instrument pointers. It is a great merit of the de Broglie-Bohm picture to force us to consider this fact. If you make axioms, rather than definitions and theorems, about the 'measurement' of anything else, then you commit redundancy and risk inconsistency. [5, p. 166]

In Bohmian mechanics, the standard quantum observables, represented by self-adjoint operators, indeed arise from an analysis of quantum experiments, as "definitions and theorems." If the experiment happens to be "measurement-like," and the outcomes of the experiment are calibrated by an assignment of numerical values to the different pointer orientations, then the induced probability distributions of these results will be given by the familiar quantum measurement postulates, involving squares of probability amplitudes and eigenvectors of the relevant quantum observable—i.e., by the spectral measure of a self-adjoint operator $A$ associated with the experiment. In this case we speak, in orthodox quantum theory, of a "measurement of the observable $A$."

In this chapter we also discuss the widespread idea that in a realistic quantum theory all quantum observables should possess actual values, which is in fact impossible by the Kochen-Specker theorem. We argue that that idea was from the outset not as reasonable as it may have appeared, but rather was based on taking operators as observables too seriously—an attitude, strongly suggested by the word "observable," that we call *naive realism about operators*.

## 1.8 The Book

### Chapter 4. Quantum Philosophy: The Flight from Reason in Science

Here we pause and reconsider the rise of quantum philosophy.

### Chapter 5. Seven Steps Towards the Classical World

Classical physics is about real objects, like apples falling from trees, whose motion is governed by Newtonian laws. In standard quantum mechanics only the wave function or the results of measurements exist, and to answer the question of how the classical world can be part of the quantum world is a rather formidable task. However, this is not the case for Bohmian mechanics, which, like classical mechanics, is a theory about real objects. In Bohmian terms, the problem of the classical limit becomes very simple: *when do the Bohmian trajectories look Newtonian?* This is the main question we address in this chapter.

### Chapter 6. On the Quantum Probability Flux through Surfaces

Here we argue that the often ignored quantum probability current is fundamental for a genuine understanding of scattering phenomena and, in particular, for the statistics of the time and position of the first exit of a quantum particle from a given region, which may be simply expressed in terms of the current. This simple formula for these statistics does not appear in the standard quantum literature. A full understanding of the quantum current and the associated formula is indeed provided by Bohmian mechanics.

### Chapter 7. On the Weak Measurement of Velocity in Bohmian Mechanics

In a recent article [27], Wiseman has proposed the use of so-called weak measurements for the determination of the velocity of a quantum particle at a given position, and has shown that according to quantum mechanics the result of such a procedure is the Bohmian velocity of the particle. Although Bohmian mechanics is empirically equivalent to variants based on velocity formulas different from the Bohmian one, and although it has been proven that the velocity in Bohmian mechanics is not measurable, we argue here for the somewhat paradoxical conclusion that Wiseman's weak measurement procedure indeed constitutes a genuine measurement of velocity in Bohmian mechanics. We reconcile the apparent contradictions and elaborate on some of the different senses of measurement at play here.

### Chapter 8. Topological Factors Derived From Bohmian Mechanics

Bohmian mechanics can be formulated for $N$ identical particles—notwithstanding a fact that could be felt to contradict their indistinguishability, namely that the particle trajectories in physical space $\mathbb{R}^3$ determine "who is who" at different times, i.e., select a one-to-one association between the $N$ points at any time and the $N$ points at another time. Taking the notion of a particle seriously, as one should in Bohmian mechanics, one recognizes that the configuration space $\mathscr{Q}$ of $N$ identical particles is best regarded as the manifold of all $N$-point subsets of $\mathbb{R}^3$. This manifold has non-trivial topological properties, and in particular is multiply connected, as its fundamental (homotopy) group is isomorphic to the group of permutations of $N$ objects.

In this chapter we derive for Bohmian mechanics topological factors for quantum systems with a multiply-connected configuration space $\mathscr{Q}$. These include nonabelian factors corresponding to what we call holonomy-twisted representations of the fundamental group of $\mathscr{Q}$. As a byproduct of our analysis, we obtain an explanation, within the framework of Bohmian mechanics, of the fact that the wave function of a system of identical particles is either symmetric or anti-symmetric. Thus, Bohmian mechanics lends support to the modern view that the symmetrization postulate emerges as a topological effect, due to the non-trivial topology of the natural configuration space.

**Chapter 9. Hypersurface Bohm-Dirac Models**

This chapter deals with the issue of Lorentz invariance. We investigate some simple models—not to be taken seriously as models of the world—which nevertheless embody quantum nonlocality and are, arguably, fundamentally Lorentz invariant. More precisely, we define here a class of Lorentz invariant Bohmian quantum models for $N$ entangled but noninteracting Dirac particles. Lorentz invariance is achieved for these models through the incorporation of a suitable foliation of space-time. These models can be regarded as the extension of Bohm's model for $N$ Dirac particles, corresponding to the foliation into the equal-time hyperplanes for a distinguished Lorentz frame, to more general foliations. These models allow for a simple statistical analysis of position correlations analogous to the analysis discussed in Chap. 2.

**Chapter 10. Bohmian Mechanics and Quantum Field Theory**

In 1986 Bell proposed how to associate particle trajectories with a lattice quantum field theory, yielding what can be regarded as a $|\Psi|^2$-distributed Markov process on the appropriate configuration space. A similar process can be defined in the continuum, for more or less any regularized quantum field theory. This chapter provides a succinct exposition of such an extension.

We show that for more or less any regularized quantum field theory there is a corresponding theory of particle motion, which in particular ascribes trajectories to the electrons or whatever sort of particles the quantum field theory is about. Corresponding to the non conservation of particle number operators in the quantum field theory, the theory describes explicit creation and annihilation events: the world lines for the particles can begin and end.

The upshot of this chapter is that not only does the notion of particle not conflict with the prominence of quantum field operators but quantum field operators have a natural place in a theory whose ultimate goal it is to govern the motion of particles. One of their important roles is to relate the Hilbert space of a quantum field theory to configuration space. Quantum theory of fields or quantum theory of particles? A theory of particle motion exploiting field operators!

**Chapter 11. Quantum Spacetime without Observers**

We explore the possibility of a Bohmian approach to the problem of finding a quantum theory incorporating gravitational phenomena. The major conceptual problems of canonical quantum gravity are the problem of time and the problem of diffeomorphism invariant observables. We find that these problems are artifacts of the subjectivity and vagueness inherent in the framework of orthodox quantum theory. When we insist upon ontological clarity—the distinguishing characteristic of a Bohmian approach—these conceptual problems vanish. We also discuss the implications of a Bohmian perspective for the significance of the wave function, concluding with unbridled speculation as to why the universe should be governed by laws so apparently bizarre as those of quantum mechanics.

**Chapter 12. Reality and the Role of the Wave Function**

In our final chapter we focus on what is perhaps the most puzzling issue in the foundations of quantum mechanics, namely that of the status of the wave function of a system in a quantum universe. Is the wave function objective or subjective? Does it represent the physical state of the system or merely our information about the system? And if the former, does it provide a complete description of the system or only a partial description? We address these questions here and argue that part of the difficulty in ascertaining the status of the wave function in quantum mechanics arises from the fact that there are two different sorts of wave functions involved. The most fundamental wave function is that of the universe. From it, together with the configuration of the universe, one can define the wave function of a subsystem. We argue that the fundamental wave function, the wave function of the universe, has a law-like character.

# References

1. W. Heisenberg. *The Physicist's Conception of Nature*. Harcourt Brace, 1958. Trans. Arnold J. Pomerans.
2. L. D. Landau and E. M. Lifshitz. *Quantum Mechanics: Non-relativistic Theory*. Pergamon Press, Oxford and New York, 1958. Translated from the Russian by J. B. Sykes and J. S. Bell.
3. M. Jammer. *The Philosophy of Quantum Mechanics*. Wiley, New York, 1974.
4. W. Moore. *Schrödinger*. Cambridge University Press, New York, 1989.

5. J. S. Bell. *Speakable and Unspeakable in Quantum Mechanics*. Cambridge University Press, Cambridge, 1987.
6. P. A. Schilpp, editor. *Albert Einstein, Philosopher-Scientist*. Library of Living Philosophers, Evanston, Ill., 1949.
7. L. de Broglie. La Nouvelle Dynamique des Quanta. In *Electrons et Photons: Rapports et Discussions du Cinquième Conseil de Physique tenu à Bruxelles du 24 au 29 Octobre 1927 sous les Auspices de l'Institut International de Physique Solvay*, pages 105–132, Paris, 1928. Gauthier-Villars.
8. D. Bohm. A Suggested Interpretation of the Quantum Theory in Terms of "Hidden" Variables: Part I. *Physical Review*, 85:166–179, 1952. Reprinted in [211].
9. D. Bohm. A Suggested Interpretation of the Quantum Theory in Terms of "Hidden" Variables: Part II. *Physical Review*, 85:180–193, 1952. Reprinted in [211].
10. E. P. Wigner. Interpretation of Quantum Mechanics. In [211], 1976.
11. M. Born. Quantenmechanik der Stoßvorgänge. *Zeitschrift für Physik*, 37:863–867, 1926.
12. M. Born. Quantenmechanik der Stoßvorgänge. *Zeitschrift für Physik*, 38:803–827, 1926. English translation (Quantum Mechanics of Collision Processes) in [149].
13. E. Nelson. Derivation of the Schrödinger Equation From Newtonian Mechanics. *Physical Review*, 150:1079–1085, 1966.
14. E. Nelson. *Dynamical Theories of Brownian Motion*. Princeton University Press, Princeton, N.J., 1967.
15. L. de Broglie. Scientific Papers Presented to Max Born. In *L'Interprétation de Mécanique Ondulatoire à l'Aide d'Ondes à Régions Singulières*, page 22, New York, 1953. Hafner Publishing Company Inc.
16. D. Edge and S. Toulmin. *Quanta and Reality*. Hutchinson and Co, 1962.
17. S. Kochen and E. P. Specker. The Problem of Hidden Variables in Quantum Mechanics. *Journal of Mathematics and Mechanics*, 17:59–87, 1967.
18. J. S. Bell. On the Problem of Hidden Variables in Quantum Mechanics. *Reviews of Modern Physics*, 38:447–452, 1966. Reprinted in [211] and in [26].
19. F. D. Peat. *Infinite Potential*. Addison-Wesley, 1997.
20. Solvay Conference. *Electrons et Photons: Rapports et Discussions du Cinquième Conseil de Physique tenu à Bruxelles du 24 au 29 Octobre 1927 sous les Auspices de l'Institut International de Physique Solvay*. Gauthier-Villars, Paris, 1928.
21. G. Bacciagaluppi and A. Valentini. *Quantum Theory at the Crossroads: Reconsidering the 1927 Solvay Conference*. Cambridge Univ. Press, 2009.
22. R. Feynman, R. B. Leighton, M. Sands. *The Feynman Lectures on Physics, I*. Addison-Wesley, New York, 1963.
23. R. Feynman. *The Character of Physical Law*. MIT Press, Cambridge, MA, 1967.
24. S. Kocsis, B. Braverman, S. Ravets, M.J. Stevens, R. P. Mirin, L. K Shalm, and A. M. Steinberg. Observing the Average Trajectories of Single Photons in a Two-Slit Interferometer. *Science*, 332:1170–1173, 2011.
25. M. F. Pusey, J. Barrett, and T. Rudolph. The Quantum State Cannot Be Interpreted Statistically. arXiv:1111.3328v1, 2011.
26. B. J. Hiley C. Philippidis, C. Dewdney. Quantum Interference and the Quantum Potential. *Il Nuovo Cimento B*, 52:15–28, 1979.
27. H. M. Wiseman. Grounding Bohmian Mechanics in Weak Values and Bayesianism. *New Journal of Physics*, 9:165, 2007.

# Part I
# Quantum Equilibrium

# Chapter 2
# Quantum Equilibrium and the Origin of Absolute Uncertainty

## 2.1 Introduction

> I am, in fact, rather firmly convinced that the essentially statistical character of contemporary quantum theory is solely to be ascribed to the fact that this (theory) operates with an incomplete description of physical systems. (Einstein, in [1], p. 666)

What is randomness? probability? certainty? knowledge? These are old and difficult questions, and we shall not focus on them here. Nonetheless, we shall obtain sharp, striking conclusions concerning the relationship between these concepts.

Our primary concern in this chapter lies with the status and origin of randomness in quantum theory. According to the quantum formalism, measurements performed on a quantum system with definite wave function $\psi$ typically yield random results. Moreover, even the specification of the wave function of the composite system including the apparatus for performing the measurement will not generally diminish this randomness. However, the quantum dynamics governing the evolution of the wave function over time, at least when no measurement is being performed, and given, say, by Schrödinger's equation, is completely deterministic. Thus, insofar as the particular physical processes which we call measurements are governed by the same fundamental physical laws that govern all other processes,[1] one is naturally led to the hypothesis that the origin of the randomness in the results of quantum measurements lies in random initial conditions, in our ignorance of the complete description of the system of interest—including the apparatus—of which we know only the wave function.

But according to orthodox quantum theory, and most nonorthodox interpretations as well, *the complete description* of a system is provided by its wave function alone, and there is no property of the system beyond its wave function (our ignorance of) which might account for the observed quantum randomness. Indeed, it used to be widely claimed, on the authority of von Neumann [2], that such properties, the

---

[1] And it is difficult to believe that this is not so; the very notion of measurement itself seems too imprecise to allow such a distinction within a fundamental theory, even if we were otherwise somehow attracted by the granting to measurement of an extraordinary status.

so-called hidden variables, are impossible, that as a matter of mathematics, averaging over ignorance cannot reproduce statistics compatible with the predictions of the quantum formalism. And this claim is even now not uncommon, despite the fact that a widely discussed counterexample, the quantum theory of David Bohm [3, 4], has existed for almost four decades.[2]

We shall call this theory, which will be "derived" and described in detail in Sect. 2.3, *Bohmian mechanics*. Bohmian mechanics is a new mechanics, a completely deterministic—but distinctly non-Newtonian—theory of particles in motion, with the wave function itself guiding this motion. (Thus the "hidden variables" for Bohmian mechanics are simply the particle positions themselves.) Moreover, while its *formulation* does not involve the notion of quantum observables, as given by self-adjoint operators—so that its relationship to the quantum formalism may at first appear somewhat obscure—it can in fact be shown that Bohmian mechanics not only accounts for quantum phenomena [4, 5, 6], but also embodies the quantum formalism itself as the very expression of its empirical import, see Chap. 3. (The analysis in the present chapter establishes agreement between Bohmian mechanics and the quantum formalism without addressing the question of how the *detailed* quantum formalism naturally emerges—how and why specific operators, such as the energy, momentum, and angular momentum operators, end up playing the roles they do, as well as why "observables" should rather generally be identified with self-adjoint operators. We shall answer some of these questions in Chap. 3, in which a general analysis of measurement from a Bohmian perspective is presented. We emphasize that the present chapter is not at all concerned directly with measurement per se, not even of positions.) That this is so is for the most part quite straightforward, but it does involve a crucial subtlety which, so far as we know, has never been dealt with in a completely satisfactory manner.

The subtlety to which we refer concerns the origin of the very randomness so characteristic of quantum phenomena. The predictions of Bohmian mechanics concerning the results of a quantum experiment can easily be seen to be precisely those of the quantum formalism, *provided* it is assumed that prior to the experiment the positions of the particles of the systems involved are randomly distributed according to Born's statistical law, i.e., according to the probability distribution given by $|\psi|^2$. And the difficulty upon which we shall focus here concerns the status—the justification and significance—of this assumption within Bohmian mechanics: not just why it should be satisfied, but also, and perhaps more important, what—in a completely deterministic theory—it could possibly mean!

In Sect. 2.3 we provide some background to Bohmian mechanics, describing its relationship to other approaches to quantum mechanics and how in fact it emerges from an analysis of these alternatives. This section, which presents a rather personal perspective on these matters, will play no role in the detailed analysis of the later sections and may be skipped on a first reading of this chapter.

---

[2] For an analysis of why von Neumann's and related "impossibility proofs" are not nearly so physically relevant as frequently imagined, see Bell's article [7]. (See also the celebrated article of Bell [8] for an "impossibility proof" which does have physical significance. See as well [9]). For a recent, and comprehensive, account of Bohm's ideas see [10].

## 2.1 Introduction

The crucial concepts in our analysis of Bohmian mechanics are those of *effective wave function* (Sect. 2.5) and *quantum equilibrium* (Sects. 2.4, 2.6, 2.13, and 2.14). The latter is a concept analogous to, but quite distinct from, thermodynamic equilibrium. In particular, quantum equilibrium provides us with a precise and natural notion of *typicality* (Sect. 2.7), a concept which frequently arises in the analysis of "large systems" and of the "long time behavior" of systems of any size. For a universe governed by Bohmian mechanics it is of course true that, given the initial wave function and the initial positions of all particles, *everything* is completely determined and nothing whatsoever is actually random. Nonetheless, we show that typical initial configurations, for the universe as a whole, evolve in such a way as to give rise to the *appearance* of randomness, with *empirical distributions* (Sects. 2.7 and 2.10) in agreement with the predictions of the quantum formalism. (Sects. 2.8–2.10 should perhaps be skipped at first reading.)

From a general perspective, perhaps the most noteworthy consequence of our analysis concerns *absolute uncertainty* (Sect. 2.11). In a universe governed by Bohmian mechanics there are sharp, precise, and irreducible limitations on the possibility of obtaining knowledge, limitations which can in no way be diminished through technological progress leading to better means of measurement.

This absolute uncertainty is in precise agreement with Heisenberg's uncertainty principle. But while Heisenberg used uncertainty to argue for the meaninglessness of particle trajectories, we find that, with Bohmian mechanics, absolute uncertainty arises as a necessity, emerging as a remarkably clean and simple consequence of the existence of trajectories. Thus quantum uncertainty, regarded as an experimental fact, is *explained* by Bohmian mechanics, rather than *explained away* as it is in orthodox quantum theory.

Our analysis covers all of nonrelativistic quantum mechanics. However, since our concern here is mainly conceptual, we shall for concreteness and simplicity consider only particles without spin, and shall ignore indistinguishability and the exclusion principle. Spin and permutation symmetry arise naturally in Bohmian mechanics [3, 7, 11, 12], and an analysis *explicitly* taking them into account would differ from the one given here in no essential way (see Chaps. 3 and 8).

In fact, our analysis really depends only on rather general qualitative features of the structure of abstract quantum theory, not on the details of any specific quantum theory—such as nonrelativistic quantum mechanics or a quantum field theory. In particular, the analysis does not require a particle ontology; a field ontology, for example, would do just as well.

Our analysis is, however, fundamentally nonrelativistic. It may well be the case that a fully relativistic generalization of the kind of physics explored here requires new concepts [13, 14, 15, 16]—if not new mathematical structures. But if one has not first understood the nonrelativistic case, one could hardly know where to begin for the relativistic one.

Perhaps this chapter should be read in the following spirit: In order to grasp the essence of Quantum Theory, one must first completely understand *at least one* quantum theory.

## 2.2 Reality and the Role of the Wave Function

> For each measurement one is required to ascribe to the $\psi$-function a characteristic, quite sudden change, which *depends on the measurement result obtained,* and so *cannot be forseen*; from which alone it is already quite clear that this second kind of change of the $\psi$-function has nothing whatever in common with its orderly development *between* two measurements. The abrupt change by measurement...is the most interesting point of the entire theory....For *this* reason one can *not* put the $\psi$-function directly in place of...the physical thing...because in the realism point of view observation is a natural process like any other and cannot *per se* bring about an interruption of the orderly flow of natural events. (Schrödinger [17])

The conventional wisdom that the wave function provides a complete description of a quantum system is certainly an attractive possibility: other things being equal, monism—the view that there is but one kind of reality—is perhaps more alluring than pluralism. But the problem of the origin of quantum randomness, described at the beginning of this chapter, already suggests that other things are not, in fact, equal.

Moreover, wave function monism suffers from another serious defect, to which the problem of randomness is closely related: Schrödinger's evolution tends to produce spreading over configuration space, so that the wave function $\psi$ of a macroscopic system will typically evolve to one supported by distinct, and vastly different, macroscopic configurations, to a grotesque macroscopic superposition, even if $\psi$ were originally quite prosaic. This is precisely what happens during a measurement, over the course of which the wave function describing the measurement process will become a superposition of components corresponding to the various apparatus readings to which the quantum formalism assigns nonvanishing probability. And the difficulty with this conception, of a world *completely* described by such an exotic wave function, is not even so much that it is extravagantly bizarre, but rather that this conception—or better our place in it, as well as that of the random events which the quantum formalism is supposed to govern—is exceedingly obscure. (What we have just described is often presented more colorfully as the paradox of Schrödinger's cat [17]).

What has just been said supports, not the impossibility of wave function monism, but rather its incompatibility with the Schrödinger evolution. And the allure of wave function monism is so strong that most interpretations of quantum mechanics in fact involve the abrogation of Schrödinger's equation. This abrogation is often merely implicit and, indeed, is often presented as if it were compatible with the quantum dynamics. This is the case, for example, when the measurement postulates, regarded as embodying "collapse of the wave packet," are simply combined with Schrödinger's equation in the formulation of quantum theory. The "measurement problem" is merely an expression of this inconsistency.

There have been several recent proposals—for example, by Wigner [18], by Leggett [19], by Stapp [16], by Weinberg [20] and by Penrose [21]—suggesting explicitly that the quantum evolution is not of universal validity, that under suitable conditions, encompassing those which prevail during measurements, the evolution of the wave function is not governed by Schrödinger's equation (see also [22]). A

## 2.2 Reality and the Role of the Wave Function

common suggestion is that the quantum dynamics should be replaced by some sort of "nonlinear" (possibly nondeterministic) modification, to which, on the microscopic level, it is but an extremely good approximation. One of the most concrete proposals along these lines is that of Ghirardi, Rimini, and Weber (GRW) [23].

The theory of GRW modifies Schrödinger's equation by the incorporation of a random "quantum jump," to a macroscopically localized wave function. As an explanation of the origin of quantum randomness it is thus not very illuminating, accounting, as it does, for the randomness in a rather ad hoc manner, essentially by fiat. Nonetheless this theory should be commended for its precision, and for the light it sheds on the relationship between Lorentz invariance and nonlocality (see [14]).

A related, but more serious, objection to proposals for the modification of Schrödinger's equation is the following: The quantum evolution embodies a deep mathematical beauty, which proclaims "Do not tamper! Don't degrade my integrity!" Thus, in view of the fact that (the relativistic extension of) Schrödinger's equation, or, better, the quantum theory, in which it plays so prominent a role, has been verified to a remarkable—and unprecedented—degree, these proposals for the modification of the quantum dynamics appear at best dubious, based as they are on purely conceptual, philosophical considerations.

But is wave function monism really so compelling a conception that we must struggle to retain it in the face of the formidable difficulties it entails? Certainly not! In fact, we shall argue that even if there were no such difficulties, even in the case of "other things being equal," a strong case can be made for the superiority of pluralism.

According to (pre-quantum-mechanical) scientific precedent, when new mathematically abstract theoretical entities are introduced into a theory, the physical significance of these entities, their very meaning insofar as physics is concerned, arises from their dynamical role, from the role they play in (governing) the evolution of the more primitive—more familiar and less abstract—entities or dynamical variables. For example, in classical electrodynamics the *meaning* of the electromagnetic field derives solely from the Lorentz force equation, i.e., from the field's role in governing the evolution of the positions of charged particles, through the specification of the forces, acting upon these particles, to which the field gives rise; while in general relativity a similar statement can be made for the gravitational metric tensor. That this should be so is rather obvious: Why would these abstractions be introduced in the first place, if not for their relevance to the behavior of *something else*, which somehow already has physical significance?

Indeed, it should perhaps be thought astonishing that the wave function was not also introduced in this way—insofar as it is a field on configuration space rather than on physical space, the wave function is an abstraction of even higher order than the electromagnetic field.

But, in fact, it was! The concept of the wave function originated in 1924 with de Broglie [24], who—intrigued by Einstein's idea of the "Gespensterfeld"—proposed that just as electromagnetic waves are somehow associated with particles, the photons, so should material particles, in particular electrons, be accompanied by waves. He conceived of these waves as "pilot waves," somehow governing the motion of the associated particles in a manner which he only later, in the late 1920s, made explicit

[25]. However, under an onslaught of criticism by Pauli, he soon abandoned his pilot wave theory, only to return to it more than two decades later, after his ideas had been rediscovered, extended, and vastly refined by Bohm [3, 4].

Moreover, in a paper written shortly after Schrödinger invented wave mechanics, Born too explored the hypothesis that the wave function might be a "guiding field" for the motion of the electron [26, 27]. As consequences of this hypothesis, Born was led in this paper both to his statistical interpretation of the wave function and to the creation of scattering theory. Born did not explicitly specify a guiding law, but he did insist that the wave function should somehow determine the motion of the electron only statistically, that deterministic guiding is impossible. And, like de Broglie, he later quickly abandoned the guiding field hypothesis, in large measure owing to the unsympathetic reception of Heisenberg, who insisted that physical theories be formulated directly in terms of observable quantities, like spectral lines and intensities, rather than in terms of microscopic trajectories.

The Copenhagen interpretation of quantum mechanics can itself be regarded as giving the wave function a role in the behavior of something else, namely of certain macroscopic objects, called "measurement instruments," during "quantum measurements" [28, 29]. Indeed, the most modest attitude one could adopt towards quantum theory would appear to be that of regarding it as a phenomenological formalism, roughly analogous to the thermodynamic formalism, for the description of certain *macroscopic* regularities. But it should nonetheless strike the reader as somewhat odd that the wave function, which appears to be the fundamental theoretical entity of the fundamental theory of what we normally regard as microscopic physics, should be assigned a role on the level of the macroscopic, itself an imprecise notion, and specifically in terms, even less precise, of measurements, rather than on the microscopic level.

Be that as it may, the modest position just described is not a stable one: It raises the question of how this phenomenological formalism arises from the behavior of the microscopic constituents of the macroscopic objects with which it is concerned. Indeed, this very question, in the context of the thermodynamic formalism, led to the development of statistical mechanics by Boltzmann and Gibbs, and, with some help from Einstein, eventually to the (almost) universal acceptance of the atomic hypothesis.

Of course, the Copenhagen interpretation is not quite so modest. It goes further, insisting upon the *impossibility* of just such an explanation of the (origin of the) quantum formalism. On behalf of this claim—which is really quite astounding in that it raises to a universal level the personal failure of a generation of physicists to find a satisfactory *objective* description of microscopic processes—the arguments which have been presented are not, in view of the rather dramatic conclusions that they are intended to establish, as compelling as might have been expected. Nonetheless, the very acceptance of these arguments by several generations of physicists should lead us to expect that, if not impossible, it should at best be extraordinarily difficult to account for the quantum formalism in objective microscopic terms.

Exhortations to the contrary notwithstanding, suppose that we do seek a microscopic origin for the quantum formalism, and that we do this by trying to find a

## 2.3 Bohmian Mechanics

role on the microscopic level for the wave function, relating it to the behavior of something else. How are we to proceed? A modest proposal: First try the obvious! Then proceed to the less obvious and, as is likely to be necessary, eventually to the not-the-least-bit-obvious. We shall implement this proposal here, and shall show that we need nothing but the obvious! (Insofar as nonrelativistic quantum mechanics is concerned.)

What we regard as the obvious choice of primitive ontology—the basic kinds of entities that are to be the building blocks of everything else (except, of course, the wave function)—should by now be clear: Particles, described by their positions in space, changing with time—some of which, owing to the dynamical laws governing their evolution, perhaps combine to form the familiar macroscopic objects of daily experience.

However, the *specific* role the wave function should play in governing the motion of the particles is perhaps not so clear, but for this, too, we shall find that there is a rather obvious choice, which when combined with Schrödinger's equation becomes Bohmian mechanics. (That an abstraction such as the wave function, for a many-particle system a field that is not on physical space but on configuration space, should be a fundamental theoretical entity in such a theory appears quite natural—as a compact expression of dynamical principles governing an evolution of *configurations*.)[3]

## 2.3 Bohmian Mechanics

> ...in physics the only observations we must consider are position observations, if only the positions of instrument pointers. It is a great merit of the de Broglie-Bohm picture to force us to consider this fact. If you make axioms, rather than definitions and theorems, about the 'measurement' of anything else, then you commit redundancy and risk inconsistency. (Bell [30])

Consider a quantum system of $N$ particles, with masses $m_1, \ldots, m_N$ and position coordinates $\mathbf{q}_1, \ldots, \mathbf{q}_N$, whose wave function $\psi = \psi(\mathbf{q}_1, \ldots, \mathbf{q}_N, t)$ satisfies Schrödinger's equation

$$i\hbar \frac{\partial \psi}{\partial t} = -\sum_{k=1}^{N} \frac{\hbar^2}{2m_k} \Delta_k \psi + V\psi, \qquad (2.1)$$

where $\Delta_k = \boldsymbol{\nabla}_k \cdot \boldsymbol{\nabla}_k = \partial/\partial \mathbf{q}_k$ and $V = V(\mathbf{q}_1, \ldots, \mathbf{q}_N)$ is the potential energy of the system.

Suppose that the wave function $\psi$ does not provide a complete description of the system, that the most basic ingredient of the description of the state at a given

---

[3] However, *with wave function monism,* without such a role and, indeed, without particle positions from which to form configurations, how can we make sense of a field on the *space of configurations*? We might well ask *"What configurations?"* (And the wave function really is on configuration space—it is in this representation that quantum mechanics assumes its simplest form!)

time $t$ is provided by the positions $\mathbf{Q}_1, \ldots, \mathbf{Q}_N$ of its particles at that time, and that the wave function governs the *evolution* of (the positions of) these particles. (Note that we use $q = (\mathbf{q}_1, \ldots, \mathbf{q}_N)$ as the *generic* configuration space variable, which, to avoid confusion, we distinguish from the *actual* configuration of the particles, for which we usually use capitals.)

Insofar as first derivatives are simpler than higher derivatives, the simplest possibility would appear to be that the wave function determine the *velocities* $\mathbf{v}_1^\psi, \ldots, \mathbf{v}_N^\psi$ of all the particles. Here $\mathbf{v}_k^\psi \equiv \mathbf{v}_k^\psi(\mathbf{q}_1, \ldots, \mathbf{q}_N)$ is a velocity vector field, on *configuration space*, for the $k$-th particle, i.e.,

$$\frac{d\mathbf{Q}_k}{dt} = \mathbf{v}_k^\psi(\mathbf{Q}_1, \ldots, \mathbf{Q}_N) \tag{2.2}$$

Since (2.1) and (2.2) are first order differential equations, it would then follow that the state of the system is indeed given by $\psi$ and $Q \equiv (\mathbf{Q}_1, \ldots, \mathbf{Q}_N)$—the specification of these variables at any time would determine them at all times.

Since two wave functions of which one is a nonzero constant multiple of the other should be physically equivalent, we demand that $\mathbf{v}_k^\psi$ be homogeneous of degree 0 as a function of $\psi$,

$$\mathbf{v}_k^{c\psi} = \mathbf{v}_k^\psi \tag{2.3}$$

for any constant $c \neq 0$.

In order to arrive at a form for $\mathbf{v}_k^\psi$ we shall use symmetry as our main guide. Consider first a single free particle of mass $m$, whose wave function $\psi(\mathbf{q})$ satisfies the free Schrödinger equation

$$i\hbar \frac{\partial \psi}{\partial t} = -\frac{\hbar^2}{2m} \Delta \psi. \tag{2.4}$$

We wish to choose $\mathbf{v}^\psi$ in such a way that the system of equations given by (2.4) and

$$\frac{d\mathbf{Q}}{dt} = \mathbf{v}^\psi(\mathbf{Q}) \tag{2.5}$$

is Galilean and time-reversal invariant. (Note that a first-order (Aristotelian) Galilean invariant theory of particle motion may *appear* to be an oxymoron.) Rotation invariance, with the requirement that $\mathbf{v}^\psi$ be homogeneous of degree 0, yields the form

$$\mathbf{v}^\psi = \alpha \frac{\nabla \psi}{\psi},$$

where $\alpha$ is a constant scalar, as the simplest possibility.

This form will not in general be real, so that we should perhaps take real or imaginary parts. Time-reversal is implemented on $\psi$ by the involution $\psi \to \psi^*$ of complex conjugation, which renders Schrödinger's equation time reversal invariant. If the full system, including (2.5), is also to be time-reversal invariant, we must thus

## 2.3 Bohmian Mechanics

have that

$$\mathbf{v}^{\psi^*} = -\mathbf{v}^\psi, \tag{2.6}$$

which selects the form

$$\mathbf{v}^\psi = \alpha \operatorname{Im} \frac{\nabla \psi}{\psi} \tag{2.7}$$

with $\alpha$ real.

Moreover the constant $\alpha$ is determined by requiring full Galilean invariance: Since $\mathbf{v}^\psi$ must transform like a velocity under boosts, which are implemented on wave functions by $\psi \mapsto \exp\left[(im/\hbar)\mathbf{v}_0 \cdot \mathbf{q}\right]\psi$, invariance under boosts requires that $\alpha = \hbar/m$, so that (2.7) becomes

$$\mathbf{v}^\psi = \frac{\hbar}{m} \operatorname{Im} \frac{\nabla \psi}{\psi}. \tag{2.8}$$

For a general $N$-particle system, with general potential energy $V$, we define the velocity vector field by requiring (2.8) for each particle, i.e., by letting

$$\mathbf{v}_k^\psi = \frac{\hbar}{m_k} \operatorname{Im} \frac{\nabla_k \psi}{\psi}, \tag{2.9}$$

so that (2.2) becomes

$$\frac{d\mathbf{Q}_k}{dt} = \frac{\hbar}{m_k} \operatorname{Im} \frac{\nabla_k \psi}{\psi}(\mathbf{Q}_1, \ldots, \mathbf{Q}_N) \tag{2.10}$$

We've arrived at *Bohmian mechanics*: for our system of $N$ particles the state is given by

$$(Q, \psi) \tag{2.11}$$

and the evolution by

$$\begin{aligned}\frac{d\mathbf{Q}_k}{dt} &= \frac{\hbar}{m_k} \operatorname{Im} \frac{\nabla_k \psi}{\psi}(\mathbf{Q}_1, \ldots, \mathbf{Q}_N) \\ i\hbar \frac{\partial \psi}{\partial t} &= -\sum_{k=1}^{N} \frac{\hbar^2}{2m_k} \Delta_k \psi + V\psi.\end{aligned} \tag{2.12}$$

We note that Bohmian mechanics is time-reversal invariant, and that it is Galilean invariant whenever $V$ has this property, e.g., when $V$ is the sum of a pair interaction of the usual form,

$$V(\mathbf{q}_1, \ldots, \mathbf{q}_N) = \sum_{i<j} \phi(|\mathbf{q}_i - \mathbf{q}_j|). \tag{2.13}$$

However, our analysis will not depend on the form of $V$.

Note also that Bohmian mechanics depends only upon the Riemannian structure $g = (g_{ij}) = (m_i \delta_{ij})$ defined by the masses of the particles: In terms of this Riemannian structure, the evolution Eqs. (2.1 and 2.10) of Bohmian mechanics become

$$\frac{dQ}{dt} = \hbar \, \text{Im} \, \frac{\text{grad } \psi}{\psi}(Q)$$
$$i\hbar \frac{\partial \psi}{\partial t} = -\frac{\hbar^2}{2} \Delta \psi + V \psi, \qquad (2.14)$$

where $Q = (\mathbf{Q}_1, \ldots, \mathbf{Q}_N)$ is the configuration, and $\Delta$ and grad are, respectively, the Laplace-Beltrami operator and the gradient on the configuration space equipped with this Riemannian structure. (For more detail, see Chap. 8.)

While Bohmian mechanics shares Schrödinger's equation with the usual quantum formalism, it might appear that they have little else in common. After all, the former is a theory of particles in motion, albeit of an apparently highly nonclassical, non-Newtonian character; while the observational content of the latter derives from a calculus of noncommuting "observables," usually regarded as implying radical epistemological innovations. Indeed, if the coefficient in the first equation of (2.12) were other than $\hbar/m_k$, i.e., for general constants $\alpha_k$, the corresponding theory would have little else in common with the quantum formalism. But for the particular choice of $\alpha_k$, of the coefficient in (2.12), which defines Bohmian mechanics, the quantum formalism itself emerges as a phenomenological consequence of this theory.

What makes the choice $\alpha_k = \hbar/m_k$ special—apart from Galilean invariance, which plays little or no role in the remainder of this chapter—is that with this value, the probability distribution on configuration space given by $|\psi(q)|^2$ possesses the property of equivariance, a concept to which we now turn.

Note well that $\psi$ on the right hand side of (2.2) or (2.10) is a solution to Schrödinger's equation (2.1) and is thus time-dependent, $\psi = \psi(t)$. It follows that the vector field $\mathbf{v}_k^\psi$, the right hand side of (2.10), will in general be (explicitly) time-dependent. Therefore, given a solution $\psi$ to Schrödinger's equation, we cannot in general expect the evolution on configuration space defined by (2.10) to possess a stationary probability distribution, an object which very frequently plays an important role in the analysis of a dynamical system.

However, the distribution given by $|\psi(q)|^2$ plays a role similar to that of—and for all practical purposes is just as good as—a stationary one: Under the evolution $\rho(q, t)$ of probability densities, of ensemble densities, arising from (2.10), given by the continuity equation

$$\frac{\partial \rho}{\partial t} + \text{div}\,(\rho v^\psi) = 0 \qquad (2.15)$$

with $v^\psi = (\mathbf{v}_1^\psi, \ldots, \mathbf{v}_N^\psi)$ the configuration space velocity arising from $\psi$ and div the divergence on configuration space, the density $\rho = |\psi|^2$ is stationary *relative to $\psi$*, i.e., $\rho(t)$ retains its form as a functional of $\psi(t)$. In other words,

*if $\rho(q, t_0) = |\psi(q, t_0)|^2$ at some time $t_0$, then $\rho(q, t) = |\psi(q, t)|^2$ for all $t$.* (2.16)

## 2.3 Bohmian Mechanics

We say that such a distribution is *equivariant*.[4]
To see that $|\psi|^2$ is, in fact, equivariant observe that

$$J^\psi = |\psi|^2 v^\psi \tag{2.17}$$

where $J^\psi = (\mathbf{J}_1^\psi, \ldots, \mathbf{J}_N^\psi)$ is the quantum probability current,

$$\mathbf{J}_k^\psi = \frac{\hbar}{2im_k}(\psi^* \boldsymbol{\nabla}_k \psi - \psi \boldsymbol{\nabla}_k \psi^*), \tag{2.18}$$

which obeys the quantum continuity equation

$$\frac{\partial |\psi|^2}{\partial t} + \text{div}(J^\psi) = 0 \tag{2.19}$$

as a consequence of Schrödinger's equation (2.1). Thus $\rho(q,t) = |\psi(q,t)|^2$ satisfies (2.15).

Now consider a quantum measurement, involving an interaction between a system "under observation" and an apparatus which performs the "observation." Let $\psi$ be the wave function and $q = (q_{sys}, q_{app})$ the configuration of the composite system of system and apparatus. Suppose that prior to the measurement, at time $t_i$, $q$ is random, with probability distribution given by $\rho(q, t_i) = |\psi(q, t_i)|^2$. When the measurement has been completed, at time $t_f$, the configuration at this time will, of course, still be random, as will typically be the outcome of the measurement, as given by appropriate apparatus variables, for example, by the orientation of a pointer on a dial or by the pattern of ink marks on paper. Moreover, by equivariance, the distribution of the configuration q at time $t_f$ will be given by $\rho(q, t_f) = |\psi(q, t_f)|^2$, in agreement with the prediction of the quantum formalism for the distribution of $q$ at this time. In particular, Bohmian mechanics and the quantum formalism then agree on the statistics for the outcome of the measurement.[5]

---

[4] More generally, and more precisely, we say that a functional $\psi \to \mu^\psi$, from wave functions to finite measures on configuration space, is *equivariant* if the diagram

$$\begin{array}{ccc} \psi & \longrightarrow & \mu^\psi \\ U_t \downarrow & & \downarrow F_t^\psi \\ \psi_t & \longrightarrow & \mu^{\psi_t} \end{array}$$

is commutative, where $U_t = \exp[-(i/\hbar)tH]$, with Hamiltonian $H = -\Sigma_{k=1}^N (\hbar^2/2m_k)\Delta_k \psi + V\psi$, is the solution map for Schrödinger's equation and $F_t^\psi$ is the solution map for the natural evolution on measures which arises from (2.10), with initial wave function $\psi$. ($F_t^\psi(\mu)$ is the measure to which $\mu$ evolves in $t$ units of time when the initial wave function is $\psi$).

[5] This argument appears to leave open the possibility of disagreement when the outcome of the measurement is not configurationally grounded, i.e., when the apparatus variables which express this outcome are not functions of $q_{app}$. However, the reader should recall Bohr's insistence that the outcome of a measurement be describable in classical terms, as well as note that results of measurements must always be at least *potentially* grounded configurationally, in the sense that we

## 2.4 The Problem of Quantum Equilibrium

> Then for instantaneous macroscopic configurations the pilot-wave theory gives the same distribution as the orthodox theory, insofar as the latter is unambiguous. However, this question arises: what is the good of *either* theory, giving distributions over a hypothetical ensemble (of worlds!) when we have only one world. (Bell [31])

Suppose a system has wave function $\psi$. We shall call the probability distribution on configuration space given by $\rho = |\psi|^2$ the *quantum equilibrium* distribution. And we shall say that a system is *in quantum equilibrium* when its coordinates are "randomly distributed" according to the quantum equilibrium distribution. As we have seen, when a system and apparatus are in quantum equilibrium the results of "measurement" arising from the interaction between system and apparatus will conform with the predictions of the quantum formalism for such a measurement.

More precisely(!), we say that a system is in quantum equilibrium when the quantum equilibrium distribution is appropriate for its description. It is a major goal of this chapter to explain what exactly this might mean and to show that, indeed, when understood properly, it is *typically* the case that systems are in quantum equilibrium. In other words, our goal here is to clarify and justify the *quantum equilibrium hypothesis*:

When a system has wave function $\psi$, the distribution $\rho$ of its coordinates satisfies

$$\rho = |\psi|^2. \tag{2.20}$$

We shall do this in the later sections of this chapter. In the rest of this section we will elaborate on the *problem* of quantum equilibrium.

From a dynamical systems perspective, it would appear natural to attempt to justify (2.20) using such notions as "convergence to equilibrium," "mixing," or "ergodicity"—suitably generalized. And if it were in fact necessary to establish such properties for Bohmian mechanics in order to justify the quantum equilibrium hypothesis, we could not reasonably expect to succeed, at least not with any degree of rigor. The problem of establishing good ergodic properties for nontrivial dynamical systems is extremely difficult, even for highly simplified, less than realistic, models.

It might seem that Bohmian mechanics rather trivially *fails* to possess good ergodic properties, if one considers the motion arising from the standard energy eigenstates of familiar systems. However, quantum systems attain such simple wave functions only through complex interactions, for example with an apparatus during a measurement or preparation procedure, during which time they are not governed by a *simple* wave function. Thus the question of the ergodic properties of Bohmian mechanics refers to the motion under generic, more complex, wave functions.

We shall show, however, that establishing such properties is neither necessary nor sufficient for our purposes: That it is not necessary follows from the analysis in the

---

can arrange that they be recorded in configurational terms *without affecting the result*. Otherwise we could hardly regard the process leading to the original result as a completed measurement.

## 2.4 The Problem of Quantum Equilibrium

later sections of this chapter, and that it would not be sufficient follows from the discussion to which we now turn.

The reader may wonder why the quantum equilibrium hypothesis should present any difficulty at all. Why can we not regard it as an additional postulate, on say initial conditions (in analogy with equilibrium statistical mechanics, where the Gibbs distribution is often uncritically accepted as axiomatic)? Then, by equivariance, it will be preserved by the dynamics, so that we obtain the quantum equilibrium hypothesis for all times. In fact, when all is said and done, we shall find that this is an adequate description of the situation *provided the quantum equilibrium hypothesis is interpreted in the appropriate way*. But for the quantum equilibrium hypothesis as so far formulated, such an account would be grossly inadequate.

Note first that the quantum equilibrium hypothesis relates objects belonging to rather different conceptual categories: The right hand side of (2.20) refers to a dynamical object, which from the perspective of Bohmian mechanics is of a thoroughly objective character; while the left refers to a probability distribution—an object whose *physical* significance remains mildly obscure and moderately controversial, and which often is regarded as having a strongly subjective aspect. Thus, some explanation or justification is called for.

One very serious difficulty with (2.20) is that it *seems* to be demonstrably false in a great many situations. For example, the wave function—of system and apparatus—after a measurement (arising from Schrödinger's equation) is supported by the set of all configurations corresponding to the *possible* outcomes of the measurement, while the probability distribution at this time is supported only by those configurations corresponding to the *actual* outcome, e.g., given by a specific pointer position, a main point of measurement being to obtain the information upon which this probability distribution is grounded.

This difficulty is closely related to an ambiguity in the domain of physical applicability of Bohmian mechanics. In order to avoid inconsistency we must regard Bohmian mechanics as describing the entire universe, i.e., our system should consist of all particles in the universe: The behavior of parts of the universe, of subsystems of interest, must arise from the behavior of the whole, evolving according to Bohmian mechanics. It turns out, as we shall show, that subsystems are themselves, in fact, frequently governed by Bohmian mechanics. But if we *postulate* that subsystems must obey Bohmian mechanics, we "commit redundancy and risk inconsistency."

Note also that the very nature of our concerns—the origin and justification of (local) randomness—forces us to consider the universal level: Local systems are not (always and are never entirely) isolated. Recall that cosmological considerations similarly arise in connection with the problem of the origin of irreversibility (see R. Penrose [32]).

Thus, strictly speaking, for Bohmian mechanics only the universe has a wave function, since the *complete* state of an $N$ particle universe at any time is given by *its* wave function $\psi$ and the configuration $Q = (\mathbf{Q}_1, \ldots, \mathbf{Q}_N)$ of its particles. (The notion of the wave function of a subsystem will be the concern of the next section. However, for a smoother and more straightforward presentation, see Sect. 12.2.) Therefore the right hand side of the quantum equilibrium hypothesis (2.20) is also obscure as soon as it refers to a system smaller than the entire universe—and the systems to which (2.20) is normally applied are very small indeed, typically microscopic.

Suppose, as suggested earlier, we consider (2.20) for the entire universe. Then the right hand side is clear, but the left is completely obscure: Focus on (2.20) for THE INITIAL TIME. What physical significance can be assigned to a probability distribution on the initial configurations for the entire universe? What can be the relevance to physics of such an ensemble of universes? After all, we have at our disposal only the particular, actual universe of which we are a part. Thus, even if we could make sense of the right hand side of (2.20), and in such a way that (2.20) remains a consequence of the quantum equilibrium hypothesis at THE INITIAL TIME, we would still be far from our goal, appearances to the contrary notwithstanding.

Since the inadequacy of the quantum equilibrium hypothesis regarded as describing an ensemble of universes is a crucial point, we wish to elaborate. For each choice of initial universal wave function $\psi$ and configuration $Q$, a "history"—past, present, and future—is completely determined. In particular, the results of all experiments, including quantum measurements, are determined.

Consider an ensemble of universes initially satisfying (2.20), and suppose that it can be shown that for this ensemble the outcome of a particular experiment is randomly distributed with distribution given by the quantum formalism. This would tell us only that if we were to repeat the *very same* experiment—whatever this might mean—many times, sampling from our ensemble of universes, we would obtain the desired distribution. But this is both impossible and devoid of physical significance: While we *can* perform many *similar* experiments, differing, however, at the very least, by location or time, we cannot perform the very same experiment more than once.

What we need to know about, if we are to make contact with physics, is *empirical distributions*—actual relative frequencies within an ensemble of actual events—arising from repetitions of similar experiments, performed at different places or times, within a single sample of the universe—the one we are in. In other words, what is physically relevant is not sampling across an ensemble of universes—across (initial) $Q$'s—but sampling across space and time within a single universe, corresponding to a fixed (initial) $Q$ (and $\psi$).

Thus, to demonstrate the compatibility of Bohmian mechanics with the predictions of the quantum formalism, we must show that for at least some choice of initial universal $\psi$ and $Q$, the evolution (2.12) leads to an apparently random pattern of events, with empirical distribution given by the quantum formalism. In fact, we show much more.

We prove that for *every* initial $\psi$, this agreement with the predictions of the quantum formalism is obtained for *typical*—i.e., for the overwhelming majority of—choices of initial $Q$. And the sense of typicality here is with respect to the only mathematically natural—because equivariant—candidate at hand, namely, quantum equilibrium.

Thus, on the universal level, the physical significance of quantum equilibrium is as a measure of typicality, and the ultimate justification of the quantum equilibrium hypothesis is, as we shall show, in terms of the statistical behavior arising from a typical initial configuration.

According to the usual understanding of the quantum formalism, when a system has wave function $\psi$, (2.20) is satisfied *regardless of whatever additional information we might have*. When we claim to have established agreement between Bohmian mechanics and the predictions of the quantum formalism, we mean to include this statement among those predictions. We are thus claiming to have established that in a universe governed by Bohmian mechanics it is in principle impossible to know more about the configuration of any subsystem than what is expressed by (2.20)—despite the fact that for Bohmian mechanics the actual configuration is an *objective* property, beyond the wave function.

This may appear to be an astonishing claim, particularly since it refers to knowledge, a concept both vague and problematical, in an essential way. More astonishing still is this: This uncertainty, of an absolute and precise character, emerges with complete ease, the structure of Bohmian mechanics being such that it allows for the formulation and clean demonstration of statistical statements of a purely *objective* character which nonetheless imply our claims concerning the irreducible limitations on possible knowledge *whatever this "knowledge" may precisely mean, and however we might attempt to obtain this knowledge*, provided it is consistent with Bohmian mechanics. We shall therefore call this limitation on what can be known *absolute uncertainty*.

## 2.5 The Effective Wave Function

*No one can understand this theory until he is willing to think of $\psi$ as a real objective field rather than just a 'probability amplitude.' Even though it propagates not in 3-space but in 3N-space.* (Bell [31])

We now commence our more detailed analysis of the behavior of an $N$-particle nonrelativistic universe governed by Bohmian mechanics, focusing in this section on the notion of the effective wave function of a subsystem. We begin with some notation.

We shall use $\Psi$ as the variable for the universal wave function, reserving $\psi$ for the effective wave function of a subsystem, the definition and clarification of which is the aim of this section. By $\Psi_t = \Psi_t(q)$ we shall denote the universal wave function at time $t$. We shall denote the configuration of the universe at time $t$ by $Q_t$.

We remind the reader that according to Bohmian mechanics the state $(Q_t, \Psi_t)$ of the universe at time $t$ evolves via

$$\frac{dQ_t}{dt} = v^{\Psi_t}(Q_t)$$
$$i\hbar \frac{\partial \Psi_t}{\partial t} = -\sum_{k=1}^{N} \frac{\hbar^2}{2m_k} \Delta_k \Psi_t + V\Psi_t, \qquad (2.21)$$

where $v^{\Psi} = (\mathbf{v}_1^{\Psi}, \ldots, \mathbf{v}_N^{\Psi})$ with $\mathbf{v}_k^{\Psi}$ defined by (2.9).

For any given subsystem of particles we obtain a splitting

$$q = (x, y), \qquad (2.22)$$

with $x$ the generic variable for the configuration of the subsystem and $y$ the generic variable for the configuration of the complementary subsystem, formed by the particles not in the given subsystem. We shall call the given subsystem the $x$-*system*, and we shall sometimes call its complement—the $y$-*system*—the *environment* of the $x$-system.[6]

Of course, for any splitting (2.22) we have a splitting

$$Q = (X, Y) \qquad (2.23)$$

for the actual configuration. And for the wave function $\Psi$ we may write $\Psi = \Psi(x, y)$.

Frequently the subsystem of interest naturally decomposes into smaller subsystems. For example, we may have

$$x = (x_{sys}, x_{app}), \qquad (2.24)$$

for the composite formed by system and apparatus, or

$$x = (x_1, \ldots, x_M), \qquad (2.25)$$

for the composite formed from $M$ disjoint subsystems. And, of course, any of the $x_i$ in (2.25) could be of the form (2.24).

Consider now a subsystem with associated splitting (2.22). We wish to explore the circumstances under which we may reasonably regard this subsystem as "itself having a wave function." This will serve as motivation for our definition of the effective wave function of this subsystem. To this end, suppose first that the universal wave function factorizes so that

$$\Psi(x, y) = \psi(x)\Phi(y). \qquad (2.26)$$

Then we obtain the splitting

$$v^\Psi = (v^\psi, v^\Phi), \qquad (2.27)$$

and, in particular, we have that

$$\frac{dX}{dt} = v^\psi(X) \qquad (2.28)$$

---

[6] While we have in mind the situation in which the $x$-system consists of a set of particles selected by their labels, what we say would not be (much) affected if the $x$-system consisted, say, of all particles in a given region. In fact the splitting (2.22) could be more general than one based upon what we would normally regard as a division into complementary systems of particles; for example, the $x$-system might include the center of mass of some collection of particles, while the $y$-system includes the relative coordinates for this collection.

## 2.5 The Effective Wave Function

for as long as (2.26) is satisfied. Moreover, to the extent that the interaction between the $x$-system and its environment can be ignored, i.e., that the Hamiltonian

$$H = -\sum_{k=1}^{N} \frac{\hbar^2}{2m_k} \Delta_k + V \tag{2.29}$$

in (2.21) can be regarded as being of the form

$$H = H^{(x)} + H^{(y)} \tag{2.30}$$

where $H^{(x)}$ and $H^{(y)}$ are the contributions to $H$ arising from terms involving only the particle coordinates of the $x$-system, respectively, the $y$-system,[7] the form (2.26) is preserved by the evolution, with $\psi$, in particular, evolving via

$$i\hbar \frac{d\psi}{dt} = H^{(x)}\psi. \tag{2.31}$$

It must be emphasized, however, that the factorization (2.26) is extremely unphysical. After all, interactions between system and environment, which tend to destroy the factorization (2.26), are commonplace. In particular, they occur whenever a measurement is performed on the $x$-system. Thus, the universal wave function $\Psi$ should now be of an extremely complex form, involving intricate "quantum correlations" between $x$-system and $y$-system, however simple it may have been originally!

Note, however, that if

$$\Psi = \Psi^{(1)} + \Psi^{(2)} \tag{2.32}$$

with the wave functions on the right having (approximately[8]). disjoint supports, then (approximately)

$$v^{\Psi}(Q) = v^{\Psi^{(i)}}(Q) \tag{2.33}$$

for $Q$ in the support of $\Psi^{(i)}$. Of course, by mere linearity, if $\Psi$ is of the form (2.32) at some time $\tau$, it will be of the same form

$$\Psi_t = \Psi_t^{(1)} + \Psi_t^{(2)} \tag{2.34}$$

---

[7] The sense of the approximation expressed by (2.30) is somewhat delicate. In particular, (2.30) should not be regarded as a condition on $H$ (or $V$) so much as a condition on (the supports of) the factors $\psi$ and $\Phi$ of the wave function $\Psi$ whose evolution is governed by $H$; namely, that these supports be sufficiently well separated that all contributions to $V$ involving both particle coordinates in the support of $\psi$ and particle coordinates in the support of $\Phi$ are so small that they can be neglected when $H$ is applied to such a $\Psi$.

[8] In an appropriate sense, of course. Note in this regard that the simplest metrics $d$ on the projective space of rays $\{c\Psi\}$ are of the form $d(\Psi, \Psi') = \|\frac{\nabla \Psi}{\Psi} - \frac{\nabla \Psi'}{\Psi'}\|$, where "$\| \ \|$" is a norm on the space of complex vector fields on configuration space. Moreover the metric $d$ is preserved by the space-time symmetries (when "$\| \ \|$" is translation and rotation invariant)

for all t, where $\Psi_t^{(i)}$ is the solution agreeing with $\Psi^{(i)}$ at time $\tau$ of the second equation of (2.21). Moreover, if the supports of $\Psi^{(1)}$ and $\Psi^{(2)}$ are "sufficiently disjoint" at this time, we should expect the approximate disjointness of these supports, and hence the approximate validity of (2.33), to persist for a "substantial" amount of time.

Finally, we note that according to orthodox quantum measurement theory [2, 28, 33, 34], after a measurement, or preparation, has been performed on a quantum system, the wave function for the composite formed by system and apparatus is of the form

$$\sum_\alpha \psi_\alpha \otimes \phi_\alpha \qquad (2.35)$$

with the different $\phi_\alpha$ supported by the macroscopically distinct (sets of) configurations corresponding to the various possible outcomes of the measurement, e.g., given by apparatus pointer positions. Of course, for Bohmian mechanics the terms of (2.35) are not all on the same footing: one of them, and only one, is selected, or more precisely supported, by the outcome—corresponding, say, to $\alpha_0$—which *actually* occurs. To emphasize this we may write (2.35) in the form

$$\psi \otimes \phi + \Psi^\perp \qquad (2.36)$$

where $\psi = \psi_{\alpha_0}, \phi = \phi_{\alpha_0}$, and $\Psi^\perp = \sum_{\alpha \neq \alpha_0} \psi_\alpha \otimes \phi_\alpha$.

Motivated by these observations, we say that a subsystem, with associated splitting (2.22), has *effective wave function* $\psi$ (at a given time) if the universal wave function $\Psi = \Psi(x, y)$ and the actual configuration $Q = (X, Y)$ (at that time) satisfy

$$\Psi(x, y) = \psi(x)\Phi(y) + \Psi^\perp(x, y) \qquad (2.37)$$

with $\Phi$ and $\Psi^\perp$ having macroscopically disjoint $y$-supports, and

$$Y \in \operatorname{supp} \Phi. \qquad (2.38)$$

Here, by the macroscopic disjointness of the $y$-supports of $\Phi$ and $\Psi^\perp$ we mean not only that their supports are disjoint but that there is a macroscopic function of $y$—think, say, of the orientation of a pointer—whose values for $y$ in the support of $\Phi$ differ by a *macroscopic* amount from its values for $y$ in the support of $\Psi^\perp$.

Readers familiar with quantum measurement theory should convince themselves (see (2.35 and 2.36)) that our definition of effective wave function coincides with the usual practice of the quantum formalism in ascribing wave functions to systems *whenever the latter does assign a wave function*. In particular, whenever a system has a wave function for orthodox quantum theory, it has an effective wave function for Bohmian mechanics.[9] However, there may well be situations in which a system has an effective wave function according to Bohmian mechanics, but the standard quantum

---

[9] Note that the $x$-system will not have an effective wave function—even approximately—when, for example, it belongs to a larger microscopic system whose effective wave function does not factorize in the appropriate way. Note also that the *larger* the environment of the $x$-system, the *greater* is the potential for the existence of an effective wave function for this system, owing in effect to the

## 2.5 The Effective Wave Function

formalism has nothing to say. (We say "may well be" because the usual quantum formalism is too imprecise and too controversial insofar as these questions—for which "collapse of the wave packet" must in some ill-defined manner be invoked—are concerned to allow for a more definite statement.) Readers who are not familiar with quantum measurement theory can—as a consequence of our later analysis—simply replace whatever vague notion they may have of the wave function of a system with the more precise notion of effective wave function.

Despite the slight vagueness in the definition of effective wave function, arising from its reference to the imprecise notion of the macroscopic, the effective wave function, when it exists, is unambiguous. In fact, it is given by the *conditional wave function* (we identify wave functions related by a nonzero constant factor)

$$\psi(x) = \Psi(x, Y), \tag{2.39}$$

which, moreover, is (almost) always defined (assuming continuity, which, of course, we must). In fact, the main result of this chapter, concerning the *statistical* properties of subsystems, remains valid when the notion of effective wave function is replaced by the completely precise, and less restrictive, formulation provided by the conditional wave function (2.39).[10]

It follows from (2.39) that when the after-measurement wave function of system and apparatus has the form (2.35), the conditional wave function of the system is one of the wave functions $\psi_\alpha$, namely the one corresponding to the outcome that actually occurs, with $\alpha$ such that the actual configuration of the apparatus is in the support of $\phi_\alpha$. Connecting this to the quantum formalism, when $\psi_\alpha$ is the projection of the initial system wave function onto the subspace of the eigenstates of a measured observable corresponding to $\alpha$, this corresponds to the usual collapse rule of quantum mechanics.

Note that by virtue of the first equation of (2.21), the velocity vector field for the $x$-system is generated by its conditional wave function. However, the conditional wave function will not in general evolve (even approximately) according to Schrödinger's equation, even when the $x$-system is dynamically decoupled from its environment. Thus (2.39) by itself lacks the central *dynamical* implications, as suggested by the preliminary discussion, of our definition (2.37, 2.38). And it is of course from these dynamical implications that the wave function of a system derives much of its physical significance.[11]

---

greater abundance of "measurement-like" interactions with a larger environment (see, for example, Point 20 of the Appendix and the references therein).

[10] We therefore need not be too concerned here by the fact that our definition is also somewhat unrealistic, in the sense that in situations where we would in practice say that a system has wave function $\psi$, the terms on the right hand side of (2.37) are only approximately disjoint, or, what amounts to the same thing, the first term on the right is only approximately of the product form, though to an enormously good degree of approximation.

[11] In this regard note the following: Let $W^Y(x) = V_I(x, Y)$, where $V_I$ is the contribution to $V$ arising from the terms which represent interactions between the $x$-system and the $y$-system, i.e., $H = H^x + H^{(y)} + V_I$. Suppose that $W^Y$ does not depend upon $Y$ for $Y$ in the support of $\Phi$, $W^Y = W$ for $Y \in \text{supp } \Phi$. Then the effective wave function $\psi$ satisfies $i\hbar(d\psi/dt) = (H^{(x)} + W)\psi$. The

Note well that the notion of effective wave function, or conditional wave function, is made possible by the existence of the *actual* configuration $Q = (X, Y)$ as well as $\Psi$! (In particular, the effective—or conditional—wave function is *objective,* while a related notion in Everett's Many-Worlds or Relative State interpretation of quantum theory [35] is merely *relative*. For an incisive critique of the Many-Worlds interpretation, as well as a detailed comparison with Bohmian mechanics, see Bell [31, 36].) Note also that the conditional wave function is the function of $x$ most naturally arising from $\Psi$ and $Y$.[12] (For more on the notion of conditional wave function, see Sects. 3.2.2 and 12.2.)

We emphasize that the effective wave function—as well as the conditional wave function—is, like any honest to goodness attribute or objective property, a functional of state description, here a function-valued functional of $\Psi$ and $Q = (X, Y)$ which depends on $Q$ only through $Y$. We shall sometimes write

$$\psi = \psi^{Y,\Psi} \tag{2.40}$$

to emphasize this relationship. For the conditional or effective wave function at time $t$ we shall sometimes write

$$\psi_t = \psi^{Y_t,\Psi_t} \equiv \psi_t^{Y_t}, \tag{2.41}$$

suppressing the dependence upon $\Psi$.

Note that though we speak of $\psi$ as a property of the $x$-system, it depends not upon the coordinates of the $x$-system but only upon the environment, a distinctly peculiar situation from a classical perspective. In fact, it is precisely because of this that the effective wave function behaves like a degree of freedom for the $x$-system which is independent of its configuration $X$.

Consider now a composite $x = (x_1, \ldots, x_M)$ of *microscopic* subsystems, with $M$ not too large, i.e., not "macroscopically large." Suppose that (simultaneously) each $x_i$-system has effective wave function $\psi_i$. Then the $x$-system has effective wave function

$$\psi(x) = \psi_1(x_1)\psi_2(x_2) \cdots \psi_M(x_M), \tag{2.42}$$

in agreement with the quantum formalism.[13] To see this, note that for each $i$ we have

---

reader should think, for example, of a gas confined by the walls of a box, or of a particle moving among obstacles. The interaction of the gas or the particle with the walls or the obstacles—which after all are part of the environment—is expressed thru $W$.

[12] For particles with spin our definition (2.37, 2.38) needs no essential modification. However, (2.39) would have to be replaced by $\Psi(x, Y) = \psi(x) \otimes \Phi$, where "$\otimes$" here denotes the tensor product over the spin degrees of freedom. In particular, for particles with spin, a subsystem need not have even a conditional wave function. But it will always have a conditional density matrix, see [37].

[13] As far as the quantum formalism is concerned, recall that from a purely operational perspective, whatever procedure simultaneously prepares each system in the corresponding quantum state is a preparation of the product state for the composite. Moreover, an analysis of such a simultaneous preparation in terms of quantum measurement theory would, of course, lead to the same conclusion. Note also that if the $x$-system is described by a density matrix whose reduced density matrix for

## 2.5 The Effective Wave Function

that

$$\Psi = \psi_i(x_i)\Phi_i(y_i) + \Psi_i^\perp(x_i, y_i) \tag{2.43}$$

with $\Phi_i$ and $\Psi_i^\perp$ having macroscopically disjoint $y_i$-supports and hence, because the $x_i$-systems are microscopic, having disjoint $y$-supports as well.[14] Moreover,

$$Y \in \text{supp } \Phi_1 \cap \text{supp } \Phi_2 \cap \cdots \cap \text{supp } \Phi_M, \tag{2.44}$$

and for all such $Y$ we have

$$\Psi(x_1, \ldots, x_M, Y) = \psi_i(x_i)\Phi_i(\hat{x}_i, Y) \tag{2.45}$$

for all i, where $\hat{x}_i = (x_1, \ldots, x_M)$ with $x_i$ missing. It follows by separation of variables, writing

$$\Psi(x, Y) = \psi_1(x_1) \cdots \psi_M(x_M)\Phi(x, Y) \tag{2.46}$$

and dividing by $\prod_i \psi_i$, that for $Y$ satisfying (2.44)

$$\Psi(x, Y) = \psi_1(x_1) \cdots \psi_M(x_M)\Phi(Y) \tag{2.47}$$

and, indeed, that the $x$-system has an effective wave function, given by the product (2.42).

Note that this result would not in general be valid for conditional wave functions. In fact, the derivation of (2.42), which is used for the equal-time analysis of Sect. 2.7, is the only place where more than (2.39) is required for our results, and even here only the more precise consequence (2.45) is needed. Moreover, our more general, multitime analysis (see Sects. 2.8–2.10) does not appeal to (2.42) and requires only (2.39).

We wish to point out that while the qualifications under which we have established (2.42) are so mild that in practice they exclude almost nothing, (2.42) is nonetheless valid in much greater generality. In fact, whenever it is "known" that the subsystems have the $\psi_i$ as their respective effective wave functions—by investigators, by devices, or by any records or traces whatsoever—*insofar as this "knowledge" is grounded in the environment of the composite system,* i.e., is reflected in $y$, (2.42) follows without further qualification.

Nonetheless, in order better to appreciate the significance of the qualification "microscopic" for (2.42), the reader should consider the following unrealistic but

---

each $x_i$-system is given by the wave function $\psi_i$, then this density matrix is itself, in fact, given by the corresponding product wave function.

[14] It is at this point that the condition that $M$ not be "too large"—so large that $x$ can be used to form a macroscopic variable—becomes relevant. And while the problematical situation which worries us here may seem far fetched, it is not as far fetched as it initially might appear to be. It may be that SQUIDs, superconducting quantum interference devices, can be regarded as giving rise to a situation just like the one with which we are concerned, in which lots of microscopic systems have, say, the same effective wave function, but the composite does not have the corresponding product as effective wave function. See, however, the comment following the proof of (2.42).

instructive example: Consider a pair of macroscopic systems with the composite system having effective wave function $\psi(x) = \psi_L(x_1)\psi_L(x_2)+\psi_R(x_1)\psi_R(x_2)$, where $\psi_L$ is a wave function supported by configurations in which a macroscopic coordinate is "on the left," and similarly for $\psi_R$. Suppose that $X_1$ and $X_2$ are "on the left." Then each system has effective wave function $\psi_L$.

What wave function would the quantum formalism assign to, say, system 1 in the previous example? Though we can imagine many responses, we believe that the best answer is, perhaps, that while the quantum formalism is for all practical purposes unambiguous, we are concerned here with one of those "impractical purposes" for which the usual quantum formalism is not sufficiently precise to allow us to make any definite statement on its behalf. In this regard, see Bell [38].

We shall henceforth often say "wave function" instead of "effective wave function."

## 2.6 The Fundamental Conditional Probability Formula

> The intellectual attractiveness of a mathematical argument, as well as the considerable mental labor involved in following it, makes mathematics a powerful tool of intellectual prestidigitation—a glittering deception in which some are entrapped, and some, alas, entrappers. Thus, for instance, the delicious ingenuity of the Birkhoff ergodic theorem has created the general impression that it must play a central role in the foundations of statistical mechanics.... The Birkhoff theorem does us the service of establishing its own inability to be more than a questionably relevant superstructure upon [the] hypothesis [of absolute continuity]. (Schwartz [39])

We are ready to begin the detailed analysis of the quantum equilibrium hypothesis (2.20). We shall find that by employing, purely as a mathematical device, the quantum equilibrium distribution on the universal scale, at, say, THE INITIAL TIME, we obtain the quantum equilibrium hypothesis in the sense of empirical distributions for all scales at all times. The key ingredient in the analysis is an elementary conditional probability formula.

Let us now denote the initial universal wave function by $\Psi_0$ and the initial universal configuration by $Q$, and for definiteness let us take THE INITIAL TIME to be $t = 0$. For the purposes of our analysis we shall regard $\Psi_0$ as fixed and $Q$ as random. More precisely, for given fixed $\Psi_0$ we equip the space $\mathscr{Q} = \{Q\}$ of initial configurations with the quantum equilibrium probability distribution $\mathbb{P}(dQ) = \mathbb{P}_{\Psi_0}(dQ) = |\Psi_0(Q)|^2 dQ$. $Q_t$ is then a random variable on the probability space $\{\mathscr{Q}, \mathbb{P}\}$, since it is determined via (2.21) by the initial condition given by $Q_0 = Q$ and $\Psi_0$. Thus, for any subsystem, with associated splitting (2.22), $X_t$, $Y_t$, and $\psi_t$ are also random variables on $\{\mathscr{Q}, \mathbb{P}\}$, where $Q_t = (X_t, Y_t)$ is the splitting of $Q_t$ arising from (2.22), and $\psi_t$ is the (conditional) wave function of the $x$-system at time $t$ (see Eq. 2.41).[15]

---

[15] The reader may wonder why we don't also treat $\Psi_0$ as random. First of all, we don't have to—we are able to establish our results for *every* initial $\Psi_0$, without having to invoke in any way any randomness in $\Psi_0$. Moreover, if it had proven necessary to invoke randomness in $\Psi_0$, the results

## 2.6 The Fundamental Conditional Probability Formula

We wish again to emphasize that, taking into account the discussion in Sect. 2.4, we regard the quantum equilibrium distribution $\mathbb{P}$, at least for the time being, solely as a mathematical device, facilitating the extraction of *empirical* statistical regularities from Bohmian mechanics(in a manner roughly analogous to the use of ergodicity in deriving the *pointwise* behavior of time averages for dynamical systems), and otherwise *devoid of physical significance*. (However, as a *consequence* of our analysis, the reader, if he so wishes, can safely also regard $\mathbb{P}$ as providing a measure of *subjective* probability for the initial configuration $Q$. After all, $\mathbb{P}$ could in fact be *somebody's* subjective probability for $Q$.)

Note that by equivariance the distribution of the random variable $Q_t$ is given by $|\Psi_t|^2$. It thus follows directly from (2.37), and even more directly from (2.39), that for the conditional probability distribution of the configuration of a subsystem, given the configuration of its environment, we have the *fundamental conditional probability formula*[16]

$$\mathbb{P}(X_t \in dx | Y_t) = |\psi_t(x)|^2 dx, \qquad (2.48)$$

where $\psi_t = \psi_t^{Y_t}$ is the (conditional) wave function of the subsystem at time $t$. In particular, this conditional distribution on the configuration of a subsystem depends on the configuration of its environment only through its wave function—an object of quite independent dynamical significance. In other words, $X_t$ and $Y_t$ are conditionally independent given $\psi_t$. The entire *empirical* statistical content of Bohmian mechanics flows from (2.48) with remarkable ease.

We wish to emphasize that (2.48) involves conditioning on the detailed microscopic configuration of the environment—far more information than could ever be remotely accessible. Thus (2.48) is extremely strong. Note that it implies in particular that

$$\mathbb{P}(X_t \in dx | \psi_t) = |\psi_t(x)|^2 dx, \qquad (2.49)$$

which involves conditioning on what we would be minimally expected to know if we were testing Born's statistical law (2.20). However, it would be very peculiar to know *only* this—to know no more than the wave function of the system of interest. But (2.48) suggests—and we shall show, see Sect. 2.11—that whatever additional information we might have can be of no relevance whatsoever to the possible value of $X_t$.[17]

---

so obtained would be of dubious physical significance, since to account for the nonequilibrium character of our world, the initial wave function must be a nonequilibrium, i.e., "atypical," wave function. See the discussion in Sects. 2.12–2.14.

[16] $\psi$ is to be understood as normalized whenever we write $|\psi|^2$.

[17] It immediately follows from (2.48) that for random $\Psi_0$ we have that

$$\mathbb{P}(X_t \in dx | Y_t, \Psi_0) = |\psi_t(x)|^2 dx,$$

where now $\mathbb{P}(dQ, d\Psi_0) = |\Psi_0(Q)|^2 dQ \, \mu(d\Psi_0)$ with $\mu$ any probability measure whatsoever on initial wave functions. Moreover (2.49) remains valid.

## 2.7 Empirical Distributions

> ...a single configuration of the world will show statistical distributions over its different parts. Suppose, for example, this world contains an actual ensemble of similar experimental set-ups....it follows from the theory that the 'typical' world will approximately realize quantum mechanical distributions over such approximately independent components. The role of the hypothetical ensemble is precisely to permit definition of the word 'typical.' (Bell [31])

In this section we present the simplest application of (2.48), to the empirical distribution on configurations arising from a large collection of subsystems, all of which have the "same" wave function at a common time. This is the situation relevant to an equal-time test of Born's statistical law. In practice the subsystems in our collection would be widely separated, perhaps even in different laboratories.

Consider $M$ subsystems, with configurations $x_1, \ldots, x_M$, where $x_i$ are coordinates relative to a frame of reference convenient for the $i$-th subsystem. Suppose that with respect to these coordinates each subsystem has at time $t$ the same wave function $\psi$, with the composite $x = (x_1, \ldots, x_M)$ having the corresponding product

$$\psi_t(x) = \psi(x_1) \cdots \psi(x_M) \tag{2.50}$$

as its wave function at that time. Then applying the fundamental conditional probability formula to the $x$-system, we obtain

$$\mathbb{P}(X_t \in dx \mid Y_t = Y) = |\psi(x_1)|^2 \cdots |\psi(x_M)|^2 \, dx_1 \cdots dx_M, \tag{2.51}$$

where $Y_t = Y$ is the configuration of the environment at this time. In other words, we find that relative to the conditional probability distribution $\mathbb{P}_t^Y(dQ) \equiv \mathbb{P}(dQ|Y_t = Y)$ given the configuration of the environment of the composite system at time $t$, the (actual) coordinates $X_1, \ldots, X_M$ of the subsystems at this time form a collection of independent random variables, identically distributed, with common distribution $\rho_{qe} = |\psi|^2$.

In any test of the quantum equilibrium hypothesis (2.20), it is the *empirical distribution*

$$\rho_{emp}(z) = \frac{1}{M} \sum_{i=1}^{M} \delta(z - X_i) \tag{2.52}$$

of $(X_1, \ldots, X_M)$ which is *directly* observed—so that the operational significance of the quantum equilibrium hypothesis is that $\rho_{emp}$ be (approximately) given by $\rho_{qe}$. Notice that $\rho_{emp}$ is a (distribution-valued) random variable on $(\mathcal{Q}, \mathbb{P})$, and that $\rho_{emp}(\Gamma) \equiv \int_\Gamma \rho_{emp}(z) \, dz$ is the relative frequency in our ensemble of subsystems of the event "$X_i \in \Gamma$".

It now follows from the weak law of large numbers that when the number $M$ of subsystems is large, $\rho_{emp}$ is very close to $\rho_{qe}$ for ($\mathbb{P}_t^Y$-)most initial configurations $Q \in \mathcal{Q}_t^Y \equiv \{Q \in \mathcal{Q} \mid Y_t = Y\}$, the *fiber* of $\mathcal{Q}$ for which $Y_t = Y$: For any bounded function $f(z)$, and any $\varepsilon > 0$, let the *"agreement set"* $\mathbf{A}(M, f, \varepsilon, t) \subset \mathcal{Q}_t^Y$ be the set

## 2.7 Empirical Distributions

of initial configurations $Q \in \mathcal{Q}_t^Y$ for which

$$\|\rho_{emp} - \rho_{qe}\|_f \equiv \left| \int \left( \rho_{emp}(z) - \rho_{qe}(z) \right) f(z)\, dz \right|$$
$$= \left| \frac{1}{M} \sum_{i=1}^{M} f(X_i) - \int f(z)\, |\psi(z)|^2\, dz \right| \quad (2.53)$$
$$\leq \varepsilon.$$

(We suppress the dependence of **A** upon Y and on the subsystems under consideration.) Then by the weak law of large numbers

$$\mathbb{P}_t^Y(\mathbf{A}(M, f, \varepsilon, t)) = 1 - \delta(M, f, \varepsilon) \quad (2.54)$$

where $\delta \to 0$ as $M \to \infty$.

For a single function $f$, $\|\cdot\|_f$ cannot provide a very good measure of closeness. Therefore, consider any finite collection $\mathbf{f} = (f_\alpha)$ of bounded functions, corresponding for example to a coarse graining of value space, and let

$$\mathbf{A}(M, \mathbf{f}, \varepsilon, t) \equiv \cap_\alpha \mathbf{A}(M, f_\alpha, \varepsilon, t)$$
$$\equiv \left\{ Q \in \mathcal{Q}_t^Y \,\Big|\, \|\rho_{emp} - \rho_{qe}\|_\mathbf{f} \equiv \sup_\alpha \|\rho_{emp} - \rho_{qe}\|_{f_\alpha} \leq \varepsilon \right\}. \quad (2.55)$$

It follows from (2.54) that

$$\mathbb{P}_t^Y(\mathbf{A}(M, \mathbf{f}, \varepsilon, t)) = 1 - \delta(M, \mathbf{f}, \varepsilon) \quad (2.56)$$

where $\delta(M, \mathbf{f}, \varepsilon) \leq \sum_\alpha \delta(M, f_\alpha, \varepsilon)$.

The empirical distribution $\rho_{emp}$ does not probe in a significant way the joint distribution (2.51), i.e., the independence, of $X_1, \ldots, X_M$—the law of large numbers is valid under conditions far more general than independence. To explore independence one might employ pair functions $f(X_i, X_j)$, or functions of several variables, in a manner analogous to that of the preceding analysis. Rather than proceeding in this way, we merely note—more generally—the following:

For any decision regarding the joint distribution of the $X_i$, we have at our disposal only the values which happen to occur. On the basis of some feature of these values, we must arrive at a (possibly rather tentative) conclusion. With any such feature we may associate a subset $\mathcal{T}$ of the space $\mathbb{R}^{DM} = \{(x_1, \ldots, x_M)\}$ of possible joint values, where $D = \dim(X_i)$ is the dimension of our subsystems.

Let $\mathcal{T} \subset \mathbb{R}^{DM}$ be a *statistical test* for the hypothesis that $X_1, \ldots, X_M$ are independent, with distribution $|\psi|^2$. This means that the failure to occur of the event $(X_1, \ldots X_M) \in \mathcal{T}$ can be regarded as a strong indication that $X_1, \ldots, X_M$ are not generated by such a joint distribution; in other words, it means that

$$\mathsf{P}(\mathcal{T}) = 1 - \delta(\mathcal{T}) \quad (2.57)$$

with $\delta \ll 1$, where $\mathsf{P}(dx_1, \ldots, dx_M) = |\psi(x_1)|^2 \cdots |\psi(x_M)|^2 \, dx_1 \cdots dx_M$ is the joint distribution under examination. $1 - \delta(\mathscr{T})$ is a measure of the reliability of the test $\mathscr{T}$.

Let

$$\mathbf{A}(\mathscr{T}, t) = \{Q \in \mathscr{Q}_t^Y \mid X_t \equiv (x_1, \ldots, x_M) \in \mathscr{T}\} \tag{2.58}$$

Then, trivially,

$$\mathbb{P}_t^Y(\mathbf{A}(\mathscr{T}, t)) = 1 - \delta(\mathscr{T}); \tag{2.59}$$

i.e., the $\mathbb{P}_t^Y$-size of the set of initial configurations in $Q_t^Y$ for which the test is passed matches precisely the reliability of the test. (We remind the reader that the *existence* of useful tests, analogous to, but more general than, the one defined for example by (2.53), is a consequence of the weak law of large numbers.) In particular, the size of $M$ required for $\delta$ in (2.56) to be "sufficiently" small is precisely the size required for the corresponding test

$$\mathscr{T} = \left\{ (x_1, \ldots, x_M) \in \mathbb{R}^{DM} \ \middle| \ \sup_\alpha \left| \frac{1}{M} \sum_{i=1}^M f_\alpha(x_i) - \int f_\alpha(z) |\psi(z)|^2 \, dz \right| \leq \varepsilon \right\} \tag{2.60}$$

to be "sufficiently" reliable (see Point 12 of the Appendix).

Equations (2.54, 2.56, and 2.59) are valid only for $Y$ as described, i.e., when the $x$-system has (conditional) wave function $\psi_t \equiv \psi^{Y, \Psi_t}$ of the form (2.50), with which we are primarily concerned. We remark, however, that for a general $Y$ these equations remain valid, provided the agreement sets which appear in them are sensibly defined in terms of the conditional distribution $\mathsf{P}_t^Y(dx) = |\psi^{Y, \Psi_t}(x)|^2 \, dx$ of $X_t$ given $Y_t = Y$. For example, we may let

$$\mathbf{A}(Y, t) = \{Q \in \mathscr{Q}_t^Y \mid X_t \in \mathscr{T}(\mathsf{P}_t^Y)\}, \tag{2.61}$$

where, for any distribution $\mathsf{P}$ (on $\mathbb{R}^{DM}$), $\mathscr{T} = \mathscr{T}(\mathsf{P})$ is a test for $\mathsf{P}$, satisfying (2.57) with $\delta(\mathscr{T}) \ll 1$.

In terms of such *conditioned agreement sets* $\mathbf{A}(Y, t)$, we may define an *unconditioned agreeement set* $\mathbf{A}(t)$ by requiring that

$$\mathbf{A}(t) \cap \mathscr{Q}_t^Y = \mathbf{A}(Y, t); \tag{2.62}$$

directly in terms of the tests $\mathscr{T}$,

$$\mathbf{A}(t) = \{Q \in \mathscr{Q} \mid X_t \in \mathscr{T}(\mathsf{P}_t^{Y_t})\}. \tag{2.63}$$

Corresponding to Eq. (2.54, 2.56, and 2.59) we then have that

$$\mathbb{P}(\mathbf{A}(t)) = 1 - \delta(t) \tag{2.64}$$

## 2.7 Empirical Distributions

where

$$\delta(t) = \int \delta(Y_t, t) \, d\mathbb{P} \ll 1 \qquad (2.65)$$

with $\delta(Y, t) \equiv \delta(\mathscr{T}(\mathsf{P}_t^Y))$.

Having said this, we wish to emphasize that Eqs. (2.54, 2.56, and 2.59) (for a general $Y$), expressing the "largeness" of the conditioned agreement sets, are much stronger and much more relevant than the Eqs. (2.64, 2.65) which we have just obtained: The original equations demand that the *disagreement set* $\mathbf{B}(t) = \mathbf{A}(t)^c \equiv \mathscr{Q} \setminus \mathbf{A}(t)$ be "small," not just for *"most"* fibers $\mathscr{Q}_t^Y$ corresponding to the possible environments $Y$ at time $t$, but for *all* such fibers. Insofar as the actual environment $Y_t$ at time $t$ might be rather special—for example, because it describes a world containing (human) life—the fact that "disagreement" has "insignificant probability" for *every* environment, regardless of how special, is quite important.[18] Indeed, it is the crucial element in our analysis of absolute uncertainty in Sect. 2.11.

We may summarize the conclusion at which we have so far arrived with the assertion that for Bohmian mechanics *typical* initial configurations lead to empirical statistics at time t which are governed by the quantum formalism (see the last paragraph of Sect. 2.3). Typicality is to be here understood in the sense of quantum equilibrium: something is true for *typical* initial configurations if the set of initial configurations for which it is false is small in the sense provided by the quantum equilibrium distribution $\mathbb{P}$ (and the appropriate conditional quantum equilibrium distributions $\mathbb{P}_t^Y$ arising from $\mathbb{P}$).

We wish to emphasize the role of equivariance in our analysis. Notice that Eq. (2.55, 2.56) would remain valid—with $\delta$ small—if, for example, $\rho_{qe}$ were replaced by $|\psi|^4$, *provided* the sense $\mathbb{P}$ of typicality were given, not by $|\Psi|^4$ (which is not equivariant), but by the density to which $|\Psi_t|^4$ would (backwards) evolve as the time decreases from $t$ to THE INITIAL TIME 0. This distribution, this sense of typicality, would presumably be extravagantly complicated and exceedingly artificial.

More important, it would depend upon the time $t$ under consideration, while equivariance provides a notion of typicality that works for all $t$. In fact, because of this time independence of typicality for quantum equilibrium, we immediately obtain the typicality of joint agreement for a not-too-large collection of times $t_1, \ldots, t_J$

$$\mathbb{P}\left(\cup_j \mathbf{B}(t_j)\right) \ll 1, \qquad (2.66)$$

as well as the typicality of joint agreement at *most* times of a collection of any size. We shall not go into this in more detail here because equivariance in fact yields results far more powerful than these, covering the empirical distribution for configurations $X_1, \ldots, X_M$ referring to times $t_1, \ldots, t_M$ which may all be different, to which we now turn. We shall find that in exploring this general situation, further novelties of the quantum domain emerge.

---

[18] Note, in particular, that for any condition $\mathscr{C}$ on environments implying, among other things, that the wave function of the $x$-system at time $t$ is of the form (2.50), we have the same statement of the "smallness" of the disagreement set with respect to the conditional distribution given $Y_t \in \mathscr{C}$.

## 2.8 Multitime Experiments: the Problem

In the previous section we analyzed the joint distribution of the simultaneous configurations $X_1, \ldots, X_M$ of $M$ (distinct and disjoint) subsystems, each of which has the same wave function $\psi$. We would now like to consider the more general, and more realistic, situation in which $X_1, \ldots, X_M$ refer to any $M$ subsystems, some or all of which might in fact be the same, at respective times $t_1, \ldots, t_M$, which might all be different. And we would again like to conclude that suitably conditioned, $X_1, \ldots, X_M$ are independent, each with distribution given by $|\psi|^2$; this would imply, precisely as in Sect. 2.7, the corresponding results about empirical distributions and tests.

We shall find, however, that this multitime situation requires considerably more care than we have so far needed; in particular, what we might think at first glance we would like to be true, in fact turns out to be in general false!

To begin to appreciate the difficulty, consider configurations $X_1$ and $X_2$ referring to the *same* system but at different times $t_1 < t_2$, and suppose this system has wave function $\psi$ at both of these times. Can we conclude that $X_1$ and $X_2$ are independent? Of course not! For example, if the system is suitably isolated between the times $t_1$ and $t_2$, so that its configuration undergoes an autonomous evolution, then $X_2$ will in fact be a function of $X_1$; in the simplest case, when the wave function $\psi$ is a ground state, we will in fact have that $X_2 = X_1$.

What has just been described is not, however, an instance of disagreement with the quantum formalism, which concerns only the results of *observation*—and in the previous example observation would destroy the isolation upon which the strong correlation between $X_1$ and $X_2$ was based. Moreover, the particular difficulty just described is easily remedied by taking "observation" into account. However, it is perhaps worth noting that for the equal-time analysis it was not necessary in any way to take observation directly into account to obtain agreement with the quantum formalism—$X_1, \ldots, X_M$ had the distribution given by the quantum formalism regardless of whether these variables were observed.

A much more serious, and subtle, difficulty arises from the fact that the wave function $\psi_t$ of a system at time $t$ is itself a random variable (see (2.41)), while we wish to consider situations in which our systems each have the same (non-random) wave function $\psi$. In the equal-time case this consideration led to no difficulty—and was barely noticed—since $\psi_t$ is *nonrandom relative* to the environment $Y_t$ upon which we there conditioned. For the multitime case, however, it is at first glance by no means clear how we should capture the stipulation that our systems each have wave function $\psi$.

One possibility would be to treat this stipulation as further conditioning, i.e., to consider the conditional distribution of $X_1, \ldots, X_M$ given, among other things, that the wave functions $\psi_{t_i}$ of our respective systems at the respective times $t_1, \ldots, t_M$ satisfy $\psi_{t_i} = \psi$ for all $i$. This would be a bad idea! The conditioning just described can affect the distribution of the configurations $X_1, \ldots, X_M$ in surprising, and uncontrollable, ways.

For example, suppose that when the result of an observation of $X_1$ is "favorable," the happy experimenter proceeds somehow to prepare the second system in state $\psi$

2.9 Random Systems

at time $t_2$, while if the result is "unfavorable," the depressed experimenter requires some extra time to recuperate, and prepares the second system in state $\psi$ at time $t_2' > t_2$. In this situation $X_1$ need not be independent of $\psi_{t_2}$, so that conditioning on $\psi_{t_2}$ may bias the distribution of $X_1$.

Moreover, we believe that this example is not nearly so artificial as it may at first appear. In the real world, of which the experimenters and their equipment are a part, which experiments get performed where and when can, and typically will, be correlated with the results of previous experiments, with each other, and with any number of other factors, such as, for example, the weather, which we would not normally take into account. Therefore, stochastic conditioning can be a very tricky business here, yielding conditional distributions of a surprising, and thoroughly unwanted, character.

What has just been said suggests that our multitime formulation is, while nonetheless inadequate, also perhaps not as general as we might want. The times at which our experiments are performed, and indeed the subsystems upon which they are performed, may themselves be random, and a more general formulation, like the one we shall give, should take this into account. However, we wish to emphasize that, as we shall see, the primary value of such a "random system" formulation is not increased generality. Rather, it is first of all simply the case that, strictly speaking, the systems upon which experiments get performed are, in fact, themselves random—not just the results, or the state of the system, but the time of the experiment as well as the specific system, the particular collection of particles, upon which we focus and act. Furthermore, when we properly take *this* into account, the difficulty we have been discussing vanishes!

## 2.9 Random Systems

Consider a pair $\sigma = (\pi, T)$, where $T \in \mathbb{R}$ (with $T \geq 0$ if THE INITIAL TIME is 0) and $\pi$ is a splitting

$$q = (x, y) \equiv (\pi q, \pi^\perp q) \tag{2.67}$$

(see Sect. 2.5); we identify $\pi$ with the projection $\mathcal{Q} \equiv \mathbb{R}^{3N} \to \mathbb{R}^{3m}$ onto the configuration of the ($m$-particle) $x$-system, with the components of $x \equiv \pi q$ ordered, say, as in $q$. $\pi$ comes together with $\pi^\perp$, the complementary projection, onto the coordinates of the environment (also ordered as in $q$). Thus we may identify $\pi$ with the subset of $\{1, \ldots, N\}$ corresponding to the particles of the $x$-system. $\sigma$ specifies a subsystem at a given time, for example, the system upon which we experiment and the time at which the experiment begins. (If indistinguishability were taken into account, our identification of $\pi$ would have to be modified accordingly. We might then associate it, for example, with a subset of $\mathbb{R}^3$. See footnote 6.)

Now allow both $T$ and $\pi$ to be random, i.e., allow $T$ to be a real-valued, and $\pi$ to be a projection-valued, function on the space $\mathcal{Q}$ of initial configurations. ($\pi$ may thus be identified with a random subset of $\{1,\ldots,N\}$.) For $\sigma = (\pi, T)$ we write

$$X_\sigma = \pi Q_T \tag{2.68}$$

for the configuration of the system and

$$Y_\sigma = \pi^\perp Q_T \tag{2.69}$$

for the configuration of its environment.[19]

We say that a pair

$$\sigma = (\pi, T), \tag{2.71}$$

consisting of a random projection and a random time as described, is a *random system* provided

$$\{\sigma = \sigma_0\} \in \mathscr{F}(Y_{\sigma_0}) \tag{2.72}$$

for any (nonrandom) $\sigma_0 = (\pi_0, t)$.[20] Here we use the notation $\mathscr{A} \in \mathscr{F}(W_1, W_2, \ldots)$ to convey that $I_\mathscr{A}$, the indicator function of the event $\mathscr{A} \subset \mathscr{Q}$, is a function of $W_1, W_2, \ldots$. [More precisely, $\mathscr{F}(W_1, W_2, \ldots)$ denotes the sigma-algebra generated by the random variables $W_1, W_2, \ldots$.]

We emphasize that for a random system $\sigma$, the configuration $X_\sigma$ ($Y_\sigma$) of the system (of its environment) is *doubly random*—$\sigma$ is itself random, and for a given value $\sigma_0$ of $\sigma$, $X_{\sigma_0}$ ($Y_{\sigma_0}$) is, of course, still random.

The condition (2.72) says that the value of a random system, i.e., the identity of the particular subsystem and time that it happens to specify, is reflected in its environment. In practice, this value is expressed by the state of the experimenters, their devices and records, and whatever other features of the environment form the basis of its *selection*. It is for this reason that we usually fail to notice that our systems are random: relative to "ourselves," which we naturally don't think of as random,

---

[19] More explicitly, when $\pi$ and $T$ are random, $X_\sigma$ is the random variable

$$X_\sigma(Q) = \pi(Q)\left(Q_{T(Q)}\right) \tag{2.70}$$

and similarly for $Y_\sigma$.

[20] The condition (2.72), which is formally what we need, technically suffers from "measure-0 defects"—since a random time $T$ will typically be a continuous random variable, the event $\{\sigma = \sigma_0\}$ will typically have measure 0, while conditional probabilities, for which (2.72) is formally utilized, are strictly defined only up to sets of measure 0. This defect can be eliminated by replacing (2.72) by the condition that for any $t$ there exist a number $\varepsilon_0(t) > 0$ such that

$$\{\pi = \pi_0, \, t - \varepsilon \leq T \leq t\} \in \mathscr{F}(Y_{(\pi_0, t)}) \tag{2.73}$$

for all $0 < \varepsilon < \varepsilon_0(t)$, using which our formal analysis becomes rigorous via standard continuity-density arguments. (Of course, if time were discrete no such technicalities would arise.)

## 2.9 Random Systems

they are completely determined. Notice also that (2.72) fits nicely with the notion of the wave function of a subsystem, as expressed, e.g., by (2.39).[21]

We shall write $\psi_\sigma$ for the (effective or conditional) wave function of the random system $\sigma$—given $Q \in \mathcal{Q}$, the wave function at time $T(Q)$ of the system defined by $\pi(Q)$. Using the notation of Eq. (2.41), we have that

$$\psi_\sigma = \psi_{T,\pi}^{Y_\sigma}, \qquad (2.75)$$

where the subscript $\pi$ makes explicit the dependence of $\psi_t^Y$ upon the splitting $q = (x, y)$. Note that $\psi_\sigma$ is a functional of both $\sigma$ and $Y_\sigma$.

The crucial ingredient in our multitime analysis is the observation that the fundamental conditional probability formula (2.48) remains valid for random systems: For any random system [the conditioning here on $\sigma$ can of course be removed if $\sigma \in \mathcal{F}(Y_\sigma)$ or, more generally, if $\psi_\sigma \in \mathcal{F}(Y_\sigma)$, e.g., if $\psi_\sigma = \psi$ is constant, i.e., nonrandom]

$$\mathbb{P}(X_\sigma \in dx | Y_\sigma, \sigma) = |\psi_\sigma(x)|^2 dx, \qquad (2.76)$$

which can in a sense be regarded as the most compact expression of the entire quantum formalism. To see this note that for any value $\sigma_0 = (\pi_0, t)$ of $\sigma$, we have that on $\{\sigma = \sigma_0\}$

$$\begin{aligned}\mathbb{P}(X_\sigma \in dx | Y_\sigma, \sigma) &= \mathbb{P}(X_\sigma \in dx | Y_\sigma, \sigma = \sigma_0) \\ &= \mathbb{P}(X_{\sigma_0} \in dx | Y_{\sigma_0}, \sigma = \sigma_0) \\ &= \mathbb{P}(X_{\sigma_0} \in dx | Y_{\sigma_0}) \equiv \mathbb{P}(X_t \in dx | Y_t) \\ &= |\psi_t(x)|^2 dx \equiv |\psi_{\sigma_0}(x)|^2 dx \\ &= |\psi_\sigma(x)|^2 dx, \end{aligned} \qquad (2.77)$$

where we have used (2.48) and (2.72), as well as the obvious fact that $X_\sigma$, $Y_\sigma$, and $\psi_\sigma$ agree respectively with $X_{\sigma_0}(\equiv X_t)$, $Y_{\sigma_0}(\equiv Y_t)$, and $\psi_{\sigma_0}(\equiv \psi_t)$ on $\{\sigma = \sigma_0\}$. (The reader familiar with stochastic processes should note the similarity between (2.72) and (2.76) on the one hand, and the notions of stopping time and the strong Markov property from Markov process theory. Indeed, (2.48) can be regarded as a kind of Markov property, in relation to which (2.76) then becomes a strong Markov property.)

---

[21] While the preceding informal description may not appear to discriminate between (2.72) and the perhaps equally natural condition

$$\sigma \in \mathcal{F}(Y_\sigma),$$

which we may formally write as

$$\{\sigma = \sigma_0\} \in \mathcal{F}(Y_\sigma), \qquad (2.74)$$

a careful reading should convey (2.72). The conditions (2.72) and (2.74) are not, in fact, equivalent, nor even comparable. In practice both are satisfied, the validity of (2.74) deriving mainly from the existence of "clocks." We have defined the notion of random system using only (2.72) because this is what turns out to be relevant for our analysis. (Note also that, trivially, $\sigma \in \mathcal{F}(Y_\sigma, \sigma)$.)

## 2.10 Multitime Distributions

...every atomic phenomenon is closed in the sense that its observation is based on registrations obtained by means of suitable amplification devices with irreversible functioning such as, for example, permanent marks on the photographic plate...the quantum-mechanical formalism permits well-defined applications only to such closed phenomena... (Bohr [40], pp. 73 and 90)

Now consider a sequence $\sigma_i = (\pi_i, T_i)$, $i = 1, \ldots, M$, of random systems, ordered so that (with probability 1)

$$T_1 \leq T_2 \leq \cdots \leq T_M. \tag{2.78}$$

We write $X_i$ for $X_{\sigma_i}$, $Y_i$ for $Y_{\sigma_i}$, and let

$$\mathscr{F}_i = \mathscr{F}(Y_{\sigma_i}, \sigma_i). \tag{2.79}$$

Suppose that for the wave function of the $i$-th system we have

$$\psi_{\sigma_i} = \psi_i \tag{2.80}$$

where $\psi_i$ is *nonrandom*, i.e., (with probability 1) the random wave function $\psi_{\sigma_i}$ is the specific wave function $\psi_i$. This will be the case if the requirement that the $i$-th system have wave function $\psi_i$ forms part of the basis of selection for this system, i.e., for $\sigma_i$— for example, if the $i$-th experiment, by prior decision, must be preceded by a successful preparation of the state $\psi_i$.

Finally, suppose that

$$X_i \in \mathscr{F}_j \quad \text{for all } i < j, \tag{2.81}$$

i.e., for all $i < j$ $X_i$ is a function of $Y_j$ and $\sigma_j$. This will hold, for example, if, with probability 1, each $X_i$ is measured—if the $i$-th measurement has not been completed, and the result "recorded," prior to time $T_j$, then the $i$-th system, together with the apparatus which measures it, must still be isolated at time $T_j$, from $\sigma_j$ as well as from the rest of its environment, remaining so until the completion of this measurement.

Notice that since $\psi_j$ is nonrandom, it follows from (2.81) and the fundamental conditional probability formula (2.76) that

$$\begin{aligned}\mathbb{P}(X_j \in dx_j | X_1, \ldots, X_{j-1}) &= \mathbb{P}(X_j \in dx_j | Y_j, \sigma_j) \\ &= |\psi_j(x_j)|^2 dx_j.\end{aligned} \tag{2.82}$$

Thus

$$\begin{aligned}\mathbb{P}\left(X_i \in dx_i, i \leq j\right) &= \mathbb{P}\left(X_i \in dx_i, i \leq j-1\right) \mathbb{P}\left(X_j \in dx_j | X_1 = x_1, \ldots, X_{j-1} = x_{j-1}\right) \\ &= \mathbb{P}\left(X_i \in dx_i, i \leq j-1\right) |\psi_j(x_j)|^2 dx_j \\ &= |\psi_1(x_1)|^2 \cdots |\psi_j(x_j)|^2 dx_1 \cdots dx_j,\end{aligned} \tag{2.83}$$

## 2.10 Multitime Distributions

and

$X_1, \ldots, X_M$ are independent, with each $X_i$ having distribution given by $|\psi_i|^2$. 
(2.84)

As it stands (2.84) is mildly useless, since the probability distribution $\mathbb{P}$ with respect to which it is formulated does not take into account any "prior" information, some of which we might imagine to be relevant to the outcomes of our sequence of experiments. Therefore, it is significant that our entire random system analysis (including 2.78, 280, and 2.81) can be relativized to any set $\mathscr{M} \subset \mathscr{Q}$—i.e., we may replace $(\mathscr{Q}, \mathbb{P})$ by $(\mathscr{M}, \mathbb{P}^{\mathscr{M}})$ where $\mathbb{P}^{\mathscr{M}}(dQ) = \mathbb{P}(dQ|\mathscr{M})$—without essential modification, *provided* the random systems $\sigma$ under consideration satisfy

$$\mathscr{M} \in \mathscr{F}(Y_\sigma, \sigma).$$
(2.85)

In particular, (2.84) is valid even with respect to $\mathbb{P}^{\mathscr{M}}$ provided that for all $i$

$$\mathscr{M} \in \mathscr{F}_i.$$
(2.86)

We might think of $\mathscr{M}$ as reflecting the "macroscopic state" at a time prior to all of our experiments, though one might argue about whether (2.86) would then be satisfied. Be that as it may, any event $\mathscr{M}$ describing any sort of prior information to which we could conceivably have access would be expected to satisfy (2.86), particularly if this information were recorded.

Now suppose that $\psi_i = \psi$ for all $i$. Then the joint distribution of $X_1, \ldots, X_M$ with respect to $\mathbb{P}^{\mathscr{M}}$ is precisely the same as in the equal time situation of Sect. 2.7.[22] Since the analysis there depended only upon this joint distribution, we may draw the same conclusions concerning empirical distributions and tests as before. We thus find for our sequence of experiments that *typical* initial configurations—typical with respect to $\mathbb{P}$ or $\mathbb{P}^{\mathscr{M}}$—yield empirical statistics governed by the quantum formalism.

Perhaps this claimed agreement with the quantum formalism requires elaboration. We have been explicitly concerned here only with the statistics governing the outcomes of *position* measurements. Now we were also concerned only with configurations in our equal-time analysis of Sect. 2.7. But our results there directly implied agreement with the quantum formalism for the results of measurements of any observable:

Our statistical conclusions there were valid regardless of whether or not the configurations—the $X_i$—were "measured." Thus, for the equal time case the joint distribution of any functions $Z_i = f_i(X_i)$ of the configurations must be inherited from the distribution of the $X_i$ themselves. In particular, by considering subsystems of the form (2.24), where the apparatus "measures the observable"—i.e., self-adjoint operator—$\hat{Z}_i$, with wave functions $\hat{\psi}_i = \psi_i \otimes \phi_i$ where $\phi_i$ is the initial(ized) wave function of the $i$-th apparatus, letting $Z_i$ be the outcome of this "measurement of $\hat{Z}_i$"

---

[22] Notice that equal-time experiments are covered by our multitime analysis—all the $T_i$ can be identical—and in this case (2.81) is automatically satisfied. However, for our earlier equal-time results it was necessary that $\psi$ be the effective wave function, while here conditional is sufficient.

and using what we know about the joint distribution of the $X_i$, it follows that the $Z_i$ are independent, and, as in the last paragraph of Sect. 2.3, that each $Z_i$ must have the distribution provided by the quantum formalism, namely, that given by the spectral measure $\rho_{\psi_i}^{\hat{Z}_i}(dz)$ for $\hat{Z}_i$ in the state $\psi_i$. (For a detailed account of how this comes about see [4, 6], and Chap. 3.)

The corresponding result for the multitime case does not, in fact, follow from (2.84). The latter does require that the configurations be "measured," and a "measurement of $\hat{Z}_i$" need not involve, and indeed may be incompatible with, a "measurement" of $X_i$.

But, while it does not follow from the *result* for the $X_i$, the corresponding result for "general measurements" does, in fact, follow from the *analysis* for the $X_i$. We need merely suppose for the $Z_i$ what we did for the $X_i$, namely, that

$$Z_i \in \mathscr{F}_j \quad \text{for all } i < j, \tag{2.87}$$

to conclude, for the sequence of outcomes $Z_i$ of "measurements of observables" $\hat{Z}_i$ in states $\psi_i$, that (with respect to $\mathbb{P}^{\mathscr{M}}$ for $\mathscr{M}$ satisfying (2.85))

$Z_1, \ldots, Z_M$ are independent, with each $Z_i$ having distribution given by $\rho_{\hat{Z}_i}^{\psi_i}$,
(2.88)

from which the usual conclusions concerning empirical distributions and tests follow immediately.[23]

We emphasize that the assumptions (2.81, 2.87, and 2.86) are minimal. They demand merely that facts about results and initial experimental conditions not be "forgotten." Thus they are hardly assumptions at all, but almost the very conditions essential to enable us, at the conclusion of our sequence of experiments, to talk in an informed manner about the experimental conditions and results and compare these with theory.

Moreover, it is not hard to see that if these conditions are relaxed, the "predictions" should not be expected to agree with those of the quantum formalism.[24] This is a striking illustration of the way in which Bohmian mechanics does not *merely* agree with the quantum formalism, but, eliminating ambiguities, illuminates, clarifies, and sharpens it.[25]

---

[23] That $Z_i = f_i(X_i)$ will in fact *be* the outcome of what would normally be considered a measurement of $\hat{Z}_i$ can be expected only if $\psi_i$ is the effective wave function of the $i$-th system, and not merely the conditional wave function: The functional form of $Z_i$ is based upon the evolution of a system initially with effective wave function $\psi_i$ interacting with a suitable apparatus but otherwise isolated. However, the conclusion (2.88) for $Z_i = f_i(X_i)$ is valid even for $\psi_i$ merely the conditional wave function, though in this case $Z_i$ may have little connection with what is actually observed.

[24] Note that by selectively "forgetting" results we can dramatically alter the statistics of those that we have not "forgotten."

[25] The analysis we have presented does not allow for the possibility that with nonvanishing probability $T_i = \infty$, i.e., the conditions for the selection of $\sigma_i$ are never satisfied. Our results extend to this case provided that $(X, \ldots, X_i$ and $\{T_{i+1} < \infty\}$ are conditionally independent given $\{T_i < \infty\}$ for all $i = 1, \ldots, M - 1$, in which case our results are valid given $\{T_M < \infty\}$. Note that without the aforementioned conditional independence our results would not be expected to hold: Suppose,

## 2.11 Absolute Uncertainty

That the quantum equilibrium hypothesis $\rho = |\psi|^2$ conveys the *most detailed* knowledge *possible* concerning the present configuration of a subsystem (of which the "observer" or "knower" is not a part—see Point 23 of the Appendix), what we have called *absolute uncertainty*, is implicit in the results of Sect. 2.7 and 2.10.[26] The key observation relevant to this conclusion is this: Whatever we may reasonably mean by knowledge, information, or certainty—and what precisely these do mean is not at all an easy question—it simply must be the case that the experimenters, their measuring devices, their records, and whatever other factors may form the basis for, or representation of, what could conceivably be regarded as knowledge of, or information concerning, the systems under investigation, must be a part of or grounded in the environment of these systems.

The possession by experimenters of such information must thus be reflected in *correlations* between the system properties to which this information refers and the features of the environment which express or represent this information. We have shown, however, that *given* its wave function there can be no correlation between (the configuration of) a system and (that of) its environment, even if the full microscopic environment $Y$—itself grossly more than what we could conceivably have access to—is taken into account.

Because we consider absolute uncertainty to be a very important conclusion, with significance extending beyond the conceptual foundations of quantum theory, we shall elaborate on how our results, for both the equal-time and the general multitime cases, entail this conclusion. The crucial point is that the possession of knowledge or information implies the existence of certain features of the environment, an environmentally based selection criterion, such that systems selected on the basis of this criterion satisfy the conditions expressed by this information. (For example, when a measuring device registers, or the associated computer printout records, that "$|X| < 1$", it should in fact be more or less the case that $|X| < 1$.)

Suppose that our $M$ systems of Sect. 2.7 have been chosen on the basis of some features of the environment, say by selection from an ensemble of $M'$ systems, also of the form considered there. The selection criterion can be based upon any property

---

for example, that if the initial results are "unfavorable," the depressed experimenter destroys humankind, and systems no longer get prepared properly. Thus, conditioning on $\{T_M < \infty\}$ yields a "biased" sample. The preceding points to perhaps a different, albeit rather minor, ambiguity in the quantum formalism, of which Bohmian mechanics again forces one to take note, and in so doing to rectify.

[26] Note, however, that as far as knowledge of the past is concerned, it is possible to do a good deal better than what would be permitted by absolute uncertainty for knowledge of the present: Having prepared our subsystem in a specific (not-too-localized) quantum state, with known wave function $\psi$, we may proceed to measure the configuration $X$ of this system, thereby obtaining detailed knowledge of both its wave function and its configuration for some *past* time. But note well that the determination of the configuration may—indeed, as we show, must—lead to an appropriate "collapse" of $\psi$, and hence our knowledge of the (present) configuration will be compatible with $\rho = |\psi|^2$ for the *present* wave function. (Note also that for quantum orthodoxy as well it is sometimes argued that knowledge of the past need not be constrained by the uncertainty principle.)

of the environment $Y_t = Y$ of the original (preselection) ensemble. (We allow for a rather arbitrary selection criterion, though in practice selection would of course be quite constrained. In particular, a realistic selection criterion should, perhaps, be the "same" for each system; i.e., whether or not the $i$-th system is selected should depend, for all $i$, upon the same property of $Y$ relative to this system. However, we need here no such constraints.)

Since, with respect to $\mathbb{P}_t^Y$, the configurations of the systems of our original ensemble were independent, with each having distribution given by $|\psi|^2$, and since our selection criterion is based solely upon the environment $Y$ of the original ensemble and in no way directly on the values of the configurations themselves, it follows that the configurations $X_1, \ldots, X_M$ of our selected subsystems have precisely the same distribution (also relative to $\mathbb{P}_t^Y$) as the original ensemble. Thus, for typical initial universal configurations, the empirical distribution of configurations across our selected ensemble will be given (approximately) by $|\psi|^2$, just as for the original ensemble. It follows that, whatever else it may be, our selection criterion cannot be based upon what we could plausibly regard as *information* concerning system configurations (more detailed than what is already expressed by $|\psi|^2$).

For the general case of multitime experiments as described in Sect. 2.10, the analysis is perhaps even simpler. In fact, for this case there is really nothing to do, beyond observing that any (environmentally based) selection criterion, whatever it may be, can be incorporated into the definition of our random systems, as part of the basis for their selection. It thus follows from the results of Sect. 2.10 that no such criterion can be regarded as reflecting any information, beyond $|\psi|^2$, about the configurations of these systems. Therefore, no devices whatsoever, based on any present or future technology, will provide us with the corresponding knowledge. *In a Bohmian universe such knowledge is absolutely unattainable!*[27]

We emphasize that we do not claim that knowledge of the detailed configuration of a system is impossible, a claim that would be manifestly false. We maintain only that—as a consequence of the fact that the configuration $X$ of a system and the configuration $Y$ of its environment are conditionally independent given its wave

---

[27] The reader concerned that we have overlooked the possibility that information may sometimes be grounded in non-configurational features of the environment, for example in velocity patterns, should consider the following (recall as well footnote 5):

1. Knowledge and information are, in fact, almost always, if not always, configurationally grounded. Examples are hardly necessary here, but we mention one—synaptic connections in the brain.
2. Dynamically relevant differences between environments, e.g., velocity differences, which are not instantaneously correlated with configurational differences quickly generate them anyway. And we need not be concerned with differences which are not dynamically relevant!
3. Knowledge and information must be communicable if they are to be of any social relevance; their content must be stable under communication. But communication typically produces configurational representations, e.g., pressure patterns in sound waves.
4. In any case, in view of the effective product form (2.37), when a system has an effective wave function, the configuration $Y$ provides an exhaustive description of the state of its environment (aside from the universal wave function $\Psi$—and through it $\Phi$—which for convenience of exposition we are regarding as given—see also footnotes 15 and 17).

function $\psi$—*all such knowledge must be mediated by* $\psi$. And we emphasize that a major reason for the not insignificant length of our argument, as presented in Sects. 2.6–2.11, was the necessity to extract from the aforementioned conditional independence analogous conclusions concerning empirical correlations.

From our conclusion that when a system has wave function $\psi$ we cannot know more about its configuration $X$ than what is expressed by $|\psi|^2$, it follows trivially that *knowledge* that its wave function is $\psi$ similarly constrains our knowledge of the configuration. It also trivially follows that detailed knowledge of $X$, for example that $X \in I$ for a given set of values $I$, entails detailed conclusions concerning the wave function, for example that the (conditional) wave function of the system is supported by $I$ (and even if the system does not have an effective wave function, we have that any density matrix describing the system must also be "supported" by $I$).

Finally, in order to further sharpen the character of our absolute uncertainty, one more point must be made. We have focused here primarily on the *statistical* aspect of the wave function of a system. But any "absolute uncertainty" based solely upon the fact that knowledge of the configuration $X$ of a system must be mediated by (knowledge of) some "object," in the sense that the distribution of $X$ can be expressed simply in terms of that "object," may be sorely lacking in substance if the "object" is *merely* statistical. In such a case, knowledge of the "object" need amount to nothing more than knowledge that $X$ has the distribution so expressed.

What lends substance to the "absolute uncertainty" in Bohmian mechanics—and justifies our use of that phrase—is the fact that the relevant "object," the wave function $\psi$, plays a dual role: it has, in addition to its statistical aspect, also a dynamical one, as expressed, e.g., in Eq. (2.28 and 2.31). Thus, knowledge of the wave function of a system, which sharply constrains our knowledge of its configuration, is knowledge of something in its own right, something "real," and not merely knowledge that the configuration has distribution $|\psi|^2$.

Moreover, the *detailed* character of this dynamical aspect is such that a wave function with narrow support quickly spreads, owing to the dispersion in Schrödinger's equation, to one with broad support, a change which generates a similar change in the distribution of the configuration. It follows that the unavoidable price we must pay for sharp knowledge of the present configuration of a system is at best hazy knowledge of its future configuration, i.e., of its "effective velocity." In particular, our absolute uncertainty embodies absolute unpredictability. More generally, the usual uncertainty relations for noncommuting "observables" become a corollary of the quantum equilibrium hypothesis $\rho = |\psi|^2$ as soon as the dynamical role of the wave function is taken into account; a detailed analysis can be found in [4, 6], and Chap. 3.

## 2.12 Knowledge and Nonequilibrium

The alert reader may be troubled that we have established results about randomness and uncertainty, results of a flavor often associated with "chaos" and "strong ergodic properties," without having to invoke any of the hard estimates and delicate analysis usually required to establish such properties. Indeed, our analysis neither used nor referred to any such properties. How can this be?

The short answer is quantum equilibrium, with all that the notion of equilibrium entails and conveys, an answer upon which we shall elaborate in the next section. Here we would like merely to observe that what is truly remarkable is not absolute uncertainty, irreducible limitations on what we *can* know, but rather that it is possible to know anything at all!

We take (the possibility of) knowledge, our information gathering and storing abilities, too much for granted. (And we conclude all too readily that the unknowable is unreal.) Of course, it is not at all surprising that we should do so, in view of the essential role such abilities play in our existence and survival. But that there should arise stable systems embodying (what can reasonably be regarded as) such abilities is a perhaps astonishing fact about the way our universe works, about the laws of nature!

The point is that we, the knowers, are separate and distinct from the things about which we know, and know in marvelous detail. How can there be, between completely disjoint entities, sufficiently strong correlations to allow for a representation in one of these entities of detailed features of the other? Indeed, such correlations are absent in thermodynamic equilibrium. With respect to (any of the distributions describing) global thermodynamic equilibrium, disjoint systems are more or less independent, and systems are more or less independent of their environments, facts incompatible with the existence of knowledge or information.

What renders knowledge at all possible is nonequilibrium. In fact, rather trivially, the very existence of the devices and records, not to mention brains, yielding or embodying any sort of information is impossible under global equilibrium. And, according to Heisenberg, "every act of observation is by its very nature an irreversible process" [41] (p. 138), and thus fundamentally nonequilibrium.

Thus, the very notion of quantum equilibrium, of equilibrium of configurations relative to the wave function, already suggests the unknowability of these configurations beyond the wave function. Our results merely provide a firm foundation for this suggestion. What is, however, striking is the simplicity of the analysis and how absolute and clean are the conclusions.

Insofar as equilibrium is associated with the impossibility of knowledge, equilibrium alone does not provide an adequate perspective on our analysis. In particular, our results say perhaps little of physical relevance unless *some* knowledge is possible, e.g., of the wave function of a particular system, or of the results of observations. But for this *nonequilibrium* is essential.

## 2.13 Quantum Equilibrium and Thermodynamic (non)Equilibrium

[In] a complete physical description, the statistical quantum theory would...take an approximately analogous position to the statistical mechanics within the framework of classical mechanics. (Einstein, in [1], p. 672)

## 2.13 Quantum Equilibrium and Thermodynamic (non)Equilibrium

We would like now to place quantum equilibrium within a broader context by comparing it with classical thermodynamic equilibrium.

According to the quantum equilibrium hypothesis, when a system has wave function $\psi$, the distribution $\rho$ of its configuration is given by

$$\rho = |\psi|^2. \tag{2.89}$$

Similarly, the Gibbs postulate of statistical mechanics asserts that for a system at temperature $T$, the distribution $\rho$ of its phase space point is given by

$$\rho = \frac{e^{-H/kT}}{Z}, \tag{2.90}$$

where $H$ is the classical Hamiltonian of the system (including, say, the "wall potential"), $k$ is Boltzmann's constant, and $Z$, the partition function, is a normalization.

In addition, we found that (2.89) assumed sharp mathematical form when understood as expressing the conditional probability formula (2.48). Equation (2.90) is perhaps also best regarded as a conditional probability formula, for the distribution of the phase point of the system given that of its environment—after all, the Hamiltonian $H$ typically involves interactions with the environment, and the temperature $T$ (like the wave function) can be regarded as a function of (the state of) the *environment*. (How otherwise would we know the temperature?) Furthermore, for a rigorous analysis of equilibrium distributions in the thermodynamic limit—i.e., of (the idealization given by) global thermodynamic equilibrium—the equations of Dobrushin and Lanford-Ruelle [42, 43], stipulating that (2.90)—regarded as expressing such a conditional distribution—be satisfied for all subsystems, often play a defining role.[28]

Moreover, what we have just described is only a part of a deeper and broader analogy, between the scheme

classical mechanics $\Longrightarrow$ equilibrium statistical mechanics $\Longrightarrow$ thermodynamics,
$$\tag{2.91}$$

which outlines the (classical) connection between the microscopic level of description and a phenomenological formalism on the macroscopic level; and the scheme

Bohmian mechanics $\Longrightarrow$ quantum equilibrium: statistical mechanics relative to the wave function $\Longrightarrow$ the quantum formalism,
$$\tag{2.92}$$

which outlines the (quantum) connection between the microscopic level and another phenomenological formalism—the quantum measurement formalism. We began this

---

[28] However, for a universe which, like ours, is not in global thermodynamic equilibrium, there is presumably no probability distribution on initial phase points with respect to which the probabilities (2.90), for all subsystems which happen to be "in thermodynamic equilibrium" and all times, are the conditional probabilities given the environments of the subsystems. In other words, roughly speaking, (2.90) is not equivariant (See Krylov [44], as well as the discussion after (2.92)).

section by comparing only the middle components of (2.91) and (2.92), but it is in fact the full schemes which are roughly analogous.

In particular, note that the middle of both schemes concerns the equilibrium distribution for the complete state description of the structure on the left with respect to the state for the structure on the right—the macrostate, as described by temperature (or energy) and, say, volume; or the quantum state, specified by the wave function. However, the quantum formalism does not live entirely on the macroscopic level, since the wave function for, say, an atom is best regarded as inhabiting (mainly) the microscopic level, at least for Bohmian mechanics.

The second arrow of (2.91) is, of course, associated primarily with the work of Gibbs [45]; the corresponding arrow of (2.92), upon which we have not focused here, will be the subject of Chap. 3. (See also [4, 6]). We have here focused on the first arrow of (2.92), i.e., on deriving the quantum equilibrium hypothesis from Bohmian mechanics. The corresponding arrow of (2.91) remains an active area of research, though it does not appear likely that a comprehensive rigorous analysis will be forthcoming any time soon. Conventional wisdom to the contrary notwithstanding, the problem of the rigorous justification, from first principles, of the use of the "standard ensembles," i.e., of the derivation of randomness governed by detailed probabilities, is far more difficult for classical thermodynamic equilibrium than for quantum theory!

How can this be? How is it possible so easily to derive the quantum equilibrium hypothesis from first principles (i.e., from Bohmian mechanics), while the corresponding result for thermodynamics—the rigorous derivation of the Gibbs postulate from first principles—is so very difficult? The answer, we believe, is that "pure equilibrium" is easy, while nonequilibrium, even a little bit, is hard. In our nonequilibrium universe, systems which happen to be in thermodynamic equilibrium are surrounded by, and arose from, (thermodynamic) nonequilibrium. Thus with thermodynamic equilibrium we are dealing with *islands of equilibrium in a sea of nonequilibrium*. But with quantum equilibrium we are in effect dealing with a *global* equilibrium, albeit relative to the wave function.

What makes nonequilibrium so very difficult is the fact that for nontrivial dynamics it is extremely hard to get a handle on the evolution of nonequilibrium ensembles adequate to permit us to conclude much of anything concerning the present distribution that would arise from a given nonequilibrium distribution in the (distant) past. To establish "convergence to equilibrium" for times $t \to \infty$ (mixing) is itself extremely difficult, but even this would be of little physical relevance, since we generally deal with, and can survive only during, times much earlier than the epoch of global thermodynamic equilibrium.

We should perhaps elaborate on why global equilibrium is so easy. A key aspect of equilibrium is, of course, stationarity—or equivariance. But how can this be sufficient for our purposes? Mere stationarity is not normally sufficient in a dynamical system analysis to conclude that typical behavior embodies randomness governed by the stationary distribution. Such "almost everywhere"-type assertions usually require the ergodicity of the dynamics. Why did we not find it necessary to establish some sort of ergodicity?

## 2.14 Global Equilibrium Beneath Nonequilibrium

The answer, we believe, lies in another critical aspect of the notion of equilibrium, shared by the schemes (2.91) and (2.92), and arising from the fact that both schemes are concerned with large "systems," with the thermodynamic limit as it were. In equilibrium, whether quantum or thermodynamic, most configurations or phase points are "macroscopically similar": quantities given by suitable spatial averages—e.g., density, energy density, or velocity fluctuations for thermodynamic equilibrium, and empirical correlations for quantum equilibrium—are more or less constant over the state space, in a sense defined by the equilibrium distribution. To say that a system is in equilibrium is then to say that its configuration or phase point is typical, in the sense that the values of these spatial averages are typical.

Now while the individual subsystems with which we have been concerned may be microscopic, our analysis, in fact, is effectively a "large system analysis." This is manifest in the equal-time analysis of Sect. 2.7, and for the general, multitime analysis it is implicit in our measurability conditions (2.81) and (2.85), which are plausible only for a universe having a large number of degrees of freedom. Thus, just as for a system *already in thermodynamic equilibrium,* we have no need for the ergodicity of the dynamics—just "stationarity"—since the kind of behavior we wish to establish occurs for a huge set of initial configurations, the "overwhelming majority."

(It might also be argued that we have, in fact, established for Bohmian mechanics a kind of effective Bernoulliness, and hence an effective ergodicity. And, again, the fact that we can do this with little work comes from the "thermodynamic limit" aspect of our analysis.)

The reader should compare the impossibility of perpetual motion machines, which is associated with the scheme (2.91), with that of "knowledge machines," as expressed by absolute uncertainty, associated with the scheme (2.92). In both cases the existence of devices of a certain character is precluded by general theoretical considerations—more or less equilibrium considerations for both—rather than by a detailed analysis of the workings of the various possible devices.

## 2.14 Global Equilibrium Beneath Nonequilibrium

> But to admit things not visible to the gross creatures that we are is, in my opinion, to show a decent humility, and not just a lamentable addiction to metaphysics. (Bell [14])

The schemes (2.91) and (2.92) refer to different universes, a classical universe and a quantum (Bohmian) universe. Since our universe happens to be a quantum one, it would, perhaps, be better to consider, instead of (2.91), the analogous quantum scheme[29]

---

[29] While it can be shown that in the "macroscopic limit"

$$\text{Bohmian mechanics} \Longrightarrow \text{classical mechanics,}$$

a proper understanding of thermodynamics must be in terms of the *actual* behavior of the constituents of equilibrium systems, i.e., quantum behavior.

$$\text{Bohmian mechanics} \Longrightarrow \text{quantum statistical mechanics} \Longrightarrow \text{thermodynamics.} \tag{2.93}$$

While the second arrow of (2.93) is standard, and presumably nonproblematical, research on the first arrow has not yet reached its infancy.

Note that it would make little sense to ask for a derivation of quantum statistical mechanics from the first principles provided by *orthodox* quantum theory. The very meaning of orthodox quantum theory is so entwined with processes, such as measurements, in which thermodynamic considerations play a crucial role that it is difficult to imagine where such a derivation might begin, or, for that matter, what such a derivation could possibly mean! (And insofar as Bohmian mechanics clarifies the meaning and significance of the wave function of a system, and permits a coherent analysis of the microscopic and macroscopic domains within a common theoretical framework, it may well be that the last word has not yet been written concerning the connection represented by the second arrow.)

If nonequilibrium is an essential aspect of our universe, and if configurations are in quantum equilibrium, i.e., pure equilibrium relative to the wave function, what then is the source, in our universe, of nonequilibrium? What is it that is *not in equilibrium*? The wave function, of course—both the universal wave function $\Psi$ and, as a consequence, subsystem wave functions $\psi$. At the same time, the middle of the scheme (2.93) can be regarded as concerned with the distribution of the subsystem wave function $\psi$ for subsystems which happen to be in thermodynamic equilibrium. But by exploiting global thermodynamic *nonequilibrium* we are able to see beneath the thermodynamic-macroscopic level of description, while with global quantum equilibrium there is no quantum nonequilibrium to reveal the system configuration $X$ beneath the system wave function $\psi$.

It is important, however, not to succumb to the temptation to conclude, as does Heisenberg [41], that configurations therefore provide merely an "ideological superstructure" best left out of quantum theory; for, as we have seen, the very meaning of the wave function $\psi$ of a subsystem requires the existence of configurations, i.e., those of its environment. And when we determine the wave function of a system we do so on the basis of the configuration of the environment. Recall also that both aspects of the wave function of a subsystem, the statistical and the dynamical, cannot coherently be formulated without reference to configurations. It is therefore not at all astonishing that orthodox quantum theory, by refusing to accept configurations as part of the description of the state of a system, has led to so much conceptual confusion.

Note that the fact that thermodynamics seems to depend only upon $\psi$, and not on any contribution to the total thermodynamic entropy from the actual configuration $X$, is an immediate consequence of quantum equilibrium: For a universe in quantum equilibrium the entropy associated with configurations is maximal, i.e., constant as a functional of $\psi$, and thus plays no thermodynamic role.

A crucial feature of our quantum universe is the peaceful coexistence between global equilibrium (quantum) and nonequilibrium (thermodynamic), providing us with what we may regard as an "equilibrium laboratory," a glimpse, as it were, of pure equilibrium, with all the surprising consequences it entails. Our analysis has shown how the interplay between the corresponding levels of structure—the nonequilibrium level given by the wave function, and, beneath the level of the wave function, that of the particles, described by their positions, in equilibrium relative to the wave function—leads to the randomness and uncertainty so characteristic of quantum theory.

We have argued, and believe our analysis demonstrates, that quantum randomness can best be understood as arising from ordinary "classical" uncertainty—about what is *there* but *unknown*. The denial of the existence of this unknowable—or only partially knowable—reality leads to ambiguity, incoherence, confusion, and endless controversy. What does it gain us?

## 2.15 Appendix: Random Points

In the following remarks we expand upon concepts introduced in this chapter, placing our conclusions within a broader perspective and comparing ours with related approaches.

1. Bohmian mechanics is what emerges from Schrödinger's equation, which is said to describe the evolution of the wave function of a system of *particles*, when we take this language seriously, i.e., when we insist that *"particles" means particles*. Thus Bohmian mechanics is the minimal interpretation of nonrelativistic quantum theory, arising as it does from the assertion that a familiar word has its familiar meaning.

In particular, if Bohmian mechanics is somehow strange or unacceptable, it must be because either Schrödinger's equation, or the assertion that "particles" means particles, or their combination is strange or unacceptable. Now the assertion that "particles" means particles can hardly be regarded as in any way problematical. On the other hand, Schrödinger's equation, for a field on *configuration* space, is a genuine innovation, though one that physicists by now, of course, take quite for granted. However, as we have seen in Section 2.2, when it is appropriately combined with the assertion that "particles" means particles, its strangeness is, in fact, very much diminished.

2. Quantum mechanics is notoriously nonlocal [46], a novelty which is in no way ameliorated by Bohmian mechanics. In fact, "in this theory an explicit causal mechanism exists whereby the disposition of one piece of apparatus affects the results obtained with a distant piece" [7]. We wish to emphasize, however, that *relative to the wave function*, Bohmian mechanics is completely *local*: the nonlocality in Bohmian mechanics derives solely from the nonlocality built into the structure of standard quantum theory, as provided by a wave function on configuration space.

> That the guiding wave, in the general case, propagates not in ordinary three-space but in a multidimensional-configuration space is the origin of the notorious 'nonlocality' of quantum mechanics. It is a merit of the de Broglie-Bohm version to bring this out so explicitly that it cannot be ignored. (Bell [47])

3. A rather fortunate property of Bohmian mechanics is that the behavior of the parts—of subsystems—reflects that of the whole. Indeed, if this were not the case it would have been difficult, if not impossible, to have ever discovered the full theory. We believe that a major reason why nonlocality is so often regarded as problematical is not nonlocality per se but rather that it *suggests* the breakdown of precisely this feature.

4. Notice that the effective wave function $\psi$ is, in effect, a "collapsed" wave function. Thus our analysis implicitly explains the status and role of "collapse of the wave packet" in the quantum formalism. (See also Point 21, recalling that the Wigner formula [33] for the joint distribution of the outcomes of a sequence of quantum measurements, to which we there refer, is usually based upon collapse; see Sect. 3.3.9.)

In particular, note that the effective wave function of a subsystem evolves according to Schrödinger's equation only when this system is suitably isolated. More generally, the evolution $\psi(t)$ of the effective wave function defines a stochastic process, one which embodies collapse in just the right way—with respect to the conditional probability distribution given the (initial) configuration of the environment of the composite system which includes the apparatus, with $\psi$ the effective wave function of the system alone, i.e., not including the apparatus. For details see Chap. 3.

Note also that the very notion of the effective wave function, as well as its behavior, depends upon the location of the split between the "observed" and the "observer," i.e., between the system of interest and the rest of the world, a dependence whose importance has been emphasized by Bohr [40], by von Neumann [2], and by a great many others, see for example [28, 29, 48]. In particular, while the effective wave function will "collapse" during measurement if the apparatus is *not* included in the system, it need not, in principle, collapse if the apparatus *is* included, precisely as emphasized by von Neumann [2]. But von Neumann was left with the "measurement paradox," while with Bohmian mechanics no hint of paradox remains.

5. The fact that knowledge of the configuration of a system must be mediated by its wave function may partially account, from a Bohmian perspective, for how the physics community could identify the state of a quantum system—its complete description—with its wave function without encountering any *practical* difficulties. Indeed, the conclusion of our analysis can be partially summarized with the assertion that the wave function $\psi$ of a subsystem represents maximal information about its configuration $X$. This is primarily because of the wave function's statistical role, but its dynamical role is also relevant here. Thus it is natural, even in Bohmian mechanics, to regard the wave function as the "state" of the system.

6. It has been clear, at least since von Neumann [2], that for all practical purposes the quantum formalism, regarded in strictly operational terms, is consistent. However, it has not, at least for many (e.g., Einstein), been clear that the "full" quantum theory, regarded as including the assertion of "completeness" based upon Heisenberg's uncertainty principle—which has itself traditionally been regarded as arising from the apparent impossibility of certain measurements described in more or less *classical* terms—is also consistent. (See [49] for a recent expression of related concerns.) If

## 2.15 Appendix: Random Points

nothing else, Bohmian mechanics establishes and makes clear this consistency—even including absolute uncertainty.

Indeed, as is well known, Einstein tried for many years to devise thought experiments in which the limitations expressed by the uncertainty principle could be evaded. The reason Einstein persisted in this endeavor is presumably connected with the fact that the arguments presented by Heisenberg and Bohr against such a possibility were, to say the least, not entirely convincing, relying, as they did, on a peculiar, nearly contradictory, combination of quantum and classical "reasoning." In this regard, recall that in order to rescue (a version of) the uncertainty principle from one of Einstein's final onslaughts (see [50]), Bohr felt compelled to exploit certain effects arising from Einstein's general theory of relativity [50].

However, from the perspective of a Bohmian universe the uncertainty principle is sharp and clear. In particular, from such a perspective it makes no sense to try to devise *thought* experiments by means of which the uncertainty principle can be evaded, since this principle is a mathematical consequence of Bohmian mechanics itself. One could, of course, imagine a universe governed by different laws, in which the uncertainty principle, and a great deal else, *would* be violated, but there can be no universe governed by Bohmian mechanics—and in quantum equilibrium— which fails to embody absolute uncertainty and the uncertainty principle which it entails.

7. The notion of effective wave function developed in Sect. 2.5 should perhaps be compared with a related notion of Bohm, namely, the "active" piece of the wave function [51, 10] (see also Bohm [3]): If $\Psi$ is of the form (2.32) with the supports of $\Psi^{(1)}$ and $\Psi^{(2)}$ "sufficiently disjoint," then $\Psi^{(i)}$ is "active" if the actual configuration $Q$ is in the support of $\Psi^{(i)}$ (See (2.33) and the surrounding discussion). When this active wave function appropriately factorizes—see (2.26)—the (active) wave function of a subsystem could be defined in terms of the obvious factor.

This notion of subsystem wave function will agree with ours if, as is likely to be the case, the active and inactive pieces have suitably disjoint $y$-supports, and it will otherwise disagree. (In this regard see also Point 20.) For example, if

$$\Psi^{(i)}(x, y) = \psi^{(i)}(x)\Phi(y) \tag{2.94}$$

with $\psi^{(1)}$ and $\psi^{(2)}$ suitably disjoint (e.g., because the $x$-system is macroscopic and ...) then the "active" wave function of the $x$-system is the appropriate $\psi^{(i)}$, while using our notion the $x$-system has effective wave function $\psi^{(1)} + \psi^{(2)}$. Note, in particular, that with our notion the effective wave function of the universe is the universal wave function $\Psi$, not the active piece of $\Psi$.

Our notion of effective wave function—and not the notion based upon the active piece—has a distinctly epistemological aspect: While for both choices we have that "$\rho = |\psi|^2$," the latter will be the conditional distribution given the configuration of the environment only if $\psi$ agrees with our effective (or conditional) wave function. Moreover, whenever we can be said to "*know* that the $x$-system has wave function $\psi$," then the $x$-system indeed has effective wave function $\psi$ in our sense.

Note that while both of these choices are somewhat vague, in that they appeal to the notion of the "macroscopic"—or to some such notion—our effective wave function, when it exists, is, as we have seen, completely unambiguous. Moreover, as we have also seen, with our notion reference to something like the macroscopic is not critical. Removing such a reference—as we did in defining the notion of the conditional wave function—leads to a precise formulation which remains entirely adequate (in fact, perfect) for our purposes. But for the choice based on the active piece, removing such a reference would lead to utter vagueness.

There is, of course, no real physics contingent upon a particular choice of (notion of) "effective wave function"; rather this choice is simply a matter of convenience of expression, of how we talk most efficiently about the physics. But such considerations can be quite important!

8. Sometimes it is helpful to try to imagine how things appear to God. This is of course audacious, but, in fact, the very activity of a physicist, his attempting to find the deepest laws of nature, is nothing if not audacious. Indeed, one might even argue that the defining activity of the physicist is the search for the divine perspective.

Be that as it may, to create a universe God must first decide upon the ontology—on what there is—and then on the dynamical laws—on how what is behaves. But this alone would not be sufficient. What is missing is a particular realization, out of all possible solutions, of the dynamics—the one corresponding to the actual universe. In other words, at least for a deterministic theory, what is further required is a choice of initial conditions. And unless there is somehow a natural special choice, the simplest possibility would appear to be a completely random initial condition, with an appropriate natural measure for the description of this randomness (whatever this might mean, even given the measure). The notion of typicality so defined would, in a sense, be an essential ingredient of the theory governing this hypothetical universe.

For Bohmian mechanics, *with somehow given initial wave function* $\Psi_0$, this measure of typicality is given by the quantum equilibrium distribution $|\Psi_0|^2$. Moreover, the dynamics itself is also generated by $\Psi_0$. It seems most fitting that God should design the universe in so efficient a manner, that a single object, the wave function $\Psi_0$, should generate all the necessary (extra-ontological) ingredients.

9. Regarding the question of universal initial conditions, we should perhaps contrast the issue of the initial configuration with that of the initial wave function. Insofar as the latter is a nonequilibrium wave function, the initial wave function must correspond to low entropy—it must be very atypical, i.e., of a highly improbable character. As has been much emphasized by Penrose [32], in order to understand our nonequilibriuim world we must face the problem of why God should have chosen such improbable initial conditions as demanded by nonequilibrium. On the other hand, for the universal initial configuration—in quantum equilibrium—we of course have no such problem. On the contrary, quantum randomness itself, including even absolute uncertainty, arising as it does from quantum equilibrium, in effect requires no explanation. (Concerning the choice of initial universal wave function, see also Point 13).

10. Naive agreement with the quantum formalism *demands* the existence of a small set of bad initial configurations, corresponding to outcomes which are very unlikely

## 2.15 Appendix: Random Points

but *not* impossible. It is thus hard to see how our results could be improved upon or significantly strengthened.

More generally, for any theory with probablistic content, particularly one describing a relativistic universe, we arrive at a similar conclusion: Once we recognize that there is but one world (of relevance to us), only one actual space-time history, we must also recognize that the ultimate meaning of probability, insofar as it is employed in the formulation of the predictions of the theory, must be in terms of a specification of typicality—one such that theoretically predicted empirical distributions are typical. When all is said and done, the physical import of the theory must arise from its provision of such a notion of typical space-time histories (at the very least of "macroscopic" events), presumably specified via a probability distribution on the set of all (kinematically) possible histories. And given a theory, i.e., such a probability distribution, describing a large but finite universe, atypical space-time histories, with empirical distributions disagreeing with the theoretical predictions, are, though extremely unlikely, not impossible.

11. It is quite likely that the fiber $\mathcal{Q}_t^Y \equiv \{Q \in \mathcal{Q} \mid Y_t = Y\}$ of $\mathcal{Q}$ for which $Y_t = Y$, discussed in Sect. 2.7, is extremely small, owing to the expansive and dispersive effects of the Laplacian $\Delta$ in Schrödinger's equation. If so, it follows that any regular (continuous) $\Psi_0$ (or $|\Psi_0|^2$) should be approximately constant on $\mathcal{Q}_t^Y$ (as on any sufficiently small set of initial conditions). This would imply that $\mathbb{P}_t^Y$, the conditional measure given $\mathcal{Q}_t^Y$, should be approximately the same as the uniform distribution—Lebesgue measure—on $\mathcal{Q}_t^Y$, so that typicality defined in terms of quantum equilibrium agrees with typicality in terms of Lebesgue measure.

Now, as we have already indicated in Sect. 2.4, under more careful scrutiny this argument does not sustain its appearance of relevance. However, it may nonetheless have some heuristic value.

12. We wish to emphasize that a byproduct of our analysis, quite aside from the relevance of this analysis to the interpretation of quantum theory, is the clarification and illumination of the meaning and role of probability in a deterministic (or even nondeterministic) universe. Moreover, our analysis of statistical tests in Sect. 2.7— the very triviality of this analysis, see Eqs. (2.57 and 2.59)—sharply underlines the centrality of typicality in the elucidation of the concept of probability.

13. We should mention some examples of nonequilibrium (initial) universal wave functions:

(1) Suppose that physical space is finite, say the 3-torus $\mathbb{T}^3$ rather than $\mathbb{R}^3$, and suppose, say, that the potential energy $V = 0$. Let $\Psi_0(\mathbf{q}_1, \ldots, \mathbf{q}_N) = 1$ if all $\mathbf{q}_i \in B$, where $B \subset \mathbb{T}^3$ is a "small" region in physical space, and be otherwise 0. Then $\Psi_0$ is a nonequilibrium wave function, since an equilibrium wave function should be "spread out" over $\mathbb{T}^3$. Moreover the initial quantum equilibrium distribution on configurations is uniform over configurations of $N$ particles in $B$.

More generally, any well localized $\Psi_0$ is a nonequilibrium wave function. And if physical space is $\mathbb{R}^3$, any localized or square-integrable wave function is a nonequilibrium wave function.

(2) For a nonequilibrium wave function of a rather different character, consider the following: Take $\mathbb{T}^3$ again for physical space, but instead of considering free particles, suppose that $V$ arises from Coulomb interactions, with half of the particles having charge $+e$ and half $-e$. Now suppose that $\Psi_0$ is constant, $\Psi_0 = 1$ on $\mathbb{T}^3$. (Thus, quantum equilibrium now initially corresponds to a uniform distribution on configurations.) That this $\Psi_0$, though "spread out," is nevertheless a nonequilibrium wave function can be seen in various ways. Dynamically, the Schrödinger evolution should presumably lead to the formation of "atoms," of suitable pairing in the (support properties of the) wave function. Entropically, $\Psi_0$ is very special. An equilibrium ensemble of initial wave functions is determined by the values of the infinite set of constants of the motion given by the absolute squares of the amplitudes with respect to a basis of energy eigenfunctions. Wave functions in this ensemble are then specified by the phases of these amplitudes. A random choice of phases leads to an equilibrium wave function, which should reflect the existence of "atoms." On the other hand, the wave function $\Psi_0 = 1$ corresponds to a particular, very special choice of phases, so that "atoms cancel out."

Note also that this example is relevant to the Penrose problem mentioned in Point 9. What choice of initial wave function could be simpler—and thus in a sense more natural—than the one which is everywhere constant? And, again, while it might at first glance seem that this choice corresponds to equilibrium, the attractive (in both senses) effects of the Coulomb interaction presumably imply that this is not so!

From a classical perspective the situation is similar: The initial state in which the particles are uniformly distributed in space with velocities all 0 (or with independent Maxwellian velocities) is a nonequilibrium state. In fact, an infinite amount of entropy can be extracted from suitable clustering of the particles, arising from the great volume in momentum space liberated when pairs of oppositely charged particles get close. (Of course, for Newtonian gravitation—as well as for general relativity—this tendency to cluster is, in a sense, far stronger still.)

14. To account for (the) most (familiar) applications of the quantum formalism one rarely needs to apply (the conclusions of) our quantum equilibrium analysis to systems of the form (2.24): Randomness in the result of even a quantum measurement usually arises solely from randomness in the system, randomness in the apparatus making essentially no contribution. This is because most real-world measurements are of the scattering-detection type—and a particle (or atom ...) will be detected more or less where it is at. Think, for example, of a two-slit-type experiment, or of the purpose of a cloud chamber, or of a Stern-Gerlach measurement of spin.

15. When all is said and done, what does the incorporation of actual configurations buy us? A great deal! It accounts for:

1. randomness
2. absolute uncertainty
3. the meaning of the wave function of a (sub)system
4. collapse of the wave packet
5. coherent—indeed, familiar—(macroscopic) reality

## 2.15 Appendix: Random Points

Moreover, it makes possible an appreciation of the basic significance of the universal wave function $\Psi$, as an embodiment of *law*, which cannot be clearly discerned without a coherent ontology to be governed by some law.

16. Recall that in principle the wave function $\psi$ of a (sub)system could depend upon the universal wave function $\Psi$ and on the choice of system $\sigma = (\pi, T)$, as well as on the configuration $Y$ of the environment of this system. In practice, however, in situations in which we in fact know what $\psi$ is, it must be given by a function of $Y$ alone, not depending upon $\sigma$, nor even on $\Psi$ (for "reasonable" nonequilibrium $\Psi$). After all, what else, beyond $Y$, do we have at our disposal to take into account when we conclude that a particular system has wave function $\psi$? In particular, $\Psi$ is unknown, apart from what we can conclude about it on the basis of $Y$ (and perhaps some a priori assumptions about reasonable initial $\Psi_0$'s. But even if $\Psi_0$ were known precisely, this information would be of little use here, since solving Schrödinger's equation to obtain $\Psi$ would be out of the question!)

Thus, whatever we can in practice conclude about $\psi$ must be based upon a *universal* function—of $Y$. It would be worthwhile to explore and elucidate the details of this function, analyzing the rules we follow in obtaining knowledge and trying to understand the validity of these rules. However, such considerations are not directly relevant to our purposes in this chapter, where our goal has been primarily to establish sharp *limitations* on the possibility of knowledge rather than to analyze what renders it at all possible. We have argued that the latter problem is perhaps far more difficult than the former, and, indeed, that this is not terribly astonishing.

17. In view of the similarity between Bohmian mechanics and stochastic mechanics [11, 52, 53], for which similarity see [12, 13], all of our arguments and results can be transferred to stochastic mechanics without significant modification. More important, the motivation for stochastic mechanics is the rather plausible suggestion that quantum randomness might originate from the merging of classical dynamics with intrinsic randomness, as described by a diffusion process, and with "noise" determined by $\hbar$. Insofar as our results demonstrate how quantum randomness naturally emerges without recourse to any such "noise," they rather drastically erode the evidential basis of stochastic mechanics.

18. The analyis of Bohmian mechanics presented here is relevant to the problem of the interpretation and application of quantum theory in cosmology, specifically, to the problem of the significance of $\rho = |\psi|^2$ on the cosmological level—where there is nothing outside of the system to perform the measurements from which $\rho = |\psi|^2$ derives its very meaning in orthodox quantum theory.

19. Our random system analysis illuminates the flexibility of Bohmian mechanics: It illustrates how joint probabilities as predicted by the quantum formalism, even for configurations, may arise from measurement and bear little resemblance to the probabilities for unmeasured quantities. And our analysis highlights the mathematical features which make this possible. This flexibility could be quite important for achieving an understanding of the relativistic domain, where it may happen that quantum equilibrium prevails only on special space-time surfaces (see [13] and Chap. 9).

Our (random system) multitime analysis illustrates how this need entail no genuine obstacle to obtaining the quantum formalism. (Our argument here of course involved the natural hypersurfaces given by $\{t = \text{const.}\}$, but the only feature of these surfaces critical to our analysis was the validity of quantum equilibrium, or, more precisely, of the fundamental conditional probability formula (2.48).)

20. A notion intermediate between that of the effective wave function and that of the conditional wave function of a subsystem, a *more-general-effective wave function* which like the effective wave function is "stable," may be obtained by replacing, in the definition (2.37)–(2.38) of effective wave function, the reference to macroscopically disjoint y-supports by "sufficiently disjoint" y-supports. This notion of more-general-effective wave function is, of course, rather vague. But we wish to emphasize that the y-supports of $\Phi$ and $\Psi^\perp$ may well be sufficiently disjoint to render negligible the (effects of) future interference between the terms of (2.37)—so that if (2.38) is satisfied, $\psi$ will indeed fully function dynamically as the wave function of the $x$-system—without their having to be actually macroscopically disjoint.

In fact, owing to the interactions—expressed in Schrödinger's equation—among the many degrees of freedom, the amount of y-disjointness in the supports of $\Phi$ and $\Psi^\perp$ will typically tend to increase dramatically as time goes on, with, as in a chain reaction, more and more degrees of freedom participating in this disjointness (see [3, 19, 54, 55]; see also [28])). When the effects of this dissipation or "decoherence" are taken into account, one finds that a small amount of y-disjointness will often tend quickly to become "sufficient," indeed becoming "much more sufficient" as time goes on, and very often indeed becoming macroscopic. Moreover, if ever we are in the position of knowing that a system has more-general-effective wave function $\psi$, then $\psi$ must be its effective wave function, since our knowledge must be based on or grounded in macroscopic distinctions (if only in the eye or brain).

Concerning dissipation, we wish also to emphasize that in practice the problem is not how to arrange for it to occur but how to keep it under control, so that superpositions of (sub)system wave functions retain their coherence and thus may interfere.

21. If we relax the condition (2.80), requiring that $\psi_{\sigma_i}$ be nonrandom, and stipulate instead merely that

$$\psi_{\sigma_i} \in \mathscr{F}(Z_1, \ldots, Z_{i-1}), \tag{2.95}$$

we find that $Z_1, \ldots, Z_M$ have joint distribution given by the familiar (Wigner) formula [33] (see also Sect. 3.3.9, [2] and [56]).

22. We wish to compare (what we take to be the lessons of) Bohmian mechanics with the approach of Gell-Mann and Hartle (GMH) [57, 58]. Unhappy about the irreducible reference to the observer in the orthodox formulation of quantum theory, particularly insofar as cosmology is concerned, they propose a program to extract from the quantum formalism a "quasiclassical domain of familiar experience," which, if we understand them correctly, defines for them the basic ontology of quantum theory. This they propose to do by regarding the Wigner formula (referred to in Points 4 and 21), for the joint probabilities of the results of a sequence

## 2.15 Appendix: Random Points

of measurements of quantum observables, as describing the probabilities of objective, i.e., not-necessarily-measured, events—what they call alternative histories. Of course, owing to interference effects one quickly gets into trouble here unless one restricts *this* use of the Wigner formula to what they call alternative (approximately) *decohering* histories, for which the Wigner formula can indeed be regarded as defining (approximate) probabilities, which are additive under coarse-graining. Thus far GMH in essence reproduce the work of Griffiths [59] and [60]. But, as GMH further note, the condition of (approximate) decoherence by itself allows for far too many possibilities. They thus introduce additional conditions, such as "fullness" and "maximality," as well as propose certain (as yet tentative) measures of "classicity" to define an optimization procedure they hope will yield a more or less unique quasiclassical domain. (They also consider the possibility that there may be many quasiclassical domains, each of which would presumably define a different physical theory.)

As in our analysis of Bohmian mechanics, universal initial conditions—for GMH the initial universal wave function (or density matrix)—play a critical role. And just as in Bohmian mechanics, the wave function does not provide a complete description of the universe, but rather attains physical significance from the role it plays in generating the behavior of something else, something *physically* primitive—for GMH the quasiclassical domain.

Insofar as nonrelativistic quantum theory is concerned, a significant difference between Bohmian mechanics and the proposal of GMH is that the latter defines a research program while the former is an already existing, and sharply formulated, physical theory. And as far as relativistic quantum theory is concerned, we believe that, appearances to the contrary notwithstanding, the lesson of Bohmian mechanics is one of flexibility (see also Point 19) while the approach of GMH is rigid. In saying this we have in mind, on the one hand, that GMH insist (1) that the possible ontologies be limited by the usual quantum description, i.e., correspond to a suitable (possibly time-dependent) choice of self-adjoint operators on Hilbert space; and (2) that this ontology be constrained further by the quantum formalism, demanding that its evolution be governed by the Wigner formula—so that for them, but not for Bohmian mechanics, the consideration of decoherence indeed becomes essential, bound up with questions of ontology.

On the other hand, one lesson of Bohmian mechanics is that ontology need not be so constrained. While the quantum formalism must—and for Bohmian mechanics does—emerge in measurement-type situations, the behavior of the basic variables, describing the fundamental ontology, outside of these situations need bear no resemblance to anything suggested by the quantum formalism. (Recall, in fact, that it quite frequently happens that simple, symmetric laws on a deeper level of description lead to a less symmetric phenomenological description on a higher level.) Indeed, these basic variables, whether they describe positions, or field configurations, or what have you, need not even correspond to self-adjoint operators. That they rather trivially do in Bohmian mechanics is, in part, merely an artifact of the equivariant measure's being a strictly local functional of the wave function, which was in no way crucial to our analysis.

In particular, while dissipation or decoherence are relevant both to Bohmian mechanics and to GMH, for GMH they are crucial to the *formulation* of the theory, to the specification of an *ontology*, while for Bohmian mechanics they are relevant only on the level of *phenomenology*. And insofar as the formation of new theories is concerned, the lesson of Bohmian mechanics is to look for fundamental microscopic laws appropriate to the (or a) natural choice of ontology, rather than to let the ontology itself be dictated by some law, let alone by what is usually regarded as a macroscopic measurement formalism.

It is perhaps worth considering briefly the two-slit experiment. In Bohmian mechanics the electron, indeed, goes through one or the other of the two slits, the interference pattern arising because the arrival of the electron at the "photographic" plate reflects the interference profile of the wave function governing the motion of the electron. In particular, and this is what we wish to emphasize here, in Bohmian mechanics a spot appears somewhere on the plate because the electron arrives there; while for GMH "the electron arrives somewhere" because the spot appears there.

23. There is one situation where we may, in fact, know more about configurations than what is conveyed by the quantum equilibrium hypothesis $\rho = |\psi|^2$: when we ourselves are part of the system! See, for example, the paradox of Wigner's friend [22]. In thinking about this situation it is important to note well that, while it may be merely a matter of convention whether or not we choose to include say ourselves in the subsystem of interest, the wave function to which the quantum equilibrium hypothesis refers—that of the subsystem—depends crucially on this choice.

24. We have shown, in part here and in part in Chapter 3, how the quantum formalism emerges within a Bohmian universe in quantum equilibrium. Thus, evidence for the quantum formalism is evidence for quantum equilibrium—global quantum equilibrium. This should be contrasted with the thermodynamic situation, in which the evidence points towards pockets of thermodynamic equilibrium within global thermodynamic nonequilibrium.

The reader may wish to explore quantum nonequilibrium. What sort of behavior would emerge in a universe which is initially in quantum nonequilibrium? What phenomenological formalism or laws would govern such behavior? We happen to have no idea! We know only that such a world is not our world! Or do we?

Valentini [61, 62] has in fact suggested the possibility of searching for and exploiting quantum nonequilibrium. Nonetheless, the situation today, in 2012, with regard to these questions about quantum nonequilibrium remains pretty much as described above. In contrast with thermodynamic nonequilibrium, we have at present no idea what quantum nonequilibrium, should it exist, would look like, despite claims and arguments to the contrary.

# References

1. P. A. Schilpp, editor. *Albert Einstein, Philosopher-Scientist*. Library of Living Philosophers, Evanston, Ill., 1949.
2. J. von Neumann. *Mathematische Grundlagen der Quantenmechanik*. Springer Verlag, New York-Heidelberg-Berlin, 1932. English translation by R. T. Beyer, *Mathematical Foundations of Quantum Mechanics*. Princeton University Press, Princeton, N.J., 1955.

# References

3. D. Bohm. A Suggested Interpretation of the Quantum Theory in Terms of "Hidden" Variables: Part I. *Physical Review*, 85:166–179, 1952. Reprinted in [211].
4. D. Bohm. A Suggested Interpretation of the Quantum Theory in Terms of "Hidden" Variables: Part II. *Physical Review*, 85:180–193, 1952. Reprinted in [211].
5. D. Bohm. Proof that Probability Density Approaches $|\psi|^2$ in Causal Interpretation of Quantum Theory. *Physical Review*, 89:458–466, 1953.
6. D. Bohm and B. J. Hiley. Measurement Understood Through the Quantum Potential Approach. *Foundations of Physics*, 14:255–274, 1984.
7. J. S. Bell. On the Problem of Hidden Variables in Quantum Mechanics. *Reviews of Modern Physics*, 38:447–452, 1966. Reprinted in [211] and in [26].
8. J. S. Bell. On the Einstein Podolsky Rosen Paradox. *Physics*, 1:195–200, 1964. Reprinted in [211], and in [26].
9. J. S. Bell. Bertlmann's Socks and the Nature of Reality. *Journal de Physique C 2*, 42:41–61, 1981. Reprinted in [26].
10. D. Bohm and B. J. Hiley. *The Undivided Universe: An Ontological Intepretation of Quantum Theory*. Routledge & Kegan Paul, London, 1993.
11. E. Nelson. *Quantum Fluctuations*. Princeton University Press, Princeton, N.J., 1985.
12. S. Goldstein. Stochastic Mechanics and Quantum Theory. *Journal of Statistical Physics*, 47:645–667, 1987.
13. D. Dürr, S. Goldstein, and N. Zanghì. On a Realistic Theory for Quantum Physics. In S. Albeverio, G. Casati, U. Cattaneo, D. Merlini, and R. Mortesi, editors, *Stochastic Processes, Geometry and Physics*, pages 374–391. World Scientific, Singapore, 1990.
14. J. S. Bell. Are There Quantum Jumps? In C. W. Kilmister, editor, *Schrödinger. Centenary Celebration of a Polymath*. Cambridge University Press, Cambridge, 1987. Reprinted in [26].
15. D. Bohm and B. J. Hiley. On the Intuitive Understanding of Non-Locality as Implied by Quantum Theory. *Foundations of Physics*, 5:93–109, 1975.
16. H. P. Stapp. Light as Foundation of Being. In B. J. Hiley and F. D. Peat, editors, *Quantum Implications: Essays in Honor of David Bohm*. Routledge & Kegan Paul, London and New York, 1987.
17. E. Schrödinger. Die Gegenwärtige Situation in der Quantenmechanik. *Naturwissenschaften*, 23:807–812, 1935. English translation by J. D. Trimmer, *The Present Situation in Quantum Mechanics: A Translation of Schrödinger's "Cat Paradox" Paper*, Proceedings of the American Philosophical Society, 124:323–338, 1980. Reprinted in [211].
18. E. P. Wigner. Review of the Quantum Mechanical Measurement Problem. In P. Meystre and M. O. Scully, editors, *Quantum Optics, Experimental Gravity and Measurement Theory*, pages 43–63. Plenum, New York, 1983.
19. A. J. Leggett. Macroscopic Quantum Systems and the Quantum Theory of Measurement. *Supplement of the Progress of Theoretical Physics*, 69:80–100, 1980.
20. S. Weinberg. Precision Tests of Quantum Mechanics. *Physical Review Letters*, 62:485–488, 1989.
21. R. Penrose. Quantum Gravity and State-Vector Reduction. In R. Penrose and C. J. Isham, editors, *Quantum Concepts in Space and Time*. Oxford University Press, Oxford, 1985.
22. E. P. Wigner. Remarks on the Mind-Body Question. In I. J. Good, editor, *The Scientist Speculates*. Basic Books, New York, 1961. Reprinted in [214], and in [211].
23. G. C. Ghirardi, A. Rimini, and T. Weber. Unified Dynamics for Microscopic and Macroscopic Systems. *Physical Review D*, 34:470–491, 1986.
24. L. de Broglie. A Tentative Theory of Light Quanta. *Philosophical Magazine*, 47:446–458, 1924.
25. L. de Broglie. La Nouvelle Dynamique des Quanta. In *Electrons et Photons: Rapports et Discussions du Cinquième Conseil de Physique tenu à Bruxelles du 24 au 29 Octobre 1927 sous les Auspices de l'Institut International de Physique Solvay*, pages 105–132, Paris, 1928. Gauthier-Villars.
26. M. Born. Quantenmechanik der Stoßvorgänge. *Zeitschrift für Physik*, 37:863–867, 1926.
27. M. Born. Quantenmechanik der Stoßvorgänge. *Zeitschrift für Physik*, 38:803–827, 1926. English translation (Quantum Mechanics of Collision Processes) in [149].

28. D. Bohm. *Quantum Theory*. Prentice-Hall, Englewood Cliffs, N.J., 1951.
29. L. D. Landau and E. M. Lifshitz. *Quantum Mechanics: Non-relativistic Theory*. Pergamon Press, Oxford and New York, 1958. Translated from the Russian by J. B. Sykes and J. S. Bell.
30. J. S. Bell. On the Impossible Pilot wave. *Foundations of Physics*, 12:989–999, 1982. Reprinted in [26].
31. J. S. Bell. Quantum Mechanics for Cosmologists. In C. Isham, R. Penrose, and D. Sciama, editors, *Quantum Gravity 2*, pages 611–637. Oxford University Press, New York, 1981. Reprinted in [26].
32. R. Penrose. *The Emperor's New Mind*. Oxford University Press, New York and Oxford, 1989.
33. E. P. Wigner. The Problem of Measurement. *American Journal of Physics*, 31:6–15, 1963. Reprinted in [214], and in [211].
34. E. P. Wigner. Interpretation of Quantum Mechanics. In [211], 1976.
35. B. S. DeWitt and N. Graham, editors. *The Many-Worlds Interpretation of Quantum Mechanics*. Princeton University Press, Princeton, N.J., 1973.
36. J. S. Bell. The Measurement Theory of Everett and de Broglie's Pilot Wave. In L. de Broglie and M. Flato, editors, *Quantum Mechanics, Determinism, Causality, and Particles*, pages 11–17. Dordrecht-Holland, D. Reidel, 1976. Reprinted in [26].
37. D. Dürr, S. Goldstein, R. Tumulka, and N. Zanghì. On the Role of Density Matrices in Bohmian Mechanics. *Foundations of Physics*, 35:449–467, 2005.
38. J. S. Bell. Against "Measurement". *Physics World*, 3:33–40, 1990. Also in [157].
39. J. T. Schwartz. The Pernicious Influence of Mathematics on Science. In M. Kac, G. Rota, and J. T. Schwartz, editors, *Discrete Thoughts: Essays on Mathematics, Science, and Philosophy*, page 23. Birkhauser, Boston, 1986.
40. N. Bohr. *Atomic Physics and Human Knowledge*. Wiley, New York, 1958.
41. W. Heisenberg. *Physics and Philosophy*. Harper and Row, New York, 1958.
42. R. L. Dobrushin. The Description of a Random Field by Means of Conditional Probabilities and Conditions of its Regularity. *Theory of Probability and its Applications*, 13:197–224, 1968.
43. O. E. Lanford and D. Ruelle. Observables at Infinity and States with Short Range Correlations in Statistical Mechanics. *Communications in Mathematical Physics*, 13:194–215, 1969.
44. N. S. Krylov. *Works on the Foundations of Statistical Mechanics*. Princeton University Press, Princeton, N.J., 1979.
45. J. W. Gibbs. *Elementary Principles in Statistical Mechanics*. Yale University Press, 1902. Dover, New York, 1960.
46. E. Schrödinger. Discussion of Probability Relations Between Separated Systems. *Proceedings of the Cambridge Philosophical Society*, 31:555–563, 1935. 32: 446–452, 1936.
47. J. S. Bell. De Broglie-Bohm, Delayed-Choice Double-Slit Experiment, and Density Matrix. *International Journal of Quantum Chemistry: A Symposium*, 14:155–159, 1980. Reprinted in [26].
48. F. W. London and E. Bauer. *La Théorie de l'Observation en Mécanique Quantique*. Hermann, Paris, 1939. English translation by A. Shimony, J. A. Wheeler, W. H. Zurek, J. McGrath, and S. McLean McGrath in [211].
49. M. O. Scully and H. Walther. Quantum Optical Test of Observation and Complementarity in Quantum Mechanics. *Physical Review A*, 39:5229–5236, 1989.
50. N. Bohr. Discussion with Einstein on Epistemological Problems in Atomic Physics. In Schilpp [186], pages 199–244. Reprinted in [44], and in [211].
51. D. Bohm and B. J. Hiley. An Ontological Basis for the Quantum Theory i: Non-Relativistic Particle Systems. *Physics Reports*, 144:323–348, 1987.
52. E. Nelson. Derivation of the Schrödinger Equation From Newtonian Mechanics. *Physical Review*, 150:1079–1085, 1966.
53. E. Nelson. *Dynamical Theories of Brownian Motion*. Princeton University Press, Princeton, N.J., 1967.
54. W. H. Zurek. Environment-Induced Superselection Rules. *Physical Review D*, 26:1862–1880, 1982.
55. E. Joos and H. D. Zeh. The Emergence of Classical Properties Through Interaction with the Environment. *Zeitschrift für Physik B*, 59:223–243, 1985.

# References

56. Y. Aharonov, P. G. Bergmann, and J. L. Lebowitz. Time Symmetry in the Quantum Process of Measurement. *Physical Review B*, 134:1410–1416, 1964. Reprinted in [211].
57. M. Gell-Mann and J. B. Hartle. Quantum Mechanics in the Light of Quantum Cosmology. In W. Zurek, editor, *Complexity, Entropy, and the Physics of Information*, pages 425–458. Addison-Wesley, Reading, 1990. Also in [130].
58. M. Gell-Mann and J. B. Hartle. Alternative Decohering Histories in Quantum Mechanics. In *Proceedings of the 25th International Conference on High Energy Physics: 2-8 August, 1990, (South East Asia Theoretical Physics Association, Physical Society of Japan; Teaneck, NJ)*, volume 2, pages 1303–1310. 1991.
59. R. B. Griffiths. Consistent Histories and the Interpretation of Quantum Mechanics. *Journal of Statistical Physics*, 36:219–272, 1984.
60. R. Omnes. Logical Reformulation of Quantum Mechanics. *Journal of Statistical Physics*, 53:893–932, 1988.
61. A. Valentini. Universal Signature of Non-Quantum Systems. *Physics Letters A*, 332:187–193, 2004.
62. A. Valentini. Inflationary Cosmology as a Probe of Primordial Quantum Mechanics. *Physical Review D*, 82:063513, 2010.

# Chapter 3
# Quantum Equilibrium and the Role of Operators as Observables in Quantum Theory

## 3.1 Introduction

It is often argued that the quantum mechanical association of observables with self-adjoint operators is a straightforward generalization of the notion of classical observable, and that quantum theory should be no more conceptually problematic than classical physics *once this is appreciated*. The classical physical observables—for a system of particles, their positions $q = (\mathbf{q}_k)$, their momenta $p = (\mathbf{p}_k)$, and the functions thereof, i.e., functions on phase space—form a commutative algebra. It is generally taken to be the essence of quantization, the procedure which converts a classical theory to a quantum one, that $q$, $p$, and hence all functions $f(q, p)$ thereof are replaced by appropriate operators, on a Hilbert space (of possible wave functions) associated with the system under consideration. Thus quantization leads to a noncommutative operator algebra of "observables," the standard examples of which are provided by matrices and linear operators. Thus it seems perfectly natural that classical observables are functions on phase space and quantum observables are self-adjoint operators.

However, there is much less here than meets the eye. What should be meant by "measuring" a quantum observable, a self-adjoint operator? We think it is clear that this must be specified—without such specification it can have no meaning whatsoever. Thus we should be careful here and use safer terminology by saying that in quantum theory observables are *associated* with self-adjoint operators, since it is difficult to see what could be meant by more than an association, by an *identification* of observables, regarded as somehow having independent meaning relating to observation or measurement (if not to intrinsic "properties"), with such a mathematical abstraction as a self-adjoint operator.

We are insisting on "association" rather than identification in quantum theory, but not in classical theory, because there we begin with a rather clear notion of observable (or property) which is well-captured by the notion of a function on the phase space, the state space of *complete descriptions*. If the state of the system were observed, the value of the observable would of course be given by this function of the state $(q, p)$,

D. Dürr et al., *Quantum Physics Without Quantum Philosophy*,
DOI 10.1007/978-3-642-30690-7_3, © Springer-Verlag Berlin Heidelberg 2013

but the observable might be observed by itself, yielding only a partial specification of the state. In other words, measuring a classical observable means determining to which level surface of the corresponding function the state of the system, the phase point—which is at any time *definite* though probably unknown—belongs. In the quantum realm the analogous notion could be that of function on Hilbert space, not self-adjoint operator. But we don't measure the wave function, so that functions on Hilbert space are not physically measurable, and thus do not define "observables."

The problematical character of the way in which measurement is treated in orthodox quantum theory has been stressed by John Bell:

> The concept of 'measurement' becomes so fuzzy on reflection that it is quite surprising to have it appearing in physical theory *at the most fundamental level*. Less surprising perhaps is that mathematicians, who need only simple axioms about otherwise undefined objects, have been able to write extensive works on quantum measurement theory—which experimental physicists do not find it necessary to read. ... Does not any *analysis* of measurement require concepts more *fundamental* than measurement? And should not the fundamental theory be about these more fundamental concepts? [1]
>
> ... in physics the only observations we must consider are position observations, if only the positions of instrument pointers. It is a great merit of the de Broglie-Bohm picture to force us to consider this fact. If you make axioms, rather than definitions and theorems, about the 'measurement' of anything else, then you commit redundancy and risk inconsistency. [2]

The de Broglie-Bohm theory, Bohmian mechanics, is a physical theory for which the concept of 'measurement' does not appear at the most fundamental level—in the very formulation of the theory. It is a theory about concepts more fundamental than 'measurement,' in terms of which an analysis of measurement can be performed. In the previous chapter we have shown how probabilities for positions of particles given by $|\psi|^2$ emerge naturally from an analysis of "equilibrium" for the deterministic dynamical system defined by Bohmian mechanics, in much the same way that the Maxwellian velocity distribution emerges from an analysis of classical thermodynamic equilibrium. Our analysis entails that Born's statistical rule $\rho = |\psi|^2$ should be regarded as a local manifestation of a global equilibrium state of the universe, what we call *quantum equilibrium*, a concept analogous to, but quite distinct from, thermodynamic equilibrium: a universe in quantum equilibrium evolves so as to yield an appearance of randomness, with empirical distributions in agreement with all the predictions of the quantum formalism.

While in our earlier work we have proven, from the first principles of Bohmian mechanics, the "quantum equilibrium hypothesis" that *when a system has wave function $\psi$, the distribution $\rho$ of its configuration satisfies* $\rho = |\psi|^2$, our goal here is to show that it follows from this hypothesis, not merely that Bohmian mechanics makes the same predictions as does orthodox quantum theory for the results of any experiment, but that *the quantum formalism of operators as observables emerges naturally and simply as the very expression of the empirical import of Bohmian mechanics.*

More precisely, we shall show here that self-adjoint operators arise in association with specific experiments: insofar as the statistics for the values which result from the experiment are concerned, the notion of self-adjoint operator compactly expresses

## 3.1 Introduction

and represents the relevant data. It is the association "$\mathscr{E} \mapsto A$" between an experiment $\mathscr{E}$ and an operator $A$—an association that we shall establish in Sect. 3.2 and upon which we shall elaborate in the other sections—that is the central notion of this chapter. According to this association the notion of operator-as-observable in no way implies that anything is measured in the experiment, and certainly not the operator itself. We shall nonetheless speak of such experiments as measurements, since this terminology is unfortunately standard. When we wish to emphasize that we really mean measurement—the ascertaining of the value of a quantity—we shall often speak of *genuine* measurement.

Much of our analysis of the emergence and role of operators as observables in Bohmian mechanics, including the von Neumann-type picture of measurements at which we shall arrive, applies as well to orthodox quantum theory. Indeed, the best way to understand the status of the quantum formalism—and to better appreciate the minimality of Bohmian mechanics—is Bohr's way: What are called quantum observables obtain meaning *only* through their association with specific *experiments*. We believe that Bohr's point has not been taken to heart by most physicists, even those who regard themselves as advocates of the Copenhagen interpretation.

Indeed, it would appear that the argument provided by our analysis against taking operators too seriously as observables has even greater force from an orthodox perspective: Given the initial wave function, at least in Bohmian mechanics the outcome of the particular experiment is determined by the initial configuration of system and apparatus, while for orthodox quantum theory there is nothing in the initial state which completely determines the outcome. Indeed, we find it rather surprising that most proponents of standard quantum measurement theory, that is the von Neumann analysis of measurement [3], beginning with von Neumann, nonetheless seem to retain an uncritical identification of operators with properties. Of course, this is presumably because more urgent matters—the measurement problem and the suggestion of inconsistency and incoherence that it entails—soon force themselves upon one's attention. Moreover such difficulties perhaps make it difficult to maintain much confidence about just what *should* be concluded from the "measurement" analysis, while in Bohmian mechanics, for which no such difficulties arise, what should be concluded is rather obvious.

Moreover, a great many significant real-world experiments are simply not at all associated with operators in the usual way. Because of these and other difficulties, it has been proposed that we should go beyond operators-as-observables, to *generalized observables*, described by mathematical objects (positive-operator-valued measures, POVMs) even more abstract than operators (see, e.g., the books of Davies [4], Holevo [5] and Kraus [6]). It may seem that we would regard this development as a step in the wrong direction, since it supplies us with a new, much larger class of abstract mathematical entities about which to be naive realists. We shall, however, show that these generalized observables for Bohmian mechanics form an extremely natural class of objects to associate with experiments, and that the emergence and role these observables is merely an expression of quantum equilibrium together with the linearity of Schrödinger's evolution. It is therefore rather dubious that the occurrence of generalized observables—the simplest case of which are self-adjoint operators—can be regarded as suggesting any deep truths about reality or about epistemology.

As a byproduct of our analysis of measurement we shall obtain a criterion of measurability and use it to examine the genuine measurability of some of the properties of a physical system. In this regard, it should be stressed that measurability is theory-dependent: different theories, though empirically equivalent, may differ on what should be regarded as genuinely measurable *within* each theory. This important—though very often ignored—point was made long ago by Einstein and has been repeatedly stressed by Bell. It is best summarized by Einstein's remark [7]: *"It is the theory which decides what we can observe."*

We note in passing that measurability and reality are different issues. Indeed, for Bohmian mechanics most of what is "measurable" (in a sense that we will explain) is not real and most of what is real is not genuinely measurable. (The main exception, the position of a particle, which is both real and genuinely measurable, is, however, constrained by absolute uncertainty [8]).

In focusing here on the role of operators as observables, we don't wish to suggest that there are no other important roles played by operators in quantum theory. In particular, in addition to the familiar role played by operators as generators of symmetries and time-evolutions, we would like to mention the rather different role played by the *field operators* of quantum field theory: to link abstract Hilbert-space to space-time and structures therein, facilitating the formulation of theories describing the behavior of an indefinite number of particles [9, 10, 11].

Finally, we should mention what should be the most interesting sense of measurement for a physicist, namely the determination of the coupling constants and other parameters that define our physical theories. This has little to do with operators as observables in quantum theory and shall not be addressed here.

### 3.1.1 Notations and Conventions

$Q = (\mathbf{Q}_1, \ldots, \mathbf{Q}_N)$ denotes the actual configuration of a system of $N$ particle with positions $\mathbf{Q}_k$; $q = (\mathbf{q}_1, \ldots, \mathbf{q}_N)$ is its generic configuration. Whenever we deal with a system-apparatus composite, $x$ ($X$) will denote the generic (actual) configuration of the system and $y$ ($Y$) that of the apparatus. Sometimes we shall refer to the system as the $x$-system and the apparatus as the $y$-system. Since the apparatus should be understood as including all systems relevant to the behavior of the system in which we are interested, this notation and terminology is quite compatible with that of Sect. 3.2.2, in which $y$ refers to the environment of the $x$-system.

For a system in the state $\Psi$, $\mathsf{P}_\Psi$ will denote the quantum equilibrium measure, $\mathsf{P}_\Psi(dq) = |\Psi(q)|^2 dq$. If $Z = F(Q)$ then $\mathsf{P}_\Psi^Z$ denotes the measure induced by $F$, i.e. $\mathsf{P}_\Psi^Z = \mathsf{P}_\Psi \circ F^{-1}$.

## 3.2 Bohmian Experiments

According to Bohmian mechanics, the complete description or state of an $N$-particle system is provided by its wave function $\Psi(q,t)$, where $q = (\mathbf{q}_1, \ldots, \mathbf{q}_N) \in \mathbb{R}^{3N}$, and its configuration $Q = (\mathbf{Q}_1, \ldots, \mathbf{Q}_N) \in \mathbb{R}^{3N}$, where the $\mathbf{Q}_k$ are the positions of the particles. The wave function, which evolves according to Schrödinger's equation,

$$i\hbar \frac{\partial \Psi}{\partial t} = H\Psi, \tag{3.1}$$

choreographs the motion of the particles: these evolve according to the equation

$$\frac{d\mathbf{Q}_k}{dt} = \frac{\hbar}{m_k} \operatorname{Im} \frac{\Psi^* \nabla_k \Psi}{\Psi^* \Psi} (\mathbf{Q}_1, \ldots, \mathbf{Q}_N) \tag{3.2}$$

where $\nabla_k = \partial/\partial \mathbf{q}_k$. Note that for complex-valued $\Psi$, this is just equation (2.10) of Chap. 2, while for tensor-spinor-valued $\Psi$, it provides the guidance equation for particles with spin; see Sect. 3.2.5 below, in particular for the meaning of the numerator and denominator on the right hand side of (3.2).

In Eq. (3.1), $H$ is the usual nonrelativistic Schrödinger Hamiltonian; for spinless particles it is of the form

$$H = -\sum_{k=1}^{N} \frac{\hbar^2}{2m_k} \nabla_k^2 + V, \tag{3.3}$$

containing as parameters the masses $m_1, \ldots, m_N$ of the particles as well as the potential energy function $V$ of the system. For an $N$-particle system of nonrelativistic particles, Eqs. (3.1 and 3.2) form a complete specification of the theory. Magnetic fields[1] and spin,[2] as well as Fermi and Bose-Einstein statistics,[3] can easily be dealt with and in fact arise in a natural manner [12]–[16]. There is no need, and indeed no room, for any further *axioms*, describing either the behavior of other observables or the effects of measurement.

### *3.2.1 Equivariance and Quantum Equilibrium*

It is important to bear in mind that regardless of which observable one chooses to measure, the result of the measurement can be assumed to be given configurationally, say by some pointer orientation or by a pattern of ink marks on a piece of paper.

---

[1] When a magnetic field is present, the gradients $\nabla_k$ in the Eqs. (3.1 and 3.2) must be understood as the covariant derivatives involving the vector potential $\mathbf{A}$.

[2] See Sect. 3.2.5.

[3] For indistinguishable particles, a careful analysis of the natural configuration space, which is no longer $\mathbb{R}^{3N}$, leads to the consideration of wave functions on $\mathbb{R}^{3N}$ that are either symmetric or antisymmetric under permutations (see Chap. 8).

Then the fact that Bohmian mechanics makes the same predictions as does orthodox quantum theory for the results of any experiment—for example, a measurement of momentum or of a spin component—*provided we assume a random distribution for the configuration of the system and apparatus at the beginning of the experiment given by* $|\Psi(q)|^2$—is a more or less immediate consequence of (3.2). This is because its right hand side is the ratio of the quantum continuity current $J^\Psi$ (2.18) and the quantum probability density $\rho = |\Psi|^2 = \Psi^*\Psi$, so that the classical continuity equation

$$\frac{\partial \rho}{\partial t} + \operatorname{div} \rho\, v = 0 \qquad (3.4)$$

for the system of equations $dQ/dt = v$ defined by (3.2)—governing the evolution of the probability density $\rho$ under the motion defined by the guiding Eq. (3.2)—becomes the quantum continuity equation (2.19) for the particular choice $\rho = |\Psi|^2$. In other words, if the probability density for the configuration satisfies $\rho(q, t_0) = |\Psi(q, t_0)|^2$ at some time $t_0$, then the density to which this is carried by the motion (3.2) at any time $t$ is also given by $\rho(q, t) = |\Psi(q, t)|^2$. This is an extremely important property of any Bohmian system, as it expresses a certain compatibility between the two equations of motion defining the dynamics, which we call the *equivariance*[4] of $|\Psi|^2$.

The above assumption guaranteeing agreement between Bohmian mechanics and quantum mechanics regarding the results of any experiment is what we call the "quantum equilibrium hypothesis":

*When a system has wave function $\Psi$, its configuration $Q$ is random with probability distribution given by the measure $\mathsf{P}_\Psi(dq) = |\Psi(q)|^2 dq$.* (3.5)

When this condition is satisfied we shall say that the system is in quantum equilibrium and we shall call $\mathsf{P}_\Psi$ the quantum equilibrium distribution. While the meaning and justification of (3.5) is a delicate matter, which we have discussed at length in Chap. 2, it is important to recognize that, merely as a consequence of (3.2) and (3.5), Bohmian mechanics is a counterexample to all of the claims to the effect that a deterministic theory cannot account for quantum randomness in the familiar statistical mechanical way, as arising from averaging over ignorance: Bohmian mechanics is clearly a

---

[4] Equivariance can be formulated in very general terms: consider the transformations $U : \Psi \to U\Psi$ and $f : Q \to f(Q)$, where $U$ is a unitary transformation on $L^2(dq)$ and $f$ is a transformation on configuration space that may depend on $\Psi$. We say that the map $\Psi \mapsto \mu_\Psi$ from wave functions to measures on configuration space is equivariant with respect to $U$ and $f$ if $\mu_{U\Psi} = \mu_\Psi \circ f^{-1}$. The above argument based on the continuity Eq. (3.4) shows that $\Psi \mapsto |\Psi|^2 dq$ is equivariant with respect to $U \equiv U_t = e^{-i\frac{t}{\hbar}H}$, where $H$ is the Schrödinger Hamiltonian (3.3) and $f \equiv f_t$ is the solution map of (3.2). In this regard, it is important to observe that for a Hamiltonian $H$ which is not of Schrödinger type we shouldn't expect (3.2) to be the appropriate velocity field, that is, a field which generates an evolution in configuration space having $|\Psi|^2$ as equivariant density. For example, for $H = c\frac{\hbar}{i}\frac{\partial}{\partial q}$, where $c$ is a constant (for simplicity we are assuming configuration space to be one-dimensional), we have that $|\Psi|^2$ is equivariant *provided* the evolution of configurations is given by $dQ/dt = c$. In other words, for $U_t = e^{ct\frac{\partial}{\partial q}}$ the map $\Psi \mapsto |\Psi|^2 dq$ is equivariant if $f_t : Q \to Q + ct$.

## 3.2.2 Conditional and Effective Wave Functions

Which systems should be governed by Bohmian mechanics? An $n$-particle subsystem of an $N$-particle system ($n < N$) need not in general be governed by Bohmian mechanics, since no wave function for the subsystem need exist. This will be so even with trivial interaction potential $V$, if the wave function of the system does not properly factorize; for nontrivial $V$ the Schrödinger evolution would in any case quickly destroy such a factorization. Therefore in a universe governed by Bohmian mechanics there is a priori only one wave function, namely that of the universe, and there is a priori only one system governed by Bohmian mechanics, namely the universe itself.

Consider then an $N$-particle non relativistic universe governed by Bohmian mechanics, with (universal) wave function $\Psi$. Focus on a subsystem with configuration variables $x$, i.e., on a splitting $q = (x, y)$ where $y$ represents the configuration of the *environment* of the $x$-system. The actual particle configurations at time $t$ are accordingly denoted by $X_t$ and $Y_t$, i.e., $Q_t = (X_t, Y_t)$. Note that $\Psi_t = \Psi_t(x, y)$. How can one assign a wave function to the $x$-system? For particles without spin, i.e. for the case of a scalar-valued wave function $\Psi$ of the universe, one obvious possibility—*afforded by the existence of the actual configuration*—is given by what in Sect. 2.5 we called the *conditional* wave function of the $x$-system (Eq. (2.39))

$$\psi_t(x) = \Psi_t(x, Y_t). \tag{3.6}$$

(For the case of particles with spin, see footnote 12 of Chap. 2. We also note that a subsystem always has a conditional density matrix, see [17]. For more on the notion of conditional wave function, see Sect. 12.2.)

To get familiar with this notion consider a very simple one dimensional universe made of two particles with Hamiltonian ($\hbar = 1$)

$$H = H^{(x)} + H^{(y)} + H^{(xy)} = -\frac{1}{2}\left(\frac{\partial^2}{\partial x^2} + \frac{\partial^2}{\partial y^2}\right) + \frac{1}{2}(x - y)^2,$$

and initial wave function

$$\Psi_0 = \psi \otimes \Phi_0 \quad \text{with} \quad \psi(x) = \pi^{-\frac{1}{4}} e^{-\frac{x^2}{2}} \quad \text{and} \quad \Phi_0(y) = \pi^{-\frac{1}{4}} e^{-\frac{y^2}{2}}.$$

Then (3.1) and (3.2) are easily solved:

$$\Psi_t(x, y) = \pi^{-\frac{1}{2}}(1 + it)^{-\frac{1}{2}} e^{-\frac{1}{4}\left[(x-y)^2 + \frac{(x+y)^2}{1+2it}\right]},$$

$$X_t = a(t)X + b(t)Y \quad \text{and} \quad Y_t = b(t)X + a(t)Y,$$

where $a(t) = \frac{1}{2}[(1+t^2)^{\frac{1}{2}} + 1]$, $b(t) = \frac{1}{2}[(1+t^2)^{\frac{1}{2}} - 1]$, and $X$ and $Y$ are the initial positions of the two particles. Focus now on one of the two particles (the $x$-system) and regard the other one as its environment (the $y$-system). The conditional wave function of the $x$-system

$$\psi_t(x) = \pi^{-\frac{1}{2}} (1+it)^{-\frac{1}{2}} e^{-\frac{1}{4}\left[(x-Y_t)^2 + \frac{(x+Y_t)^2}{1+2it}\right]},$$

depends, through $Y_t$, on *both* the initial condition $Y$ for the environment *and* the initial condition $X$ for the particle. As these are random, so is the evolution of $\psi_t$, with probability law determined by $|\Psi_0|^2$. In particular, $\psi_t$ does not satisfy Schrödinger's equation for any $H^{(x)}$.

We remark that even when the $x$-system is dynamically decoupled from its environment, its conditional wave function will not in general evolve according to Schrödinger's equation. Thus the conditional wave function lacks the *dynamical* implications from which the wave function of a system derives much of its physical significance. These are, however, captured by the notion of effective wave function:

Suppose that $\Psi(x,y) = \psi(x)\Phi(y) + \Psi^\perp(x,y)$, where $\Phi$ and $\Psi^\perp$ have macroscopically disjoint $y$-supports. If $Y \in \text{supp } \Phi$ we say that $\psi$ is the effective wave function *of the $x$-system.* (3.7)

Of course, $\psi$ is also the conditional wave function since nonvanishing scalar multiples of wave functions are naturally identified.[5]

### 3.2.3 Decoherence

One might wonder why systems possess an effective wave function at all. In fact, in general they don't! For example the $x$-system will not have an effective wave function when, for instance, it belongs to a larger microscopic system whose effective wave function doesn't factorize in the appropriate way. However, the *larger* the environment of the $x$-system, the *greater* is the potential for the existence of an effective wave function for this system, owing in effect to the abundance of "measurement-like" interaction with a larger environment.[6]

---

[5] Note that in Bohmian mechanics the wave function is naturally a projective object since wave functions differing by a multiplicative constant—possibly time-dependent—are associated with the same vector field, and thus generate the same dynamics.

[6] To understand how this comes about one may suppose that initially the $y$-supports of $\Phi$ and $\Psi^\perp$ (cf. the definition above of effective wave function) are just "sufficiently" (but not macroscopically) disjoint. Then, due to the interaction with the environment, the amount of $y$-disjointness will tend to increase dramatically as time goes on, with, as in a chain reaction, more and more degrees of freedom participating in this disjointness. When the effect of this "decoherence" is taken into

## 3.2 Bohmian Experiments

We remark that it is the relative stability of the macroscopic disjointness employed in the definition of the effective wave function, arising from what are nowadays often called mechanisms of decoherence—the destruction of the coherent spreading of the wave function, the effectively irreversible flow of "phase information" into the (macroscopic) environment—which accounts for the fact that the effective wave function of a system obeys Schrödinger's equation for the system alone whenever this system is isolated. One of the best descriptions of the mechanisms of decoherence, though not the word itself, can be found in Bohm's 1952 "hidden variables" papers [13, 14].

Decoherence plays a crucial role in the very formulation of the various interpretations of quantum theory loosely called decoherence theories (Griffiths [18], Omnès [19], Leggett [20], Zurek [21], Joos and Zeh [22], Gell-Mann and Hartle [23]). In this regard we wish to emphasize, however, as did Bell in his article "Against Measurement" [24], that decoherence in no way comes to grips with the measurement problem itself, being arguably a *necessary*, but certainly not a sufficient, condition for its complete resolution. In contrast, for Bohmian mechanics decoherence is purely phenomenological—it plays no role whatsoever in the formulation (or interpretation) of the theory itself[7]—and the very notion of effective wave function accounts at once for the reduction of the wave packet in quantum measurement.

According to orthodox quantum measurement theory [3, 25, 26, 27], after a measurement, or preparation, has been performed on a quantum system, the $x$-system, the wave function for the composite formed by system and apparatus is of the form

$$\sum_\alpha \psi_\alpha \otimes \Phi_\alpha \qquad (3.8)$$

with the different $\Phi_\alpha$ supported by the macroscopically distinct (sets of) configurations corresponding to the various possible outcomes of the measurement, e.g., given by apparatus pointer orientations. Of course, for Bohmian mechanics the terms of (3.8) are not all on the same footing: one of them, and only one, is selected, or more precisely supported, by the outcome—corresponding, say, to $\alpha_0$—which *actually* occurs. To emphasize this we may write (3.8) in the form

$$\psi \otimes \Phi + \Psi^\perp$$

where $\psi = \psi_{\alpha_0}$, $\Phi = \Phi_{\alpha_0}$, and $\Psi^\perp = \sum_{\alpha \neq \alpha_0} \psi_\alpha \otimes \Phi_\alpha$. By comparison with (3.7) it follows that after the measurement the $x$-system has effective wave function $\psi_{\alpha_0}$. This

---

account, one finds that even a small amount of $y$-disjointness will often tend to become "sufficient," and quickly "more than sufficient," and finally macroscopic.

[7] However, decoherence plays an important role in the emergence of Newtonian mechanics as the description of the macroscopic regime for Bohmian mechanics, supporting a picture of a macroscopic Bohmian particle, in the classical regime, guided by a macroscopically well-localized wave packet with a macroscopically sharp momentum moving along a classical trajectory. It may, indeed, seem somewhat ironic that the gross features of our world should appear classical because of interaction with the environment and the resulting wave function entanglement, the characteristic quantum innovation.

is how *collapse* (or *reduction*) of the effective wave function to the one associated with the outcome $\alpha_0$ arises in Bohmian mechanics.

While in orthodox quantum theory the "collapse" is merely superimposed upon the unitary evolution—without a precise specification of the circumstances under which it may legitimately be invoked—we have now, in Bohmian mechanics, that the evolution of the effective wave function is actually given by a stochastic process, which consistently embodies *both* unitarity *and* collapse as appropriate. In particular, the effective wave function of a subsystem evolves according to Schrödinger's equation when this system is suitably isolated. Otherwise it "pops in and out" of existence in a random fashion, in a way determined by the continuous (but still random) evolution of the conditional wave function $\psi_t$. Moreover, it is the critical dependence on the state of the environment and the initial conditions which is responsible for the random behavior of the (conditional or effective) wave function of the system.

### 3.2.4 Wave Function and State

As an important consequence of (3.6) we have, for the conditional probability distribution of the configuration $X_t$ of a system at time $t$, given the configuration $Y_t$ of its environment, the *fundamental conditional probability formula* (2.48),

$$\mathbb{P}(X_t \in dx \mid Y_t) = |\psi_t(x)|^2 \, dx, \qquad (3.9)$$

where

$$\mathbb{P}(dQ) = |\Psi_0(Q)|^2 \, dQ,$$

with $Q = (X, Y)$ the configuration of the universe at the (initial) time $t = 0$. Formula (3.9) is the cornerstone of the analysis on the origin of randomness in Bohmian mechanics in Chap. 2. Since the right hand side of (3.9) involves only the effective wave function, it follows that *the wave function $\psi_t$ of a subsystem represents maximal information about its configuration $X_t$*. In other words, given the fact that its wave function is $\psi_t$, it is in principle impossible to know more about the configuration of a system than what is expressed by the right hand side of (3.9), even when the detailed configuration $Y_t$ of its environment is taken into account [8]

$$\mathbb{P}(X_t \in dx \mid Y_t) = \mathbb{P}(X_t \in dx \mid \psi_t) = |\psi_t(x)|^2 \, dx. \qquad (3.10)$$

The fact that the knowledge of the configuration of a system must be mediated by its wave function may partially account for the possibility of identifying the *state* of a system—its complete description—with its wave function without encountering any *practical* difficulties. This is primarily because of the wave function's statistical role, but its dynamical role is also relevant here. Thus it is natural, even in Bohmian mechanics, to regard the wave function as the "*state*" of the system. This attitude is supported by the asymmetric roles of configuration and wave function: while the *fact*

3.2 Bohmian Experiments

that the wave function is $\psi$ entails that the configuration is distributed according to $|\psi|^2$, the *fact* that the configuration is $X$ has no implications whatsoever for the wave function.[8] Indeed, such an asymmetry is grounded in the dynamical laws *and* in the initial conditions. $\psi$ is always assumed to be fixed, being usually under experimental control, while $X$ is always taken as random, according to the quantum equilibrium distribution.

When all is said and done, it is important to bear in mind that regarding $\psi$ as the "state" is only of practical value, and shouldn't obscure the more important fact that the most detailed description—*the complete state description*—is given (in Bohmian mechanics) by the wave function *and* the configuration.

### 3.2.5 The Stern-Gerlach Experiment

Information about a system does not spontaneously pop into our heads or into our (other) "measuring" instruments; rather, it is generated by an *experiment*: some physical interaction between the system of interest and these instruments, which together (if there is more than one) comprise the *apparatus* for the experiment. Moreover, this interaction is defined by, and must be analyzed in terms of, the physical theory governing the behavior of the composite formed by system and apparatus. If the apparatus is well designed, the experiment should somehow convey significant information about the system. However, we cannot hope to understand the significance of this "information"—for example, the nature of what it is, if anything, that has been measured—without some such theoretical analysis.

As an illustration of such an analysis we shall discuss the Stern-Gerlach experiment from the standpoint of Bohmian mechanics. But first we must explain how *spin* is incorporated into Bohmian mechanics: If $\Psi$ is spinor-valued, the bilinear forms appearing in the numerator and denominator of (3.2) should be understood as spinor-inner-products; e.g., for a single spin 1/2 particle the two-component wave function

$$\Psi \equiv \begin{pmatrix} \Psi_+(\mathbf{x}) \\ \Psi_-(\mathbf{x}) \end{pmatrix}$$

generates the velocity

$$\mathbf{v}^\Psi = \frac{\hbar}{m} \mathrm{Im} \frac{(\Psi, \nabla \Psi)}{(\Psi, \Psi)} \tag{3.11}$$

where $(\cdot, \cdot)$ denotes the scalar product in the spin space $\mathbb{C}^2$. The wave function evolves via (3.1), where now the Hamiltonian $H$ contains the Pauli term, for a single particle proportional to $\mathbf{B} \cdot \boldsymbol{\sigma}$, that represents the coupling between the "spin" and an external magnetic field $\mathbf{B}$; here $\boldsymbol{\sigma} = (\sigma_x, \sigma_y, \sigma_z)$ are the Pauli spin matrices which can be taken to be

---

[8] The "fact" (that the configuration is $X$) shouldn't be confused with the "knowledge of the fact": the latter does have such implications! See Sect. 2.11.

$$\sigma_x = \begin{pmatrix} 0 & 1 \\ 1 & 0 \end{pmatrix}, \quad \sigma_y = \begin{pmatrix} 0 & -i \\ i & 0 \end{pmatrix}, \quad \sigma_z = \begin{pmatrix} 1 & 0 \\ 0 & -1 \end{pmatrix}.$$

Let's now focus on a Stern-Gerlach "measurement of the operator $\sigma_z$": An inhomogeneous magnetic field **B** is established in a neighborhood of the origin, by means of a suitable arrangement of magnets. This magnetic field is oriented in the positive $z$-direction, and is increasing in this direction. We also assume that the arrangement is invariant under translations in the $x$-direction, i.e., that the geometry does not depend upon $x$-coordinate. A particle with a fairly definite momentum is directed towards the origin along the negative $y$-axis. For simplicity, we shall consider a neutral spin-1/2 particle whose wave function $\Psi$ evolves according to the Hamiltonian

$$H = -\frac{\hbar^2}{2m}\nabla^2 - \mu\boldsymbol{\sigma}\cdot\mathbf{B}, \tag{3.12}$$

where $\mu$ is a positive constant (if one wishes, one might think of a fictitious electron not feeling the Lorentz force).

The inhomogeneous field generates a vertical deflection of $\Psi$ away from the $y$-axis, which for Bohmian mechanics leads to a similar deflection of the particle trajectory according to the velocity field defined by (3.11): if its wave function $\Psi$ were initially an eigenstate of $\sigma_z$ of eigenvalue 1 or $-1$, i.e., if it were of the form

$$\Psi^{(+)} = \psi^{(+)} \otimes \Phi_0(\mathbf{x}) \quad \text{or} \quad \Psi^{(-)} = \psi^{(-)} \otimes \Phi_0(\mathbf{x})$$

where

$$\psi^{(+)} \equiv \begin{pmatrix} 1 \\ 0 \end{pmatrix} \quad \text{and} \quad \psi^{(-)} \equiv \begin{pmatrix} 0 \\ 1 \end{pmatrix} \tag{3.13}$$

then the deflection would be in the positive (negative) $z$-direction (by a rather definite angle). This limiting behavior is readily seen for $\Phi_0 = \Phi_0(z)\phi(x,y)$ and $\mathbf{B} = (0, 0, B)$, so that the $z$-motion is completely decoupled from the motion along the other two directions, and by making the standard (albeit unphysical) assumption [25], [28]

$$\frac{\partial B}{\partial z} = const > 0. \tag{3.14}$$

whence

$$\mu\boldsymbol{\sigma}\cdot\mathbf{B} = (b + az)\sigma_z$$

where $a > 0$ and $b$ are constants. Then

$$\Psi_t^{(+)} = \begin{pmatrix} \Phi_t^{(+)}(z)\phi_t(x,y) \\ 0 \end{pmatrix} \quad \text{and} \quad \Psi_t^{(-)} = \begin{pmatrix} 0 \\ \Phi_t^{(-)}(z)\phi_t(x,y) \end{pmatrix}$$

where $\Phi_t^{(\pm)}$ are the solutions of

## 3.2 Bohmian Experiments

$$i\hbar \frac{\partial \Phi_t^{(\pm)}}{\partial t} = -\frac{\hbar^2}{2m}\frac{\partial^2 \Phi_t^{(\pm)}}{\partial z^2} \mp (b + a z)\Phi_t^{(\pm)}, \quad (3.15)$$

for initial conditions $\Phi_0^{(\pm)} = \Phi_0(z)$. Since $z$ generates translations of the $z$-component of the momentum, the behavior described above follows easily. More explicitly, the limiting behavior for $t \to \infty$ readily follows by a stationary phase argument on the explicit solution[9] of (3.15). More simply, we may consider the initial Gaussian state

$$\Phi_0 = \frac{e^{(-\frac{z^2}{4d^2})}}{(2d^2\pi)^{\frac{1}{4}}}$$

for which $|\Phi_t^{\pm}(z)|^2$, the probability density of the particle being at a point of $z$-coordinate $z$, is, by the linearity of the interaction in (3.15), a Gaussian with mean and mean square deviation given respectively by

$$\bar{z}(t) = (\pm)\frac{a t^2}{2m} \qquad d(t) = d\sqrt{1 + \frac{\hbar^2 t^2}{2m^2 d^4}}. \quad (3.16)$$

For a more general initial wave function,

$$\Psi = \psi \otimes \Phi_0 \qquad \psi = \alpha \psi^{(+)} + \beta \psi^{(-)} \quad (3.17)$$

passage through the magnetic field will, by linearity, split the wave function into an upward-deflected piece (proportional to $\psi^{(+)}$) and a downward-deflected piece (proportional to $\psi^{(-)}$), with corresponding deflections of the trajectories. The outcome is registered by detectors placed in the paths of these two possible "beams." Thus of the four kinematically possible outcomes ("pointer orientations") the occurrence of no detection and of simultaneous detection can be ignored as highly unlikely, and the two relevant outcomes correspond to registration by either the upper or the lower detector. Accordingly, for a measurement of $\sigma_z$ the experiment is equipped with a "calibration" (i.e., an assignment of numerical values to the outcomes of the experiment) $\lambda_+ = 1$ for upper detection and $\lambda_- = -1$ for lower detection (while for a measurement of the $z$-component of the spin angular momentum itself the calibration is given by $\frac{1}{2}\hbar\lambda_\pm$).

Note that one can completely understand what's going on in this Stern-Gerlach experiment without invoking any putative property of the electron such as its actual

---

[9] Equation (3.15) is readily solved:

$$\Phi_t^{(\pm)}(z) = \int G^{(\pm)}(z, z'; t) \Phi_0(z')\, dz',$$

where (by the standard rules for the Green's function of linear and quadratic Hamiltonians)

$$G^{(\pm)}(z, z'; t) = \sqrt{\frac{m}{2\pi i \hbar t}} e^{\frac{i}{\hbar}\left(\frac{m}{2t}\left(z - z' - (\pm)\frac{at^2}{m}\right)^2 + \frac{(\pm)at}{2}\left(z - z' - (\pm)\frac{at^2}{m}\right) - (\pm)(az' + b)t + \frac{at^3}{3m}\right)}.$$

$z$-component of spin that is supposed to be revealed in the experiment. For a general initial wave function there is no such property. What is more, the transparency of the analysis of this experiment makes it clear that there is nothing the least bit remarkable (or for that matter "nonclassical") about the *nonexistence* of this property. But the failure to pay attention to the role of operators as observables, i.e., to precisely what we should mean when we speak of measuring operator-observables, helps create a false impression of quantum peculiarity.

### 3.2.6 A Remark on the Reality of Spin in Bohmian Mechanics

Bell has said that (for Bohmian mechanics) spin is not real. Perhaps he should better have said: *"Even* spin is not real," not merely because of all observables, it is spin which is generally regarded as quantum mechanically most paradigmatic, but also because spin is treated in orthodox quantum theory very much like position, as a "degree of freedom"—a discrete index which supplements the continuous degrees of freedom corresponding to position—in the wave function.

Be that as it may, his basic meaning is, we believe, this: Unlike position, spin is not *primitive*, i.e., no *actual* discrete degrees of freedom, analogous to the *actual* positions of the particles, are added to the state description in order to deal with "particles with spin." Roughly speaking, spin is *merely* in the wave function. At the same time, as explained in Sect. 3.2.5, "spin measurements" are completely clear, and merely reflect the way spinor wave functions are incorporated into a description of the motion of configurations.

In this regard, it might be objected that while spin may not be primitive, so that the result of our "spin measurement" will not reflect any initial primitive property of the system, nonetheless this result *is* determined by the initial configuration of the system, i.e., by the position of our electron, together with its initial wave function, and as such—as a function $X_{\sigma_z}(\mathbf{q}, \psi)$ of the state of the system—it is some property of the system and in particular it is surely real. We shall address this issue in Sects. 3.8.3 and 3.8.4.

### 3.2.7 The Framework of Discrete Experiments

We shall now consider a generic experiment. Whatever its significance, the information conveyed by the experiment is registered in the apparatus as an *output*, represented, say, by the orientation of a pointer. Moreover, when we speak of a generic experiment, we have in mind a fairly definite initial state of the apparatus, the ready state $\Phi_0 = \Phi_0(y)$, one for which the apparatus should function as intended, and in particular one in which the *pointer* has some "null" orientation, as well as a definite initial state of the system $\psi = \psi(x)$ on which the experiment is performed. Under these conditions it turns out [8] that the initial $t = 0$ wave function $\Psi_0 = \Psi_0(q)$ of

3.2 Bohmian Experiments

the composite system formed by system and apparatus, with generic configuration $q = (x, y)$, has a product form, i.e.,

$$\Psi_0 = \psi \otimes \Phi_0.$$

Such a product form is an expression of the *independence* of system and apparatus immediately before the experiment begins.[10]

For Bohmian mechanics we should expect in general, as a consequence of the quantum equilibrium hypothesis, that the outcome of the experiment—the final pointer orientation—will be random: Even if the system and apparatus initially have definite, known wave functions, so that the outcome is determined by the initial configuration of system and apparatus, this configuration is random, since the composite system is in quantum equilibrium, and the distribution of the final configuration is given by $|\Psi_T(x, y)|^2$, where $\Psi_T$ is the wave function of the system-apparatus composite at the time $t = T$ when the experiment ends, and $x$, respectively $y$, is the generic system, respectively apparatus, configuration.

Suppose now that $\Psi_T$ has the form (3.8), which roughly corresponds to assuming that the experiment admits, i.e., that the apparatus is so designed that there is, only a finite (or countable) set of possible outcomes, given, say, by the different possible macroscopically distinct pointer orientations of the apparatus and corresponding to a partition of the apparatus configuration space into macroscopically disjoint regions $G_\alpha$, $\alpha = 1, 2, \ldots$.[11] We arrive in this way at the notion of *discrete experiment*, for which the time evolution arising from the interaction of the system and apparatus from $t = 0$ to $t = T$ is given by the unitary map

$$U : \mathcal{H} \otimes \Phi_0 \to \bigoplus_\alpha \mathcal{H} \otimes \Phi_\alpha, \quad \psi \otimes \Phi_0 \mapsto \Psi_T = \sum_\alpha \psi_\alpha \otimes \Phi_\alpha \qquad (3.18)$$

where $\mathcal{H}$ is the system Hilbert space of square-integrable wave functions with the usual inner product

$$\langle \psi, \phi \rangle = \int \psi^*(x) \phi(x) \, dx.$$

and the $\Phi_\alpha$ are a *fixed* set of (normalized) apparatus states supported by the macroscopically distinct regions $G_\alpha$ of apparatus configurations.

The experiment usually comes equipped with an assignment of numerical values $\lambda_\alpha$ (or a vector of such values) to the various outcomes $\alpha$. This assignment is defined by a "calibration" function $F$ on the apparatus configuration space assuming on each region $G_\alpha$ the constant value $\lambda_\alpha$. If for simplicity we assume that these values are in one-to-one correspondence with the outcomes[12] then

---

[10] It might be argued that it is somewhat unrealistic to assume a sharp preparation of $\psi$, as well as the possibility of resetting the apparatus always in the same initial state $\Phi_0$. We shall address this issue in Sect. 3.6.

[11] Note that to assume there are only finitely, or countably, many outcomes is really no assumption at all, since the outcome should ultimately be converted to digital form, whatever its initial representation may be.

[12] We shall consider the more general case later on in Sect. 3.3.2.4.

$$p_\alpha = \int_{F^{-1}(\lambda_\alpha)} |\Psi_T(x,y)|^2 dx\, dy = \int_{G_\alpha} |\Psi_T(x,y)|^2 dx\, dy \qquad (3.19)$$

is the probability of finding $\lambda_\alpha$, for initial system wave function $\psi$. Since $\Phi_{\alpha'}(y) = 0$ for $y \in G_\alpha$ unless $\alpha = \alpha'$, we obtain

$$p_\alpha = \int dx \int_{G_\alpha} |\sum_{\alpha'} \psi_{\alpha'}(x) \Phi_{\alpha'}(y)|^2 dy = \int |\psi_\alpha(x)|^2 dx = \|\psi_\alpha\|^2. \qquad (3.20)$$

Note that when the result $\lambda_\alpha$ is obtained, the effective wave function of the system undergoes the transformation $\psi \to \psi_\alpha$.

A simple example of a discrete experiment is provided by the map

$$U : \psi \otimes \Phi_0 \mapsto \sum_\alpha c_\alpha \psi \otimes \Phi_\alpha, \qquad (3.21)$$

where the $c_\alpha$ are complex numbers such that $\sum_\alpha |c_\alpha|^2 = 1$; then $p_\alpha = |c_\alpha|^2$. Note that the experiment defined by (3.21) resembles a coin-flip more than a measurement since the outcome $\alpha$ occurs with a probability independent of $\psi$.

### 3.2.8 Reproducibility and its Consequences

Though for a generic discrete experiment there is no reason to expect the sort of "measurement-like" behavior typical of familiar quantum measurements, there are, however, special experiments whose outcomes are somewhat less random than we might have thought possible. According to Schrödinger [29]:

> The systematically arranged interaction of two systems (measuring object and measuring instrument) is called a measurement on the first system, if a directly-sensible variable feature of the second (pointer position) is always reproduced within certain error limits when the process is immediately repeated (on the same object, which in the mean time must not be exposed to additional influences).

To implement the notion of "measurement-like" experiment considered by Schrödinger, we first make some preliminary observations concerning the unitary map (3.18). Let $P_{[\Phi_\alpha]}$ be the orthogonal projection in the Hilbert space $\bigoplus_\alpha \mathcal{H} \otimes \Phi_\alpha$ onto the subspace $\mathcal{H} \otimes \Phi_\alpha$ and let $\widetilde{\mathcal{H}_\alpha}$ be the subspaces of $\mathcal{H}$ defined by

$$P_{[\Phi_\alpha]} [U(\mathcal{H} \otimes \Phi_0)] = \widetilde{\mathcal{H}_\alpha} \otimes \Phi_\alpha. \qquad (3.22)$$

(Since the vectors in $\widetilde{\mathcal{H}_\alpha}$ arise from projecting $\Psi_T = \sum_\alpha \psi_\alpha \otimes \Phi_\alpha$ onto its $\alpha$-component, $\widetilde{\mathcal{H}_\alpha}$ is the space of the "collapsed" wave functions associated with the occurrence of the outcome $\alpha$.) Then

$$U(\mathcal{H} \otimes \Phi_0) \subseteq \bigoplus_\alpha \widetilde{\mathcal{H}_\alpha} \otimes \Phi_\alpha. \qquad (3.23)$$

## 3.2 Bohmian Experiments

Note, however, that it need not be the case that $U(\mathcal{H} \otimes \Phi_0) = \bigoplus_\alpha \widetilde{\mathcal{H}_\alpha} \otimes \Phi_\alpha$, and that the spaces $\widetilde{\mathcal{H}_\alpha}$ need be neither orthogonal nor distinct; e.g., for (3.21) $\widetilde{\mathcal{H}_\alpha} = \mathcal{H}$ and $U(\mathcal{H} \otimes \Phi_0) = \mathcal{H} \otimes \sum_\alpha c_\alpha \Phi_\alpha \neq \bigoplus_\alpha \mathcal{H} \otimes \Phi_\alpha$.[13]

A "measurement-like" experiment is one which is reproducible in the sense that it will yield the same outcome as originally obtained if it is immediately repeated. (This means in particular that the apparatus must be immediately reset to its ready state, or a fresh apparatus must be employed, while the system is not tampered with so that its initial state for the repeated experiment is its final state produced by the first experiment.) Thus the experiment is *reproducible* if

$$U(\widetilde{\mathcal{H}_\alpha} \otimes \Phi_0) \subseteq \widetilde{\mathcal{H}_\alpha} \otimes \Phi_\alpha \tag{3.24}$$

or, equivalently, if there are spaces $\mathcal{H}_\alpha' \subseteq \widetilde{\mathcal{H}_\alpha}$ such that

$$U(\widetilde{\mathcal{H}_\alpha} \otimes \Phi_0) = \mathcal{H}_\alpha' \otimes \Phi_\alpha. \tag{3.25}$$

Note that it follows from the unitarity of $U$ and the orthogonality of the subspaces $\widetilde{\mathcal{H}_\alpha} \otimes \Phi_\alpha$ that the subspaces $\widetilde{\mathcal{H}_\alpha} \otimes \Phi_0$ and hence the $\widetilde{\mathcal{H}_\alpha}$ are also orthogonal. Therefore, by taking the orthogonal sum over $\alpha$ of both sides of (3.25), we obtain

$$\bigoplus_\alpha U(\widetilde{\mathcal{H}_\alpha} \otimes \Phi_0) = U\left(\bigoplus_\alpha \widetilde{\mathcal{H}_\alpha} \otimes \Phi_0\right) = \bigoplus_\alpha \mathcal{H}_\alpha' \otimes \Phi_\alpha. \tag{3.26}$$

If we now make the simplifying assumption that the subspaces $\widetilde{\mathcal{H}_\alpha}$ are finite dimensional, we have from unitarity that $\widetilde{\mathcal{H}_\alpha} = \mathcal{H}_\alpha'$, and thus, by comparing (3.23) and (3.26), that equality holds in (3.23) and that

$$\mathcal{H} = \bigoplus_\alpha \mathcal{H}_\alpha \tag{3.27}$$

with

$$U(\mathcal{H}_\alpha \otimes \Phi_0) = \mathcal{H}_\alpha \otimes \Phi_\alpha \tag{3.28}$$

for

$$\mathcal{H}_\alpha \equiv \widetilde{\mathcal{H}_\alpha} = \mathcal{H}_\alpha'.$$

Therefore if the wave function of the system is initially in $\mathcal{H}_\alpha$, outcome $\alpha$ definitely occurs and the value $\lambda_\alpha$ is thus definitely obtained (assuming again for simplicity one-to-one correspondence between outcomes and results). It then follows that for a general initial system wave function

$$\psi = \sum_\alpha P_{\mathcal{H}_\alpha} \psi,$$

---

[13] Note that if $\mathcal{H}$ has finite dimension $n$, and the number of outcomes $\alpha$ is $m$, $\dim[U(\mathcal{H} \otimes \Phi_0)] = n$, while $\dim[\bigoplus_\alpha \mathcal{H} \otimes \Phi_\alpha] = n \cdot m$.

where $P_{\mathcal{H}_a}$ is the projection in $\mathcal{H}$ onto the subspace $\mathcal{H}_\alpha$, that the outcome $\alpha$, with result $\lambda_\alpha$, is obtained with (the usual) probability

$$p_\alpha = \|P_{\mathcal{H}_a}\psi\|^2 = \langle\psi, P_{\mathcal{H}_a}\psi\rangle, \tag{3.29}$$

which follows from (3.28), (3.20), and (3.18) since $U(P_{\mathcal{H}_a}\psi \otimes \Phi_0) = \psi_\alpha \otimes \Phi_\alpha$ and hence $\|P_{\mathcal{H}_a}\psi\| = \|\psi_\alpha\|$ by unitarity. In particular, when the $\lambda_\alpha$ are real-valued, the expected value obtained is

$$\sum_\alpha p_\alpha \lambda_\alpha = \sum_\alpha \lambda_\alpha \|P_{\mathcal{H}_a}\psi\|^2 = \langle\psi, A\psi\rangle \tag{3.30}$$

where

$$A = \sum_\alpha \lambda_\alpha P_{\mathcal{H}_a} \tag{3.31}$$

is the self-adjoint operator with eigenvalues $\lambda_\alpha$ and spectral projections $P_{\mathcal{H}_a}$.

### 3.2.9 Operators as Observables

What we wish to emphasize here is that, insofar as the statistics for the values which result from the experiment are concerned,

> the relevant data for the experiment are the collection $\{\mathcal{H}_\alpha\}$ of special orthogonal subspaces, together with the corresponding calibration $\{\lambda_\alpha\}$, (3.32)

and *this data is compactly expressed and represented by the self-adjoint operator A, on the system Hilbert space $\mathcal{H}$, given by* (3.31). Thus, under the assumptions we have made, with a reproducible experiment $\mathscr{E}$ we naturally associate an operator $A = A_\mathscr{E}$, a single mathematical object, defined on the system alone, in terms of which an efficient description (3.29) of the statistics of the possible results is achieved; we shall denote this association by

$$\mathscr{E} \mapsto A. \tag{3.33}$$

If we wish we may speak of "operators as observables," and when an experiment $\mathscr{E}$ is associated with a self-adjoint operator $A$, as described above, we may say that *the experiment $\mathscr{E}$ is a "measurement" of the observable represented by the self-adjoint operator A*. If we do so, however, it is important that we appreciate that in so speaking we merely refer to what we have just derived: the role of operators in the description of certain experiments.[14]

---

[14] Operators as observables also naturally convey information about the system's wave function after the experiment. For example, for an ideal measurement, when the outcome is $\alpha$ the wave

3.2 Bohmian Experiments                                                                        97

So understood, the notion of operator-as-observable in no way implies that anything is genuinely measured in the experiment, and certainly not the operator itself! In a general experiment no system property is being measured, even if the experiment happens to be measurement-like. (Position measurements in Bohmian mechanics are of course an important exception.) What in general is going on in obtaining outcome $\alpha$ is completely straightforward and in no way suggests, or assigns any substantive meaning to, statements to the effect that, prior to the experiment, observable $A$ somehow had a value $\lambda_\alpha$—whether this be in some determinate sense or in the sense of Heisenberg's "potentiality" or some other ill-defined fuzzy sense—which is revealed, or crystallized, by the experiment. Even speaking of the observable $A$ as having value $\lambda_\alpha$ when the system's wave function is in $\mathscr{H}_\alpha$, i.e., when this wave function is an eigenstate of $A$ of eigenvalue $\lambda_\alpha$—insofar as it suggests that something peculiarly quantum is going on when the wave function is not an eigenstate whereas in fact there is nothing the least bit peculiar about the situation—perhaps does more harm than good.

It might be objected that we are claiming to arrive at the quantum formalism under somewhat unrealistic assumptions, such as, for example, reproducibility or finite dimensionality. We agree. But this objection misses the point of the exercise. The quantum formalism itself is an idealization; when applicable at all, it is only as an approximation. Beyond illuminating the role of operators as ingredients in this formalism, our point was to indicate how naturally it emerges. In this regard we must emphasize that the following question arises for quantum orthodoxy, but does not arise for Bohmian mechanics: For precisely which theory is the quantum formalism an idealization?

We shall discuss how to go beyond the idealization involved in the quantum formalism in Sect. 3.4—after having analyzed it thoroughly in Sect. 3.3. First we wish to show that many more experiments than those satisfying our assumptions can indeed be associated with operators in exactly the manner we have described.

### 3.2.10 The General Framework of Bohmian Experiments

According to (3.19) the statistics of the results of a discrete experiment are governed by the probability measure $\mathsf{P}_{\Psi_T} \circ F^{-1}$, where $\mathsf{P}_{\Psi_T}(dq) = |\Psi_T(q)|^2 dq$ is the quantum equilibrium measure. Note that discreteness of the value space of $F$ plays no role in the characterization of this measure. This suggests that we may consider a more general notion of experiment, not based on the assumption of a countable set of outcomes, but only on the *unitarity* of the operator $U$, which transforms the initial state $\psi \otimes \Phi_0$ into the final state $\Psi_T$, and on a generic *calibration function* $F$ from the configuration space of the composite system to some value space, e.g., $\mathbb{R}$, or $\mathbb{R}^m$, giving the result of the experiment as a function $F(Q_T)$ of the final configuration $Q_T$ of system and apparatus. We arrive in this way at the notion of *general experiment*

---

function of the system after the experiment is (proportional to) $P_{\mathscr{H}_\alpha} \psi$. We shall elaborate upon this in the next section.

$$\mathscr{E} \equiv \{\Phi_0, U, F\}, \tag{3.34}$$

where the unitary $U$ embodies the interaction of system and apparatus and the function $F$ could be completely general. Of course, for application to the results of real-world experiments $F$ might represent the "orientation of the apparatus pointer" or some coarse-graining thereof.

Performing $\mathscr{E}$ on a system with initial wave function $\psi$ leads to the result $Z = F(Q_T)$ and since $Q_T$ is randomly distributed according to the quantum equilibrium measure $\mathsf{P}_{\Psi_T}$, the probability distribution of $Z$ is given by the induced measure

$$\mathsf{P}_\psi^Z = \mathsf{P}_{\Psi_T} \circ F^{-1}. \tag{3.35}$$

(We have made explicit only the dependence of the measure on $\psi$, since the initial apparatus state $\Phi_0$ is of course fixed, defined by the experiment $\mathscr{E}$.) Note that this more general notion of experiment eliminates the slight vagueness arising from the imprecise notion of macroscopic upon which the notion of discrete experiment is based. Note also that the structure (3.34) conveys information about the wave function (3.6) of the system after a certain result $F(Q_T)$ is obtained.

Note, however, that this somewhat formal notion of experiment may not contain enough information to determine the detailed Bohmian dynamics, which would require specification of the Hamiltonian of the system-apparatus composite, that might not be captured by $U$. In particular, the final configuration $Q_T$ may not be determined, for given initial wave function, as a function of the initial configuration of system and apparatus. $\mathscr{E}$ does, however, determine what is relevant for our purposes about the random variable $Q_T$, namely its distribution, and hence that of $Z = F(Q_T)$.

Let us now focus on the right had side of the Eq. (3.29), which establishes the association of operators with experiments: $\langle \psi, P_{\mathscr{H}_\alpha} \psi \rangle$ is the probability that "the operator $A$ has value $\lambda_\alpha$," and according to standard quantum mechanics the statistics of the results of measuring a general self-adjoint operator $A$, not necessarily with pure point spectrum, in the (normalized) state $\psi$ are described by the probability measure

$$\Delta \mapsto \mu_\psi^A(\Delta) \equiv \langle \psi, P^A(\Delta)\psi \rangle \tag{3.36}$$

where $\Delta$ is a (Borel) set of real numbers and $P^A : \Delta \mapsto P^A(\Delta)$ is the *projection-valued-measure* (PVM) uniquely associated with $A$ by the spectral theorem. (We recall [30] that a PVM is a normalized, countably additive set function whose values are, instead of nonnegative reals, orthogonal projections on a Hilbert space $\mathscr{H}$. Any PVM $P$ on $\mathscr{H}$ determines, for any given $\psi \in \mathscr{H}$, a probability measure $\mu_\psi \equiv \mu_\psi^P : \Delta \mapsto \langle \psi, P(\Delta)\psi \rangle$ on $\mathbb{R}$. Integration against projection-valued-measure is analogous to integration against ordinary measures, so that $B \equiv \int f(\lambda) P(d\lambda)$ is well-defined, as an operator on $\mathscr{H}$. Moreover, by the spectral theorem every self-adjoint operator $A$ is of the form $A = \int \lambda P(d\lambda)$, for a unique projection-valued-measure $P = P^A$, and $\int f(\lambda) P(d\lambda) = f(A)$. )

It is then rather clear how (3.33) extends to general self-adjoint operators: *a general experiment $\mathscr{E}$ is a measurement of the self-adjoint operator $A$ if the statistics of the* *results of $\mathscr{E}$ are given by (3.36)*, i.e.,

$$\mathscr{E} \mapsto A \quad \text{if and only if} \quad \mathsf{P}^Z_\psi = \mu^A_\psi. \tag{3.37}$$

In particular, if $\mathscr{E} \mapsto A$, then the moments of the result of $\mathscr{E}$ are the moments of $A$:

$$<Z^n> = \int \lambda^n \langle \psi, P(d\lambda)\psi \rangle = \langle \psi, A^n \psi \rangle.$$

## 3.3 The Quantum Formalism

The spirit of this section will be rather different from that of the previous one. Here the focus will be on the formal structure of experiments measuring self-adjoint operators. Our aim is to show that the standard quantum formalism emerges from a *formal* analysis of the association $\mathscr{E} \mapsto A$ between operator and experiment provided by (3.37). By "formal analysis" we mean not only that the detailed physical conditions under which might $\mathscr{E} \mapsto A$ hold (e.g., reproducibility) will play no role, but also that the practical requirement that $\mathscr{E}$ be physically realizable will be of no relevance whatsoever.

Note that such a formal approach is unavoidable in order to recover the quantum formalism. In fact, within the quantum formalism one may consider measurements of arbitrary self-adjoint operators, for example, the operator $A = \hat{X}^2 \hat{P} + \hat{P} X^2$, where $\hat{X}$ and $\hat{P}$ are respectively the position and the momentum operators. However, it may very well be the case that no "real world" experiment measuring $A$ exists. Thus, in order to allow for measurements of arbitrary self-adjoint operators we shall regard (3.34) as characterizing an "*abstract experiment*"; in particular, we shall not regard the unitary map $U$ as arising necessarily from a (realizable) Schrödinger time evolution. We may also speak of virtual experiments.

In this regard one should observe that to resort to a formal analysis is indeed quite common in physics. Consider, e.g., the Hamiltonian formulation of classical mechanics that arose from an abstraction of the physical description of the world provided by Newtonian mechanics. Here we may freely speak of completely general Hamiltonians, e.g. $H(p,q) = p^6$, without being concerned about whether they are physical or not. Indeed, only very few Hamiltonians correspond to physically realizable motions!

**A Warning:** As we have stressed in the introduction and in Sect. 3.2.9, when we speak here of a measurement we don't usually mean a *genuine* measurement—an experiment revealing the pre-existing value of a quantity of interest, the measured quantity or property. (We speak in this unfortunate way because it is standard.) Genuine measurement will be discussed much later, in Sect. 3.7.

## 3.3.1 Weak Formal Measurements

The first formal notion we shall consider is that of weak formal measurement, formalizing the relevant data of an experiment measuring a self-adjoint operator:

> Any orthogonal decomposition $\mathscr{H} = \bigoplus_\alpha \mathscr{H}_\alpha$, i.e., any complete collection $\{\mathscr{H}_\alpha\}$ of mutually orthogonal subspaces, paired with any set $\{\lambda_\alpha\}$ of distinct real numbers, defines the weak formal measurement $\mathscr{M} \equiv \{(\mathscr{H}_\alpha, \lambda_\alpha)\} \equiv \{\mathscr{H}_\alpha, \lambda_\alpha\}$. (3.38)

(Compare (3.38) with (3.32) and note that now we are not assuming that the spaces $\mathscr{H}_\alpha$ are finite-dimensional.) The notion of weak formal measurement is aimed at expressing the minimal structure that all experiments (some or all of which might be virtual) measuring the same operator $A = \sum \lambda_\alpha P_{\mathscr{H}_\alpha}$ have in common ($P_{\mathscr{H}_\alpha}$ is the orthogonal projection onto the subspace $\mathscr{H}_\alpha$). Then, "to perform $\mathscr{M}$" shall mean to perform (at least virtually) any one of these experiments, i.e., any experiment such that

$$p_\alpha = \langle \psi, P_{\mathscr{H}_\alpha} \psi \rangle \qquad (3.39)$$

is the probability of obtaining the result $\lambda_\alpha$ on a system initially in the state $\psi$. (This is of course equivalent to requiring that the result $\lambda_\alpha$ is definitely obtained if and only if the initial wave function $\psi \in \mathscr{H}_\alpha$.)

Given $\mathscr{M} \equiv \{\mathscr{H}_\alpha, \lambda_\alpha\}$ consider the set function

$$P : \Delta \mapsto P(\Delta) \equiv \sum_{\lambda_\alpha \in \Delta} P_{\mathscr{H}_\alpha}, \qquad (3.40)$$

where $\Delta$ is a set of real numbers (technically, a Borel set). Then

1) $P$ is *normalized*, i.e., $P(\mathbb{R}) = I$, where $I$ is the identity operator and $\mathbb{R}$ is the real line,
2) $P(\Delta)$ is an *orthogonal projection*, i.e., $P(\Delta)^2 = P(\Delta) = P(\Delta)^*$,
3) $P$ is *countably additive*, i.e., $P(\bigcup_n \Delta_n) = \sum_n P(\Delta_n)$, for $\Delta_n$ disjoint sets.

Thus $P$ is a projection-valued-measure and therefore the notion of weak formal measurement is indeed equivalent to that of "discrete" PVM, that is, a PVM supported by a countable set $\{\lambda_\alpha\}$ of values.

More general PVMs, e.g. PVMs supported by a continuous set of values, will arise if we extend (3.38) and base the notion of weak formal measurement upon the general association (3.37) between experiments and operators. If we stipulate that

> any projection-valued-measure $P$ on $\mathscr{H}$ defines a weak formal measurement $\mathscr{M} \equiv P$,

(3.41)

## 3.3 The Quantum Formalism

then "to perform $\mathscr{M}$" shall mean to perform any experiment $\mathscr{E}$ associated with $A = \int \lambda P(d\lambda)$ in the sense of (3.37).

Note that since by the spectral theorem there is a natural one-to-one correspondence between PVMs and self-adjoint operators, we may speak equivalently of *the* operator $A = A_{\mathscr{M}}$, for given $\mathscr{M}$, or of *the* weak formal $\mathscr{M} = \mathscr{M}_A$, for given $A$. In particular, the weak formal measurement $\mathscr{M}_A$ represents the equivalence class of *all* experiments $\mathscr{E} \to A$.

### 3.3.2 Strong Formal Measurements

We wish now to classify the different experiments $\mathscr{E}$ associated with the same self-adjoint operator $A$ by taking into account the effect of $\mathscr{E}$ on the state of the system, i.e., the state transformations $\psi \to \psi_\alpha$ induced by the occurrence of the various results $\lambda_\alpha$ of $\mathscr{E}$. Accordingly, unless otherwise stated, from now on we shall assume $\mathscr{E}$ to be a discrete experiment measuring $A = \sum \lambda_\alpha P_{\mathscr{H}_\alpha}$, for which the state transformation $\psi \to \psi_\alpha$ is defined by (3.18). This leads to the notion of strong formal measurements. For the most important types of strong formal measurements, ideal, normal and standard, there is a one-to-one correspondence between $\alpha$'s and numerical results $\lambda_\alpha$.

#### 3.3.2.1 Ideal Measurements

Given a weak formal measurement of $A$, the simplest possibility for the transition $\psi \to \psi_\alpha$ is that when the result $\lambda_\alpha$ is obtained, the initial state $\psi$ is projected onto the corresponding space $\mathscr{H}_\alpha$, i.e., that

$$\psi \to \psi_\alpha = P_{\mathscr{H}_a}\psi. \tag{3.42}$$

This prescription defines uniquely the *ideal measurement*ideal of $A$. (The transformation $\psi \to \psi_\alpha$ should be regarded as defined only in the projective sense: $\psi \to \psi_\alpha$ and $\psi \to c\psi_\alpha$ ($c \neq 0$) should be regarded as the same transition.) "To perform an ideal measurement of $A$" shall then mean to perform a discrete experiment $\mathscr{E}$ whose results are statistically distributed according to (3.39) and whose state transformations (3.18) are given by (3.42).

Under an ideal measurement the wave function changes as little as possible: an initial $\psi \in \mathscr{H}_\alpha$ is unchanged by the measurement. Ideal measurements have always played a privileged role in quantum mechanics. It is the ideal measurements that are most frequently discussed in textbooks. It is for ideal measurements that the standard collapse rule is obeyed. When Dirac [31] wrote: "a measurement always causes the system to jump into an eigenstate of the dynamical variable that is being measured" he was referring to an ideal measurement.

#### 3.3.2.2 Normal Measurements

The rigid structure of ideal measurements can be weakened by requiring only that $\mathscr{H}_\alpha$ as a whole, and not the individual vectors in $\mathscr{H}_\alpha$, is unchanged by the measurement and therefore that the state transformations induced by the measurement are such that when the result $\lambda_\alpha$ is obtained the transition

$$\psi \to \psi_\alpha = U_\alpha P_{\mathscr{H}_a} \psi \qquad (3.43)$$

occurs, where the $U_\alpha$ are operators on $\mathscr{H}_\alpha$ ($U_\alpha : \mathscr{H}_\alpha \to \mathscr{H}_\alpha$). Then for any such discrete experiment $\mathscr{E}$ measuring $A$, the $U_\alpha$ can be chosen so that (3.43) agrees with (3.18), i.e., so that for $\psi \in \mathscr{H}_\alpha$, $U(\psi \otimes \Phi_0) = U_\alpha \psi \otimes \Phi_\alpha$, and hence so that $U_\alpha$ is unitary (or at least a partial isometry). Such a measurement, with unitaries $U_\alpha : \mathscr{H}_\alpha \to \mathscr{H}_\alpha$, will be called a *normal measurement* of $A$.

In contrast with an ideal measurement, a normal measurement of an operator is not uniquely determined by the operator itself: additional information is needed to determine the transitions, and this is provided by the family $\{U_\alpha\}$. Different families define different normal measurements of the same operator. Note that ideal measurements are, of course, normal (with $U_\alpha = I_\alpha \equiv$ identity on $\mathscr{H}_\alpha$), and that normal measurements with one-dimensional subspaces $\mathscr{H}_\alpha$ are necessarily ideal.

Since the transformations (3.43) leave invariant the subspaces $\mathscr{H}_\alpha$, the notion of normal measurement characterizes completely the class of reproducible measurements of self-adjoint operators. Following the terminology introduced by Pauli [32], normal measurement are sometimes called *measurements of first kind*. Normal measurements are also *quantum non demolition (QND) measurements* [33], defined as measurements such that the operators describing the induced state transformations, i.e, the operators $R_\alpha \equiv U_\alpha P_{\mathscr{H}_a}$, commute with the measured operator $A = \sum \lambda_\alpha P_{\mathscr{H}_a}$. (This condition is regarded as expressing that the measurement leaves the measured observable $A$ unperturbed).

#### 3.3.2.3 Standard Measurements

We may now drop the condition that the $\mathscr{H}_\alpha$ are left invariant by the measurement and consider the very general state transformations

$$\psi \to \psi_\alpha = T_\alpha P_{\mathscr{H}_a} \psi \qquad (3.44)$$

with operators $T_\alpha : \mathscr{H}_\alpha \to \mathscr{H}$. Then, exactly as for the case of normal measurements, it follows that $T_\alpha$ can be chosen to be unitary from $\mathscr{H}_\alpha$ onto its range $\widetilde{\mathscr{H}}_\alpha$. The subspaces $\widetilde{\mathscr{H}}_\alpha$ need be neither orthogonal nor distinct. We shall write $R_\alpha = T_\alpha P_{\mathscr{H}_a}$ for the general transition operators. With $T_\alpha$ as chosen, $R_\alpha$ is characterized by the equation $R_\alpha^* R_\alpha = P_{\mathscr{H}_a}$ (where $R_\alpha^*$ denotes the adjoint of $R_\alpha$).

The state transformations (3.44), given by unitaries $T_\alpha : \mathscr{H}_\alpha \to \widetilde{\mathscr{H}}_\alpha$, or equivalently by bounded operators $R_\alpha$ on $\mathscr{H}$ satisfying $R_\alpha^* R_\alpha = P_{\mathscr{H}_a}$, define what we shall call a *standard measurement* of $A$. Note that normal measurements are standard

## 3.3 The Quantum Formalism

measurements with $\widetilde{\mathcal{H}}_\alpha = \mathcal{H}_\alpha$ (or $\widetilde{\mathcal{H}}_\alpha \subset \mathcal{H}_\alpha$). Although standard measurements are in a sense more realistic than normal measurements (real world measurements are seldom reproducible in a strict sense), they are very rarely discussed in textbooks. We emphasize that the crucial data in a standard measurement is given by $R_\alpha$, which governs both the state transformations ($\psi \to R_a \psi$) and the probabilities ($p_\alpha = \langle \psi, P_{\mathcal{H}_a} \psi \rangle = \|R_\alpha \psi\|^2$).

We shall illustrate the main features of standard measurements by considering a very simple example: Let $\{e_0, e_1, e_2, \dots\}$, be a fixed orthonormal basis of $\mathcal{H}$ and consider the standard measurement whose results are the numbers $0, 1, 2, \dots$ and whose state transformations are defined by the operators

$$R_\alpha \equiv |e_0\rangle\langle e_\alpha| \quad \text{i.e.,} \quad R_\alpha \psi = \langle e_\alpha, \psi \rangle e_0, \quad \alpha = 0, 1, 2, \dots$$

With such $R_\alpha$'s are associated the projections $P_\alpha = R_\alpha^* R_\alpha = |e_\alpha\rangle\langle e_\alpha|$, i.e., the projections onto the one dimensional spaces $\mathcal{H}_\alpha$ spanned respectively by the vectors $e_\alpha$. Thus, this is a measurement of the operator $A = \sum_\alpha \alpha |e_\alpha\rangle\langle e_\alpha|$. Note that the spaces $\widetilde{\mathcal{H}}_\alpha$, i.e. the ranges of the $R_\alpha$'s, are all the same and equal to the space $\mathcal{H}_0$ generated by the vector $e_0$. The measurement is then not normal since $\mathcal{H}_\alpha \neq \widetilde{\mathcal{H}}_\alpha$. Finally, note that this measurement could be regarded as giving a simple model for a photo detection experiment, where any state is projected onto the "vacuum state" $e_0$ after the detection.

### 3.3.2.4 Strong Formal Measurements

We shall now relax the condition that $\alpha \mapsto \lambda_\alpha$ is one-to-one, as we would have to do for an experiment having a general calibration $\alpha \mapsto \lambda_\alpha$, which need not be invertible. This leads to (what we shall call) a *strong formal measurement*. Since this notion provides the most general formalization of the notion of a "measurement of a self-adjoint operator" that takes into account the effect of the measurement on the state of the system, we shall spell it out precisely as follows:

> Any complete (labelled) collection $\{\mathcal{H}_\alpha\}$ of mutually orthogonal subspaces, any (labelled) set $\{\lambda_\alpha\}$ of not necessarily distinct real numbers, and any (labelled) collection $\{R_\alpha\}$ of bounded operators on $\mathcal{H}$, such that $R_\alpha^* R_\alpha \equiv P_{\mathcal{H}_a}$ (the projection onto $\mathcal{H}_\alpha$), defines a strong formal measurement. (3.45)

A strong formal measurement will be compactly denoted by

$$\mathcal{M} \equiv \{(\mathcal{H}_\alpha, \lambda_\alpha, R_\alpha)\} \equiv \{\mathcal{H}_\alpha, \lambda_\alpha, R_\alpha\},$$

or even more compactly by $\mathcal{M} \equiv \{\lambda_\alpha, R_\alpha\}$ (the spaces $\mathcal{H}_\alpha$ can be extracted from the projections $P_{\mathcal{H}_a} = R_\alpha^* R_\alpha$). With $\mathcal{M}$ is associated the operator $A = \sum \lambda_\alpha P_{\mathcal{H}_a}$. Note that since the $\lambda_\alpha$ are not necessarily distinct numbers, $P_{\mathcal{H}_a}$ need not be the spectral projection $P^A(\lambda_\alpha)$ associated with $\lambda_\alpha$; in general

$$P^A(\lambda) = \sum_{\alpha:\lambda_\alpha=\lambda} P_{\mathcal{H}_\alpha},$$

i.e., it is the sum of all the $P_{\mathcal{H}_\alpha}$'s that are associated with the value $\lambda$.[15] "*To perform the measurement $\mathcal{M}$*" on a system initially in $\psi$ shall accordingly mean to perform a discrete experiment $\mathcal{E}$ such that: 1) the probability $p(\lambda)$ of getting the result $\lambda$ is governed by $A$, i.e., $p(\lambda) = \langle \psi, P^A(\lambda)\psi \rangle$, and 2) the state transformations of $\mathcal{E}$ are those prescribed by $\mathcal{M}$, i.e., $\psi \to \psi_\alpha = R_\alpha \psi$.

Observe that strong formal measurements do provide a more realistic formalization of the notion of measurement of an operator than standard measurements: the notion of discrete experiment does not imply a one-to-one correspondence between outcomes, i.e, final macroscopic configurations of the pointer, and the numerical results of the experiment.

The relationship between (weak or strong) formal measurements, self-adjoint operators, and experiments can be summarized by the following sequence of maps:

$$\mathcal{E} \mapsto \mathcal{M} \mapsto A \qquad (3.46)$$

The first map expresses that $\mathcal{M}$ (weak or strong) is a formalization of $\mathcal{E}$—it contains the "relevant data" about $\mathcal{E}$—and it will be many-to-one if $\mathcal{M}$ is a weak formal measurement[16]; the second map expresses that $\mathcal{M}$ is a formal measurement of $A$ and it will be many-to-one if $\mathcal{M}$ is (required to be) strong and one-to-one if $\mathcal{M}$ is weak. Note that $\mathcal{E} \mapsto A$ *is always many-to-one*.

### 3.3.3 From Formal Measurements to Experiments

Given a strong measurement $\mathcal{M} \equiv \{\mathcal{H}_\alpha, \lambda_\alpha, R_\alpha\}$ one may easily construct a map (3.18) defining a discrete experiment $\mathcal{E} = \mathcal{E}_\mathcal{M}$ associated with $\mathcal{M}$:

$$U: \psi \otimes \Phi_0 \mapsto \sum_\alpha (R_\alpha \psi) \otimes \Phi_\alpha \qquad (3.47)$$

The unitarity of $U$ ( from $\mathcal{H} \otimes \Phi_0$ onto the range of $U$) follows then immediately from the orthonormality of the $\{\Phi_\alpha\}$ since

$$\sum_\alpha \|R_\alpha \psi\|^2 = \sum_\alpha \langle \psi, R_\alpha^* R_\alpha \psi \rangle = \langle \psi, \sum_\alpha P_{\mathcal{H}_\alpha} \psi \rangle = \langle \psi, \psi \rangle = \|\psi\|^2 \qquad (3.48)$$

---

[15] It is for this reason that it would be pointless and inappropriate to similarly generalize weak measurements. It is only when the state transformation is taken into account that the distinction between the outcome $\alpha$ (which determines the transformation) and the result $\lambda_\alpha$ (whose probability the formal measurement is to supply) becomes relevant.

[16] There is an obvious natural unitary equivalence between the preimages $\mathcal{E}$ of a strong formal measurement $\mathcal{M}$.

3.3 The Quantum Formalism

This experiment is abstractly characterized by: (1) the finite or countable set $I$ of outcomes $\alpha$, (2) the apparatus ready state $\Phi_0$ and the set $\{\Phi_\alpha\}$ of normalized apparatus states, (3) the unitary map $U : \mathcal{H} \otimes \Phi_0 \to \bigoplus_\alpha \mathcal{H} \otimes \Phi_\alpha$ given by (3.47), (4) the calibration $\alpha \mapsto \lambda_\alpha$ assigning numerical values (or a vector of such values) to the various outcomes $\alpha$. Note that $U$ need not arise from a Schrödinger Hamiltonian governing the interaction between system and apparatus. Thus $\mathcal{E}$ should properly be regarded as an "abstract" experiment as we have already pointed out in the introduction to this section.

### 3.3.4 Von Neumann Measurements

We shall now briefly comment on the relation between our approach, based on formal measurements, and the widely used formulation of quantum measurement in terms of von Neumann measurements [3].

A *von Neumann measurement* of $A = \sum \lambda_\alpha P_{\mathcal{H}_\alpha}$ on a system initially in the state $\psi$ can be described as follows (while the nondegeneracy of the eigenvalues of $A$— i.e., that $\dim(\mathcal{H}_\alpha) = 1$—is usually assumed, we shall not do so): Assume that the (relevant) configuration space of the apparatus, whose generic configuration shall be denoted by $y$, is one-dimensional, so that its Hilbert space $\mathcal{H}_\mathcal{A} \simeq L^2(\mathbb{R})$, and that the interaction between system and apparatus is governed by the Hamiltonian

$$H = H_{vN} = \gamma A \otimes \hat{P}_y \qquad (3.49)$$

where $\hat{P}_y \equiv i\hbar \partial/\partial y$ is (minus) the momentum operator of the apparatus. Let $\Phi_0 = \Phi_0(y)$ be the ready state of the apparatus. Then for $\psi = P_{\mathcal{H}_\alpha}\psi$ one easily sees that the unitary operator $U \equiv e^{-iTH/\hbar}$ transforms the initial state $\psi_\alpha \otimes \Phi_0$ into $\psi_\alpha \otimes \Phi_\alpha$ where $\Phi_\alpha = \Phi_0(y - \lambda_\alpha \gamma T)$, so that the action of $U$ on general $\psi = \sum P_{\mathcal{H}_\alpha}\psi$ is

$$U : \psi \otimes \Phi_0 \to \sum_\alpha (P_{\mathcal{H}_\alpha}\psi) \otimes \Phi_\alpha \qquad (3.50)$$

If $\Phi_0$ has sufficiently narrow support, say around $y = 0$, the $\Phi_\alpha$ will have disjoint support around the "pointer positions" $y_\alpha = \lambda_\alpha \gamma T$, and thus will be orthogonal, so that, with calibration $F(y) = y/\gamma T$ (more precisely, $F(y) = y_\alpha/\gamma T$ for $y$ in the support of $\Phi_\alpha$), the resulting von Neumann measurement becomes a discrete experiment measuring $A$; comparing (3.50) and (3.42) we see that it is an ideal measurement of $A$.[17]

Thus, the framework of von Neumann measurements is less general than that of discrete experiments, or equivalently of strong formal measurements; at the same time, since the Hamiltonian $H_{vN}$ is not of Schrödinger type, von Neumann measure-

---

[17] It is usually required that von Neumann measurements be impulsive ($\gamma$ large, $T$ small) so that only the interaction term (3.49) contributes significantly to the total Hamiltonian over the course of the measurement.

ments are just as formal. (We note that more general von Neumann measurements of $A$ can be obtained by replacing $H_{vN}$ with more general Hamiltonians; for example, $H'_{vN} = H_0 + H_{vN}$, where $H_0$ is a self-adjoint operator on the system Hilbert space which commutes with $A$, gives rise to a *normal measurement* of $A$, with $R_\alpha = e^{-iTH_0/\hbar} P_{\mathcal{H}_a}$. Thus by proper extension of the von Neumann measurements one may arrive at a framework of measurements completely equivalent to that of strong formal measurements.)

### 3.3.5 Preparation Procedures

Before discussing further extensions of the association between experiments and operators, we shall comment on an implicit assumption apparently required for the measurement analysis to be relevant: that the system upon which measurements are to be performed can be prepared in any prescribed state $\psi$.

Firstly, we observe that the system can be prepared in a prescribed state $\psi$ by means of an appropriate standard measurement $\mathcal{M}$ performed on the system when it is initially in an unknown state $\psi'$. We have to choose $\mathcal{M} \equiv \{\mathcal{H}_\alpha, \lambda_\alpha, R_\alpha\}$ in such a way that $R_{\alpha_0} \psi' = \psi$, for some $\alpha_0$ and all $\psi'$, i.e., that $\mathrm{Ran}(R_{\alpha_0}) = \mathrm{span}(\psi)$; then from reading the result $\lambda_{\alpha_0}$ we may infer that the system has collapsed to the state $\psi$. The simplest possibility is for $\mathcal{M}$ to be an ideal measurement with at least a one-dimensional subspace $\mathcal{H}_{\alpha_0}$ that is spanned by $\psi$. Another possibility is to perform a (nonideal) standard measurement like that of the example at the end of Sect. 3.3.2.3, which can be regarded as defining a preparation procedure for the state $e_0$.

Secondly, we wish to emphasize that the existence of preparation procedures is not as crucial for relevance as it may seem. If we had only statistical knowledge about the initial state $\psi$, nothing would change in our analysis of Bohmian experiments of Sect. 3.2, and in our conclusions concerning the emergence of self-adjoint operators, except that the uncertainty about the final configuration of the pointer would originate from both quantum equilibrium and randomness in $\psi$. We shall elaborate upon this later when we discuss Bohmian experiments for initial states described by a density matrix.

### 3.3.6 Measurements of Commuting Families of Operators

As hinted in Sect. 3.2.7, the result of an experiment $\mathcal{E}$ might be more complex than we have suggested until now in Sect. 3.3: it might be given by the vector $\lambda_\alpha \equiv (\lambda_\alpha^{(1)}, \ldots, \lambda_\alpha^{(m)})$ corresponding to the orientations of $m$ pointers. For example, the apparatus itself may be a composite of $m$ devices with the possible results $\lambda_\alpha^{(i)}$ corresponding to the final state of the $i$-th device. Nothing much will change in our discussion of measurements if we now replace the numbers $\lambda_\alpha$ with the vectors

## 3.3 The Quantum Formalism

$\lambda_\alpha \equiv (\lambda_\alpha^{(1)}, \ldots, \lambda_\alpha^{(m)})$, since the dimension of the value space was not very relevant. However $\mathscr{E}$ will now be associated, not with a single self-adjoint operator, but with a commuting family of such operators. In other words, we arrive at the notion of an experiment $\mathscr{E}$ that is a *measurement of a commuting family* of self-adjoint operators,[18] namely the family

$$A \equiv \sum_\alpha \lambda_\alpha P_{\mathscr{H}_a} = \left( \sum_\alpha \lambda_\alpha^{(1)} P_{\mathscr{H}_a}, \ldots, \sum_\alpha \lambda_\alpha^{(m)} P_{\mathscr{H}_a} \right) \equiv (A_1, \ldots, A_m). \quad (3.51)$$

Then the notions of the various kinds of formal measurements—weak, ideal, normal, standard, strong—extend straightforwardly to formal measurements of commuting families of operators. In particular, for the general notion of weak formal measurement given by 3.41, $P$ becomes a PVM on $\mathbb{R}^m$, with associated operators $A_i = \int_{\mathbb{R}^m} \lambda^{(i)} P(d\lambda)$ $[\lambda = (\lambda^{(1)}, \ldots, \lambda^{(m)}) \in \mathbb{R}^m]$. And just as for PVMs on $\mathbb{R}$ and self-adjoint operators, this association in fact yields, by the spectral theorem, a one-to-one correspondence between PVMs on $\mathbb{R}^m$ and commuting families of $m$ self-adjoint operators. The PVM corresponding to the commuting family $(A_1, \ldots, A_m)$ is in fact simply the product PVM $P = P^A = P^{A_1} \times \cdots \times P^{A_m}$ given on product sets by

$$P^A(\Delta_1 \times \cdots \times \Delta_m) = P^{A_1}(\Delta_1) \cdots P^{A_m}(\Delta_m), \quad (3.52)$$

where $P^{A_1}, \ldots, P^{A_m}$ are the PVMs of $A_1, \ldots, A_m$, and $\Delta_i \subset \mathbb{R}$, with the associated probability distributions on $\mathbb{R}^m$ given by the spectral measures for $A$

$$\mu_\psi^A(\Delta) = \langle \psi, P^A(\Delta)\psi \rangle \quad (3.53)$$

for any (Borel) set $\Delta \subset \mathbb{R}^m$.

In particular, for a PVM on $\mathbb{R}^m$, corresponding to $A = (A_1, \ldots, A_m)$, the $i$-marginal distribution, i.e., the distribution of the $i$-th component $\lambda^{(i)}$, is

$$\mu_\psi^A(\mathbb{R} \times \cdots \mathbb{R} \times \Delta_i \times \mathbb{R} \times \cdots \times \mathbb{R}) = \langle \psi, P^{A_i}(\Delta_i)\psi \rangle = \mu_\psi^{A_i}(\Delta_i),$$

the spectral measure for $A_i$. Thus, by focusing on the respective pointer variables $\lambda^{(i)}$, we may regard an experiment measuring (or a weak formal measurement of) $A = (A_1, \ldots, A_m)$ as providing an experiment measuring (or a weak formal measurement of) each $A_i$, just as would be the case for a genuine measurement of $m$

---

[18] We recall some basic facts about commuting families of self-adjoint operators [3, 34, 35]. The self-adjoint operators $A_1, \ldots, A_m$ form a commuting family if they are bounded and pairwise commute, or, more generally, if this is so for their spectral projections, i.e., if $[P^{A_i}(\Delta), P^{A_j}(\Gamma)] = 0$ for all $i, j = 1, \ldots, m$ and (Borel) sets $\Delta, \Gamma \subset \mathbb{R}$. A commuting family $A \equiv (A_1, \ldots, A_m)$ of self-adjoint operators is called *complete* if every self-adjoint operator $C$ that commutes with all members of the family can be expressed as $C = g(A_1, A_2, \ldots)$ for some function $g$. The set of all such operators cannot be extended in any suitable sense (it is closed in all relevant operator topologies). For any commuting family $(A_1, \ldots, A_m)$ of self-adjoint operators there is a self-adjoint operator $B$ and measurable functions $f_i$ such that $A_i = f_i(B)$. If the family is complete, then this operator has simple (i.e., nondegenerate) spectrum.

quantities $A_1, \ldots, A_m$. Note also the following: If $\{\mathcal{H}_\alpha, \lambda_\alpha, R_\alpha\}$ is a strong formal measurement of $A = (A_1, \ldots, A_m)$, then $\{\mathcal{H}_\alpha, \lambda_\alpha^{(i)}, R_\alpha\}$ is a strong formal measurement of $A_i$, but if $\{\mathcal{H}_\alpha, \lambda_\alpha, R_\alpha\}$ is an ideal, resp. normal, resp. standard, measurement of $A$, $\{\mathcal{H}_\alpha, \lambda_\alpha^{(i)}, R_\alpha\}$ need not be ideal, resp. normal, resp. standard.

There is a crucial point to observe: the same operator may belong to different commuting families. Consider, for example, a measurement of $A = (A_1, \ldots, A_m)$ and one of $B = (B_1, \ldots, B_m)$, where $A_1 = B_1 \equiv C$. Then while both measurements provide a measurement of $C$, they could be totally different: the operators $A_i$ and $B_i$ for $i \neq 1$ need not commute and the PVMs of $A$ and $B$, as well as any corresponding experiments $\mathcal{E}_A$ and $\mathcal{E}_B$, will be in general essentially different.

To emphasize this point we shall recall a famous example, the EPRB experiment [25, 36]: A pair of spin one-half particles, prepared in a spin-singlet state

$$\psi = \frac{1}{\sqrt{2}} \left( \psi^{(+)} \otimes \psi^{(-)} + \psi^{(-)} \otimes \psi^{(+)} \right),$$

are moving freely in opposite directions. Measurements are made, say by Stern-Gerlach magnets, on selected components of the spins of the two particles. Let **a**, **b**, **c** be three different unit vectors in space, let $\boldsymbol{\sigma}_1 \equiv \boldsymbol{\sigma} \otimes I$ and let $\boldsymbol{\sigma}_2 \equiv I \otimes \boldsymbol{\sigma}$, where $\boldsymbol{\sigma} = (\sigma_x, \sigma_y, \sigma_z)$ are the Pauli matrices. Then we could measure the operator $\boldsymbol{\sigma}_1 \cdot \mathbf{a}$ by measuring either of the commuting families $(\boldsymbol{\sigma}_1 \cdot \mathbf{a}, \boldsymbol{\sigma}_2 \cdot \mathbf{b})$ and $(\boldsymbol{\sigma}_1 \cdot \mathbf{a}, \boldsymbol{\sigma}_2 \cdot \mathbf{c})$. However these measurements are different, both as weak and as strong measurements, and of course as experiments. In Bohmian mechanics the result obtained at one place at any given time will in fact depend upon the choice of the measurement simultaneously performed at the other place (i.e., on whether the spin of the other particle is measured along **b** or along **c**). However, the statistics of the results won't be affected by the choice of measurement at the other place because both choices yield measurements of the same operator and thus their results must have the same statistical distribution.

### 3.3.7 Functions of Measurements

One of the most common experimental procedures is to recalibrate the scale of an experiment $\mathcal{E}$: if $Z$ is the original result and $f$ an appropriate function, recalibration by $f$ leads to $f(Z)$ as the new result. Thus $f(\mathcal{E})$ has an obvious meaning. Moreover, if $\mathcal{E} \mapsto A$ according to (3.37) then $\mu_\psi^{f(Z)} = \mu_\psi^Z \circ f^{-1} = \mu_\psi^A \circ f^{-1}$, and

$$\mu_\psi^A \circ f^{-1}(d\lambda) = \langle \psi, P^A(f^{-1}(d\lambda))\psi \rangle = \langle \psi, P^{f(A)}(d\lambda)\psi \rangle$$

where the last equality follows from the very definition of

$$f(A) = \int f(\lambda) P^A(d\lambda) = \int \lambda P^A(f^{-1}(d\lambda))$$

provided by the spectral theorem. Thus,

## 3.3 The Quantum Formalism

$$\text{if} \quad \mu_\psi^Z = \mu_\psi^A \quad \text{then} \quad \mu_\psi^{f(Z)} = \mu_\psi^{f(A)}, \tag{3.54}$$

i.e.,

$$\text{if} \quad \mathscr{E} \mapsto A \quad \text{then} \quad f(\mathscr{E}) \mapsto f(A). \tag{3.55}$$

The notion of *function of a formal measurement* has then an unequivocal meaning: if $\mathscr{M}$ is a weak formal measurement defined by the PVM $P$ then $f(\mathscr{M})$ is the weak formal measurement defined by the PVM $P \circ f^{-1}$, so that if $\mathscr{M}$ is a measurement of $A$ then $f(\mathscr{M})$ is a measurement of $f(A)$; for a strong formal measurement $\mathscr{M} = \{\mathscr{H}_\alpha, \lambda_\alpha, R_\alpha\}$ the self-evident requirement that the recalibration not affect the wave function transitions induced by $\mathscr{M}$ leads to $f(\mathscr{M}) = \{\mathscr{H}_\alpha, f(\lambda_\alpha), R_\alpha\}$. Note that if $\mathscr{M}$ is a standard measurement, $f(\mathscr{M})$ will in general not be standard (since in general $f$ can be many-to-one).

To highlight some subtleties of the notion of function of measurement we shall discuss two examples: Suppose that $\mathscr{M}$ and $\mathscr{M}'$ are respectively measurements of the commuting families $A = (A_1, A_2)$ and $B = (B_1, B_2)$, with $A_1 A_2 = B_1 B_2 = C$. Let $f : \mathbb{R}^2 \to \mathbb{R}$, $f(\lambda_1, \lambda_2) = \lambda_1 \lambda_2$. Then both $f(\mathscr{M})$ and $f(\mathscr{M}')$ are measurement of the same self-adjoint operator $C$. Nevertheless, as strong measurements or as experiments, they could be very different: if $A_2$ and $B_2$ do not commute they will be associated with different families of spectral projections. (Even more simply, consider measurements $\mathscr{M}_x$ and $\mathscr{M}_y$ of $\sigma_x$ and $\sigma_y$ and let $f(\lambda) = \lambda^2$. Then $f(\mathscr{M}_x)$ and $f(\mathscr{M}_y)$ are measurement of $I$—so that the result must be 1)—but the two strong measurements, as well as the corresponding experiments, are completely different.)

The second example is provided by measurements designed to determine whether the operator $A = \sum \lambda_\alpha P_{\mathscr{H}_\alpha}$ (the $\lambda_\alpha$'s are distinct) has values in some given set $\Delta$. This determination can be accomplished in at least two different ways: Suppose that $\mathscr{M}$ is an ideal measurement of $A$ and let $\mathbf{1}_\Delta(\lambda)$ be the characteristic function of the set $\Delta$. Then we could perform $\mathbf{1}_\Delta(\mathscr{M})$, that is, we measure $A$ and see whether "$A \in \Delta$". But we could also perform an "*ideal determination* of $A \in \Delta$," that is, an ideal measurement of $\mathbf{1}_\Delta(A) = P^A(\Delta)$. Now, both measurements provide a "measurement of $A \in \Delta$" (i.e., of the operator $\mathbf{1}_\Delta(A)$), since in both cases the results 1 and 0 get assigned the same probabilities. However, as strong measurements, they are different: when $\mathbf{1}_\Delta(\mathscr{M})$ is performed, and the result 1 is obtained, $\psi$ undergoes the transition

$$\psi \to P_{\mathscr{H}_\alpha} \psi$$

where $\alpha$ is the outcome with $\lambda_\alpha \in \Delta$ that actually occurs. On the other hand, for an ideal measurement of $\mathbf{1}_\Delta(A)$, the occurrence of the result 1 will generate the transition

$$\psi \to P^A(\Delta)\psi = \sum_{\lambda_\alpha \in \Delta} P_{\mathscr{H}_\alpha} \psi.$$

Note that in this case the state of the system is changed as little as possible. For example, suppose that two eigenvalues, say $\lambda_{\alpha_1}, \lambda_{\alpha_2}$, belong to $\Delta$ and $\psi = \psi_{\alpha_1} + \psi_{\alpha_2}$; then determination by performing $\mathbf{1}_\Delta(\mathscr{M})$ will lead to either $\psi_{\alpha_1}$ or $\psi_{\alpha_2}$, while the ideal determination of $A \in \Delta$ will not change the state.

### 3.3.8 Measurements of Operators with Continuous Spectrum

We shall now reconsider the status of measurements of self-adjoint operators with continuous spectrum. First of all, we remark that while on the weak level such measurements arise very naturally—and, as already stressed in Sect. 3.3.1, are indeed the first to appear in Bohmian mechanics—there is no straightforward extension of the notion of strong measurement to operators with continuous spectrum.

However, for given set of real numbers $\Delta$, one may consider any determination of $A \in \Delta$, that is, any strong measurement of the spectral projection $P^A(\Delta)$. More generally, for any choice of a *simple function*

$$f(\lambda) = \sum_{i=1}^{N} c_i \, 1_{\Delta_i}(\lambda),$$

one may consider the strong measurements of $f(A)$. In particular, let $\{f^{(n)}\}$ be a sequence of simple functions converging to the identity, so that $f^{(n)}(A) \to A$, and let $\mathcal{M}_n$ be measurements of $f^{(n)}(A)$. Then $\mathcal{M}_n$ are *approximate measurements* of $A$.

Observe that the foregoing applies to operators with discrete spectrum, as well as to operators with continuous spectrum. But note that while on the weak level we always have

$$\mathcal{M}_n \to \mathcal{M},$$

where $\mathcal{M}$ is a (general) weak measurement of $A$ (in the sense of (3.41)), if $A$ has continuous spectrum $\mathcal{M}$ will not exist as a strong measurement (in any reasonable generalized sense, since this would imply the existence of a bounded-operator-valued function $R_\lambda$ on the spectrum of $A$ such that $R_\lambda^* R_\lambda \, d\lambda = P^A(d\lambda)$, which is clearly impossible). In other words, in this case there can be no actual (generalized) strong measurement that the approximate measurements $\mathcal{M}_n$ approximate—which is perfectly reasonable.

### 3.3.9 Sequential Measurements

Suppose that $n$ measurements (with for each $i$, the $\lambda_{\alpha_i}^{(i)}$ distinct)

$$\mathcal{M}_1 \equiv \{\mathcal{H}_{\alpha_1}^{(1)}, \lambda_{\alpha_1}^{(1)}, R_{\alpha_1}^{(1)}\}, \ldots, \mathcal{M}_n \equiv \{\mathcal{H}_{\alpha_n}^{(n)}, \lambda_{\alpha_n}^{(n)}, R_{\alpha_n}^{(n)}\}$$

of operators (which need not commute)

$$A_1 = \sum_{\alpha_1} \lambda_{\alpha_1}^{(1)} P_{\alpha_1}^{(1)}, \ldots, A_n = \sum_{\alpha_n} \lambda_{\alpha_n}^{(n)} P_{\alpha_n}^{(n)}$$

## 3.3 The Quantum Formalism

are successively performed on our system at times $0 < t_1 < t_2 < \cdots < t_N$. Assume that the duration of any single measurement is small with respect to the time differences $t_i - t_{i-1}$, so that the measurements can be regarded as instantaneous. If in between two successive measurements the system's wave function changes unitarily with the operators $U_t$ then, using obvious notation,

$$\text{Prob}_\psi(A_1 = \lambda^{(1)}_{\alpha_1}, \ldots, A_n = \lambda^{(n)}_{\alpha_n}) = \|R^{(n)}_{\alpha_n}(t_n) \cdots R^{(1)}_{\alpha_1}(t_1)\psi\|^2, \quad (3.56)$$

where $R^{(i)}_{\alpha_i}(t) = U_t^{-1} R^{(i)}_{\alpha_i} U_t$ and $\psi$ is the initial ($t = 0$) wave function.

To understand how (3.56) comes about consider first the case where $n = 2$ and $t_2 \approx t_1 \approx 0$. According to standard probability rules, the probability of obtaining the results $Z_1 = \lambda^{(1)}_{\alpha_1}$ for the first measurement and $Z_2 = \lambda^{(2)}_{\alpha_2}$ for the second one is the product[19]

$$\text{Prob}_\psi(Z_2 = \lambda^{(2)}_{\alpha_2} | Z_1 = \lambda^{(1)}_{\alpha_1}) \cdot \text{Prob}_\psi(Z_1 = \lambda^{(1)}_{\alpha_1})$$

where the first term is the probability of obtaining $\lambda^{(2)}_{\alpha_2}$ given that the result of the first measurement is $\lambda^{(1)}_{\alpha_1}$. Since $\mathcal{M}_1$ then transforms the wave function $\psi$ to $R^{(1)}_{\alpha_1}\psi$, the (normalized) initial wave function for $\mathcal{M}_2$ is $R^{(1)}_{\alpha_1}\psi / \|R^{(1)}_{\alpha_1}\psi\|$, this probability is equal to

$$\frac{\|R^{(2)}_{\alpha_2} R^{(1)}_{\alpha_1}\psi\|^2}{\|R^{(1)}_{\alpha_1}\psi\|^2}.$$

The second term, the probability of obtaining $\lambda^{(1)}_{\alpha_1}$, is of course $\|R^{(1)}_{\alpha_1}\psi\|^2$. Thus

$$\text{Prob}_\psi(A^{(1)} = \lambda^{(1)}_{\alpha_1}, A^{(2)} = \lambda^{(2)}_{\alpha_2}) = \|R^{(2)}_{\alpha_2} R^{(1)}_{\alpha_1}\psi\|^2$$

in this case. Note that, in agreement with the analysis of discrete experiments (see Eq. (3.20)), the probability of obtaining the results $\lambda^{(1)}_{\alpha_1}$ and $\lambda^{(2)}_{\alpha_2}$ turns out to be the square of the norm of the final system wave function associated with these results. Now, for general times $t_1$ and $t_2 - t_1$ between the preparation of $\psi$ at $t = 0$ and the performance of $\mathcal{M}_1$ and between $\mathcal{M}_1$ and $\mathcal{M}_2$, respectively, the final system wave function is $R^{(2)}_{\alpha_2} U_{t_2-t_1} R^{(1)}_{\alpha_1} U_{t_1}\psi = R^{(2)}_{\alpha_2} U_{t_2} U_{t_1}^{-1} R^{(1)}_{\alpha_1} U_{t_1}\psi$. But $\|R^{(2)}_{\alpha_2} U_{t_2} U_{t_1}^{-1} R^{(1)}_{\alpha_1} U_{t_1}\psi\| = \|U_{t_2}^{-1} R^{(2)}_{\alpha_2} U_{t_2} U_{t_1}^{-1} R^{(1)}_{\alpha_1} U_{t_1}\psi\|$, and it is easy to see, just as

---

[19] This is so because of the *conditional independence* of the outcomes of two successive measurements *given* the final conditional wave function for the first measurement. More generally, the outcome of any measurement depends only on the wave function resulting from the preceding one. For Bohmian experiments this independence is a direct consequence of (3.10). One may wonder about the status of this independence for orthodox quantum theory. We stress that while this issue might be problematical for orthodox quantum theory, it is not a problem for Bohmian mechanics: the conditional independence of two successive measurements is a consequence of the theory. (For more on this point, see [8]).) We also would like to stress that this independence assumption is in fact crucial for orthodox quantum theory. Without it, it is hard to see how one could ever be justified in invoking the quantum formalism. Any measurement we may consider will follow many earlier measurements.

for the simple case just considered, that the square of the latter is the probability for the corresponding result, whence (3.56) for $n = 2$. Iterating, i.e., by induction, we arrive at (3.56) for general $n$.

We note that when the measurements $\mathcal{M}_1, \ldots, \mathcal{M}_n$ are ideal, the operators $R_{\alpha_i}^{(i)}$ are the orthogonal projections $P_{\alpha_i}^{(i)}$, and equation (3.56) becomes the standard formula for the joint probabilities of the results of a sequence of measurements of quantum observables, usually known as Wigner's formula [26].

It is important to observe that, even for ideal measurements, the joint probabilities given by (3.56) are not in general a consistent family of joint distributions: summation in (3.56) over the outcomes of the $i$-th measurement does not yield the joint probabilities for the results of the measurements of the operators $A_1, \ldots, A_{i-1}, A_{i+1}, \ldots A_n$ performed at the times $t_1, \ldots, t_{i-1}, t_{i+1}, \ldots t_n$. (By rewriting the right hand side of (3.56) as $\langle \psi, R_{\alpha_1}^{(1)}(t_n)^* \cdots R_{\alpha_n}^{(n)}(t_n)^* R_{\alpha_n}^{(n)}(t_n) R_{\alpha_1}^{(1)}(t_1) \psi \rangle$ one easily sees that the "sum rule" will be satisfied when $i = n$ or if the operators $R_{\alpha_i}^{(i)}(t_i)$ commute. More generally, the consistency is guaranteed by the "decoherence conditions" of Griffiths, Omnès, Gell-Mann and Hartle, and Goldstein and Page [18, 23, 37].

This failure of consistency means that the marginals of the joint probabilities given by (3.56) are not themselves given by the corresponding case of the formula. This should, however, come as no surprise: Since performing the measurement $\mathcal{M}_i$ affects the state of the system, the outcome of $\mathcal{M}_{i+1}$ should in general depend on whether or not $\mathcal{M}_i$ has been performed. Note that there is nothing particularly quantum in the fact that measurements matter in this way: They matter even for genuine measurements (unlike those we have been considering, in which nothing need be genuinely measured), and even in classical physics, if the measurements are such that they affect the state of the system.

The sequences of results $\lambda_\alpha \equiv (\lambda_{\alpha_1}^{(1)}, \ldots, \lambda_{\alpha_n}^{(n)})$, the associated state transformations $R_\alpha \equiv R_{\alpha_n}^{(n)} U_{t_n - t_{n-1}} R_{\alpha_{n-1}}^{(n-1)} \cdots R_{\alpha_1}^{(1)} U_{t_1}$, and the probabilities (3.56) (i.e., given by $p_\alpha = \|R_\alpha\|^2$) define what we shall call a *sequential measurement* of $\mathcal{M}_1, \ldots \mathcal{M}_n$, which we shall denote by $\mathcal{M}_n \otimes \ldots \otimes \mathcal{M}_1$. A sequential measurement does not in general define a formal measurement, neither weak nor strong, since $R_\alpha^* R_\alpha$ need not be a projection. This fact might seem disturbing (see, e.g., [4]); we shall take up this issue in the next section.

### 3.3.10 Some Summarizing Remarks

The notion of formal measurement we have explored in this section is at the heart of the quantum formalism. It embodies the two essential ingredients of a quantum measurement: the self-adjoint operator $A$ which represents the measured observable and the set of state transformations $R_\alpha$ associated with the measured results. The operator always carries the information about the statistics of possible results. The state transformations prescribe how the state of the system changes when the measurement is performed. For ideal measurement the latter information is also provided by the operator, but in general additional structure (the $R_\alpha$'s) is required.

## 3.3 The Quantum Formalism

There are some important morals to draw. *The association between measurements and operators is many-to-one:* the same operator $A$ can be measured by many different measurements, for example ideal, or normal but not ideal. Among the possible measurements of $A$, we must consider all possible measurements of commuting families of operators that include $A$, each of which may correspond to entirely different experimental setups.

A related fact: *not all measurements are ideal measurements.*[20] No argument, physical or mathematical, suggests that ideal measurements should be regarded as "more correct" than any other type. In particular, the Wigner formula for the statistics of a sequence of ideal measurements is no more correct than the formula (3.56) for a sequence of more general measurement. Granting a privileged status to ideal measurements amounts to a drastic and arbitrary restriction on the quantum formalism *qua measurement formalism*, since many (in fact most) real world measurements would be left out.

In this regard we note that the arbitrary restriction to ideal measurements affects the research program of "decoherent" or "consistent" histories [18, 19, 23], since Wigner's formula for a sequence of ideal measurements is unquestionably at its basis. (It should be emphasized however that the special status granted to ideal measurements is probably not the main difficulty with this approach. The no-hidden-variables theorems, which we shall discuss in Sect. 3.8, show that the totality of different families of weakly decohering histories, with their respective probability formulas, is genuinely inconsistent. While such inconsistency is perfectly acceptable for a measurement formalism, it is hard to see how it can be tolerated as the basis of what is claimed to be a fundamental theory. For more on this, see [8, 38, 39].

---

[20] In this regard we observe that the vague belief in a universal collapse rule is as old, almost, as quantum mechanics. It is reflected in von Neumann's formulation of quantum mechanics [3], based on two distinct dynamical laws: a unitary evolution *between measurements*, and a nonunitary evolution *when measurements are performed*. However, von Neumann's original proposal [3] for the nonunitary evolution—that when a measurement of $A = \sum_\alpha \lambda_\alpha P_{\mathcal{H}_\alpha}$ is performed upon a system in the state given by the density matrix $W$, the state of the system after the measurement is represented by the density matrix

$$W' = \sum_\alpha \sum_\beta \langle \phi_{\alpha\beta}, W\phi_{\alpha\beta} \rangle P_{[\phi_{\alpha\beta}]}$$

where, for each $\alpha$, $\{\phi_{\alpha\beta}\}$ is a basis for $\mathcal{H}_\alpha$—does not treat the general measurement as ideal. Moreover, this expression in general depends on the choice of the basis $\{\phi_{\alpha\beta}\}$, and was thus criticized by Lüders [40], who proposed the transformation

$$W \to W' = \sum_\alpha P_{\mathcal{H}_\alpha} W P_{\mathcal{H}_\alpha},$$

as it gives a *unique* prescription. Note that for $W = P_{[\psi]}$, where $P_{[\psi]}$ is the projection onto the initial pure state $\psi$, $W' = \sum_\alpha p_\alpha P_{[\psi_\alpha]}$, where $p_\alpha = |\langle \psi, P_{\mathcal{H}_\alpha} \psi \rangle|^2$ and $\psi_\alpha = P_{\mathcal{H}_\alpha} \psi$, corresponding to an ideal measurement.

## 3.4 The Extended Quantum Formalism

As indicated in Sect. 3.2.9, the textbook quantum formalism is merely an idealization. As just stressed, not all real world measurements are ideal. In fact, in the real world the projection postulate—that when the measurement of an observable yields a specific value, the wave function of the system is replaced by its projection onto the corresponding eigenspace—is rarely obeyed. More importantly, a great many significant real-world experiments are simply not at all associated with operators in the usual way. Consider for example an electron with fairly general initial wave function, and surround the electron with a "photographic" plate, away from (the support of the wave function of) the electron, but not too far away. This setup measures the position of "escape" of the electron from the region surrounded by the plate. Notice that since in general the time of escape is random, it is not at all clear which operator should correspond to the escape position—it should not be the Heisenberg position operator at a specific time, and a Heisenberg position operator at a random time has no meaning. In fact, there is presumably no such operator, so that for the experiment just described the probabilities for the possible results cannot be expressed in the form (3.37), and in fact are not given by the spectral measure for any operator.

Time measurements, for example escape times or decay times, are particularly embarrassing for the quantum formalism. This subject remains mired in controversy, with various research groups proposing their own favorite candidates for the "time operator" while paying little attention to the proposals of the other groups. For an analysis of time measurements within the framework of Bohmian mechanics, see [41]; in this regard see also [42]–[45].

Because of these and other difficulties, it has been proposed that we should go beyond operators-as-observables, to "*generalized observables*," described by mathematical objects even more abstract than operators (see, e.g., the books of Davies [4], Holevo [5] and Kraus [6]). The basis of this generalization lies in the observation that, by the spectral theorem, the concept of self-adjoint operator is completely equivalent to that of (a normalized) projection-valued measure (PVM), an orthogonal-projection-valued additive set function, on the value space $\mathbb{R}$. Orthogonal projections are among the simplest examples of positive operators, and a natural generalization of a "quantum observable" is provided by a positive-operator-valued measure (POVM): a normalized, countably additive set function $O$ whose values are positive operators on a Hilbert space $\mathcal{H}$. When a POVM is sandwiched by a wave function it generates a probability distribution

$$\mu_\psi^O : \Delta \mapsto \mu_\psi^O(\Delta) \equiv \langle \psi, O(\Delta)\psi \rangle \tag{3.57}$$

in exactly the same manner as a PVM.

## 3.4.1 POVMs and Bohmian Experiments

From a fundamental perspective, it may seem that we would regard this generalization, to positive-operator-valued measures, as a step in the wrong direction, since it supplies us with a new, much larger class of fundamentally unneeded abstract mathematical entities far removed from the basic ingredients of Bohmian mechanics. However from the perspective of Bohmian phenomenology positive-operator-valued measures form an extremely natural class of objects—*indeed more natural than projection-valued measures*.

To see how this comes about observe that (3.18) defines a family of bounded linear operators $R_\alpha$ by

$$P_{[\Phi_\alpha]}[U(\psi \otimes \Phi_0)] = (R_\alpha \psi) \otimes \Phi_\alpha, \qquad (3.58)$$

in terms of which we may rewrite the probability (3.20) of obtaining the result $\lambda_\alpha$ (distinct) in a generic discrete experiment as

$$p_\alpha = \|\psi_\alpha\|^2 = \|R_\alpha \psi\|^2 = \langle \psi, R_\alpha^* R_\alpha \psi \rangle. \qquad (3.59)$$

By the unitarity of the overall evolution of system and apparatus we have that $\sum_\alpha \|\psi_\alpha\|^2 = \sum_\alpha \langle \psi, R_\alpha^* R_\alpha \psi \rangle = 1$ for all $\psi \in \mathcal{H}$, whence

$$\sum_\alpha R_\alpha^* R_\alpha = I. \qquad (3.60)$$

The operators $O_\alpha \equiv R_\alpha^* R_\alpha$ are obviously positive, i.e.,

$$\langle \psi, O_\alpha \psi \rangle \geq 0 \quad \text{for all} \quad \psi \in \mathcal{H} \qquad (3.61)$$

and by (3.60) sum up to the identity,

$$\sum_\alpha O_\alpha = I. \qquad (3.62)$$

Thus we may associate with a generic discrete experiment $\mathscr{E}$—with no assumptions about reproducibility or anything else, but merely *unitarity*—a POVM

$$O(\Delta) = \sum_{\lambda_\alpha \in \Delta} O_\alpha \equiv \sum_{\lambda_\alpha \in \Delta} R_\alpha^* R_\alpha, \qquad (3.63)$$

in terms of which the statistics of the results can be expressed in a compact way: the probability that the result of the experiment lies in a set $\Delta$ is given by

$$\sum_{\lambda_\alpha \in \Delta} p_\alpha = \sum_{\lambda_\alpha \in \Delta} \langle \psi, O_\alpha \psi \rangle = \langle \psi, O(\Delta) \psi \rangle. \qquad (3.64)$$

Moreover, it follows from (3.18) and (3.58) that $\mathscr{E}$ generates state transformations

$$\psi \to \psi_\alpha = R_\alpha \psi. \qquad (3.65)$$

## 3.4.2 Formal Experiments

The association between experiments and POVMs can be extended to a general experiment (3.34) in a straightforward way. In analogy with (3.37) we shall say that the POVM $O$ is associated with the experiment $\mathscr{E}$ whenever the probability distribution (3.35) of the results of $\mathscr{E}$ is equal to the probability measure (3.57) generated by $O$, i.e.,[21]

$$\mathscr{E} \mapsto O \quad \text{if and only if} \quad \mathsf{P}^Z_\psi = \mu^O_\psi, \tag{3.66}$$

We may now proceed as in Sect. 3.3 and analyze on a formal level the association (3.66) by introducing the notions of *weak* and *strong* formal experiment as the obvious generalizations of (3.41) and (3.45):

> *Any positive-operator-valued measure $O$ defines the weak formal experiment $\mathcal{E} \equiv O$. Any set $\{\lambda_\alpha\}$ of not necessarily distinct real numbers (or vectors of real numbers) paired with any collection $\{R_\alpha\}$ of bounded operators on $\mathscr{H}$ such that $\sum R^*_\alpha R_\alpha = I$ defines the strong formal experiment $\mathcal{E} \equiv \{\lambda_\alpha, R_\alpha\}$ with associated POVM (3.63) and state transformations (3.65).* (3.67)

The notion of formal experiment is a genuine extension of that of formal measurement, the latter being the special case in which $O$ is a PVM and $R^*_\alpha R_\alpha$ are the projections.

Formal experiments share with formal measurements many features. This is so because all measure-theoretic properties of projection-valued measures extend to positive-operator-valued measures. For example, just as for PVMs, integration of real functions against positive-operator-valued measure is a meaningful operation that generates self-adjoint operators: for given real (and measurable) function $f$, the operator $B = \int f(\lambda) O(d\lambda)$ is a self-adjoint operator defined, say, by its matrix elements $\langle \phi, B\psi \rangle = \int \lambda \mu_{\phi,\psi}(d\lambda)$ for all $\phi$ and $\psi$ in $\mathscr{H}$, where $\mu_{\phi,\psi}$ is the complex measure $\mu_{\phi,\psi}(d\lambda) = \langle \phi, O(d\lambda)\psi \rangle$. (We ignore the difficulties that might arise if $f$ is not bounded.) In particular, with $O$ is associated the self-adjoint operator

$$A_O \equiv \int \lambda\, O(d\lambda). \tag{3.68}$$

It is however important to observe that this association (unlike the case of PVMs, for which the spectral theorem provides the inverse) is not invertible, since the self-adjoint operator $A_O$ is always associated with the PVM provided by the spectral theorem. Thus, unlike PVMs, POVMs are not equivalent to self-adjoint operators.

---

[21] Whenever (3.66) is satisfied we may say that the experiment $\mathscr{E}$ is a measurement of the generalized observable $O$. We shall however avoid this terminology in connection with generalized observables; even when it is standard (so that we use it), i.e., when $O$ is a PVM and thus equivalent to a self-adjoint operator, it is in fact improper.

## 3.4 The Extended Quantum Formalism

In general, the operator $A_O$ will carry information only about the mean value of the statistics of the results,

$$\int \lambda \, \langle \psi, O(d\lambda)\psi \rangle = \langle \psi, A_O \psi \rangle,$$

while for the higher moments we should expect that

$$\int \lambda^n \, \langle \psi, O(d\lambda)\psi \rangle \neq \langle \psi, A_O^n \psi \rangle$$

unless $O$ is a PVM.

What we have just described is an important difference between general formal experiments and formal measurements. This and other differences originate from the fact that a POVM is a much weaker notion than a PVM. For example, a POVM $O$ on $\mathbb{R}^m$—like ordinary measures and unlike PVMs—need not be a product measure: If $O_1, \ldots, O_m$ are the *marginals* of $O$,

$$O_1(\Delta_1) = O(\Delta_1 \times \mathbb{R}^{m-1}), \ldots, O_m(\Delta_m) = O(\mathbb{R}^{m-1} \times \Delta_m),$$

the product POVM $O_1 \times \cdots \times O_m$ will be in general different from $O$. (This is trivial since any probability measure on $\mathbb{R}^m$ times the identity is a POVM.)

Another important difference between the notion of POVM and that of PVM is this: while the projections $P(\Delta)$ of a PVM, for different $\Delta$'s, commute, the operators $O(\Delta)$ of a generic POVM need not commute. An illustration of how this may naturally arise is provided by sequential measurements.

A sequential measurement (see Sect. 3.3.9) $\mathscr{M}_n \otimes \ldots \otimes \mathscr{M}_1$ is indeed a very simple example of a formal experiment that in general is not a formal measurement (see also Davies [4]). We have that

$$\mathscr{M}_n \otimes \ldots \otimes \mathscr{M}_1 = \{\lambda_\alpha, R_\alpha\}$$

where

$$\lambda_\alpha \equiv (\lambda_{\alpha_1}^{(1)}, \ldots, \lambda_{\alpha_n}^{(n)})$$

and

$$R_\alpha \equiv R_{\alpha_n}^{(n)} U_{t_n - t_{n-1}} R_{\alpha_{n-1}}^{(n-1)} \cdots R_{\alpha_1}^{(1)} . U_{t_1}.$$

Note that since $p_\alpha = \|R_\alpha \psi\|^2$, we have that

$$\sum_\alpha R_\alpha^* R_\alpha = I,$$

which also follows directly using

$$\sum_{\alpha_j} R_{\alpha_j}^{(j)*} R_{\alpha_j}^{(j)} = I, \qquad j = 1, \ldots, n$$

Now, with $\mathscr{M}_n \otimes \ldots \otimes \mathscr{M}_1$ is associated the POVM

$$O(\Delta) = \sum_{\lambda_\alpha \in \Delta} R_\alpha^* R_\alpha.$$

Note that $O(\Delta)$ and $O(\Delta')$ in general don't commute since in general $R_\alpha$ and $R_\beta$ may fail to do so.

An interesting class of POVMs for which $O(\Delta)$ and $O(\Delta')$ do commute arises in association with the notion of an *"approximate measurement"* of a self-adjoint operator: suppose that the result $Z$ of a measurement $\mathcal{M} = P^A$ of a self-adjoint operator $A$ is distorted by the addition of an independent noise $N$ with symmetric probability distribution $\eta(\lambda)$. Then the result $Z + N$ of the experiment, for initial system wave function $\psi$, is distributed according to

$$\Delta \mapsto \int_\Delta \int_\mathbb{R} \eta(\lambda - \lambda') \langle \psi, P_A(d\lambda')\psi \rangle \, d\lambda,$$

which can be rewritten as

$$\Delta \mapsto \langle \psi, \int_\Delta \eta(\lambda - A) d\lambda \, \psi \rangle.$$

Thus the result $Z + N$ is governed by the POVM

$$O(\Delta) = \int_\Delta \eta(\lambda - A) \, d\lambda. \tag{3.69}$$

The formal experiment defined by this POVM can be regarded as providing an approximate measurement of $A$. For example, let

$$\eta(\lambda) = \frac{1}{\sigma\sqrt{2\pi}} e^{-\frac{\lambda^2}{2\sigma^2}}. \tag{3.70}$$

Then for $\sigma \to 0$ the POVM (3.69) becomes the PVM of $A$ and the experiment becomes a measurement of $A$.

Concerning the POVM (3.69) we wish to make two remarks. The first is that the $O(\Delta)$'s commute since they are all functions of $A$. The second is that this POVM has a continuous density, i.e.,

$$O(d\lambda) = o(\lambda) \, d\lambda \quad \text{where} \quad o(\lambda) = \eta(\lambda - A).$$

This is another difference between POVMs and PVMs: like ordinary measures and unlike PVMs, POVMs may have a continuous density. The reason this is possible for POVMs is that, for a POVM $O$, unlike for a PVM, given $\psi \in H$, the vectors $O(\Delta)\psi$ and $O(\Delta')\psi$, for $\Delta$ and $\Delta'$ disjoint and arbitrarily small, need not be orthogonal. Otherwise, no density $o(d\lambda)$ could exist, because this would imply that there is a continuous family $\{o(\lambda)\psi\}$ of orthogonal vectors in $\mathcal{H}$.

Finally, we observe that unlike strong measurements, the notion of strong formal experiment can be extended to POVM with continuous spectrum (see Section 3.3.8). One may in fact define a strong experiment by $\mathcal{E} = \{\lambda, R_\lambda\}$, where $\lambda \mapsto R_\lambda$ is a

## 3.4 The Extended Quantum Formalism

continuous *bounded-operator-valued function* such that $\int R_\lambda^* R_\lambda \, d\lambda = I$. Then the statistics for the results of such an experiment is governed by the POVM $O(d\lambda) \equiv R_\lambda^* R_\lambda \, d\lambda$. For example, let

$$R_\lambda = \xi(\lambda - A) \quad \text{where} \quad \xi(\lambda) = \frac{1}{\sqrt{\sigma}\sqrt[4]{2\pi}} e^{-\frac{\lambda^2}{4\sigma^2}}.$$

Then $O(d\lambda) = R_\lambda^* R_\lambda \, d\lambda$ is the POVM (3.69) with $\eta$ given by (3.70). We observe that the state transformations (cf. the definition (3.6) of the conditional wave function)

$$\psi \to R_\lambda \psi = \frac{1}{\sqrt{\sigma}\sqrt[4]{2\pi}} e^{-\frac{(\lambda - A)^2}{4\sigma^2}} \psi \tag{3.71}$$

can be regarded as arising from a von Neumann interaction with Hamiltonian (3.49) (and $\gamma T = 1$) and ready state of the apparatus

$$\Phi_0(y) = \frac{1}{\sqrt{\sigma}\sqrt[4]{2\pi}} e^{-\frac{y^2}{4\sigma^2}}.$$

Experiments with state transformations (3.71), for large $\sigma$, have been considered by Aharonov and coworkers (see, e.g., Aharonov, Anandan, and Vaidman [47]) as providing "weak measurements" of operators. (The effect of the measurement on the state of the system is "small" if $\sigma$ is sufficiently large). This terminology notwithstanding, it is important to observe that such experiments are not measurements of $A$ in the sense we have discussed here. They give information about the average value of $A$, since $\int \lambda \langle \psi, R_\lambda^* R_\lambda \psi \rangle \, d\lambda = \langle \psi, A\psi \rangle$, but presumably none about its higher moments.

### 3.4.3 From Formal Experiments to Experiments

Just as with a formal measurement (see Sect. 3.3.3), with a formal experiment $\mathcal{E} \equiv \{\lambda_\alpha, R_\alpha\}$, we may associate a discrete experiment $\mathscr{E}$. The unitary map (3.18) of $\mathscr{E}$ will be given again by (3.47), i.e.,

$$U : \psi \otimes \Phi_0 \mapsto \sum_\alpha (R_\alpha \psi) \otimes \Phi_\alpha, \tag{3.72}$$

but now $R_\alpha^* R_\alpha$ of course need not be projection. The unitarity of $U$ follows immediately from the orthonormality of the $\Phi_\alpha$ using $\sum R_\alpha^* R_\alpha = I$. (Note that with a weak formal experiment $\mathcal{E} \equiv O = \{O_\alpha\}$ we may associate many inequivalent discrete experiments, defined by (3.72) with operators $R_\alpha \equiv U_\alpha \sqrt{O_\alpha}$, for *any* choice of unitary operators $U_\alpha$.)

We shall now discuss a concrete example of a discrete experiment defined by a formal experiment which will allow us to make some more further comments on the issue of reproducibility discussed in Sect. 3.2.8.

Let $\{\ldots, e_{-1}, e_0, e_1, \ldots\}$ be an orthonormal basis in the system Hilbert space $\mathcal{H}$, let $P_-, P_0, P_+$ be the orthogonal projections onto the subspaces $\widetilde{\mathcal{H}}_-, \mathcal{H}_0, \widetilde{\mathcal{H}}_+$ spanned by $\{e\}_{\alpha<0}, \{e_0\}, \{e\}_{\alpha>0}$ respectively, and let $V_+, V_-$ be the right and left shift operators,

$$V_+ e_\alpha = e_{\alpha+1}, \qquad V_- e_\alpha = e_{\alpha-1}.$$

Consider the strong formal experiment $\mathscr{E}$ with the two possible results $\lambda_\pm = \pm 1$ and associated state transformations

$$R_{\pm 1} = V_\pm (P_\pm + \frac{1}{\sqrt{2}} P_0). \tag{3.73}$$

Then the unitary $U$ of the corresponding discrete experiment $\mathscr{E}$ is given by

$$U : \psi \otimes \Phi_0 \to R_- \psi \otimes \Phi_- + R_+ \psi \otimes \Phi_+,$$

where $\Phi_0$ is the ready state of the apparatus and $\Phi_\pm$ are the apparatus states associated with the results $\pm 1$. If we now consider the action of $U$ on the basis vectors $e_\alpha$,

$$U(e_\alpha \otimes \Phi_0) = e_{\alpha+1} \otimes \Phi_+ \quad \text{for } \alpha > 0$$
$$U(e_\alpha \otimes \Phi_0) = e_{\alpha-1} \otimes \Phi_- \quad \text{for } \alpha < 0$$
$$U(e_0 \otimes \Phi_0) = \frac{1}{\sqrt{2}}(e_1 \otimes \Phi_+ + e_{-1} \otimes \Phi_-),$$

we see immediately that

$$U(\widetilde{\mathcal{H}}_\pm \otimes \Phi_0) \subset \widetilde{\mathcal{H}}_\pm \otimes \Phi_{\pm 1}.$$

Thus (3.24) is satisfied and $\mathscr{E}$ is a reproducible experiment. Note however that the POVM $O = \{O_{-1}, O_{+1}\}$ associated with (3.73),

$$O_{\pm 1} = R_{\pm 1}^* R_{\pm 1} = P_\pm + \frac{1}{2} P_0,$$

is not a PVM since the positive operators $O_{\pm 1}$ are not projections, i.e, $O_{\pm 1}^2 \neq O_{\pm 1}$. Thus $\mathscr{E}$ is not a measurement of any self-adjoint operator, which shows that without the assumption of the finite dimensionality of the subspaces $\widetilde{\mathcal{H}}_\alpha$ a reproducible discrete experiment need not be a measurement of a self-adjoint operator.

### *3.4.4 Measure-Valued Quadratic Maps*

We conclude this section with a remark about POVMs. Via (3.57) every POVM $O$ defines a "normalized quadratic map" from $\mathcal{H}$ to measures on some space (the value-space for the POVM). Moreover, every such map comes from a POVM in this

## 3.4 The Extended Quantum Formalism

way. Thus the two notions are equivalent:

(3.57) *defines a canonical one-to-one correspondence between POVMs and normalized measure-valued quadratic maps on $\mathcal{H}$.* (3.74)

To say that a measure-valued map on $\mathcal{H}$

$$\psi \mapsto \mu_\psi \tag{3.75}$$

is quadratic means that

$$\mu_\psi = B(\psi, \psi) \tag{3.76}$$

is the diagonal part of a sesquilinear map $B$, from $\mathcal{H} \times \mathcal{H}$ to the complex measures on some value space $\Lambda$. If $B(\psi, \psi)$ is a probability measure whenever $\|\psi\| = 1$, we say that the map is normalized.[22]

Proposition (3.74) is a consequences of the following considerations: For a given POVM $O$ the map $\psi \mapsto \mu_\psi^O$, where $\mu_\psi^O(\Delta) \equiv \langle \psi, O(\Delta)\psi \rangle$, is manifestly quadratic, with $B(\phi, \psi) = \langle \phi, O(\cdot)\psi \rangle$, and it is obviously normalized. Conversely, let $\psi \mapsto \mu_\psi$ be a normalized measure-valued quadratic map, corresponding to some $B$, and write $B_\Delta(\phi, \psi) = B(\phi, \psi)[\Delta]$ for the complex measure $B$ at the Borel set $\Delta$. By the Schwartz inequality, applied to the positive form $B_\Delta(\phi, \psi)$, we have that $|B_\Delta(\phi, \psi)| \leq \|\psi\|\|\phi\|$. Thus, using Riesz's lemma [30], there is a unique bounded operator $O(\Delta)$ on $\mathcal{H}$ such that

$$B_\Delta(\phi, \psi) = \langle \phi, O(\Delta)\psi \rangle.$$

Moreover, $O(\Delta)$, like $B_\Delta$, is countably additive in $\Delta$, and since $B(\psi, \psi)$ is a (positive) measure, $O$ is a positive-operator-valued measure, normalized because $B$ is.

A simple example of a normalized measure-valued quadratic map is

$$\Psi \mapsto \rho^\Psi(dq) = |\Psi|^2 dq, \tag{3.77}$$

whose associated POVM is the PVM $P^{\hat{Q}}$ for the position (configuration) operator

$$\hat{Q}\Psi(q) = q\Psi(q). \tag{3.78}$$

Note also that if the quadratic map $\mu_\psi$ corresponds to the POVM $O$, then, for any unitary $U$, the composite map $\psi \mapsto \mu_{U\psi}$ corresponds to the POVM $U^*OU$, since

---

[22] A sesquilinear map $B(\phi, \psi)$ is one that is linear in the second slot and conjugate linear in the first:

$$B(\phi, \alpha\psi_1 + \beta\psi_2) = \alpha B(\phi, \psi_1) + \beta B(\phi, \psi_2)$$
$$B(\alpha\phi_1 + \beta\phi_2, \psi) = \bar{\alpha} B(\phi_1, \psi) + \bar{\beta} B(\phi_2, \psi).$$

Clearly any such normalized $B$ can be chosen to be conjugate symmetric, $B(\psi, \phi) = \overline{B(\phi, \psi)}$, without affecting its diagonal, and it follows from polarization that any such $B$ must in fact *be* conjugate symmetric.

$\langle U\psi, O(\Delta)U\psi\rangle = \langle \psi, U^*O(\Delta)U\psi\rangle$. In particular for the map (3.77) and $U = U_T$, the composite map corresponds to the PVM $P^{\hat{Q}_T}$, with $\hat{Q}_T = U^*\hat{Q}U$, the Heisenberg position (configuration) at time $T$, since $U_T^* P^{\hat{Q}} U_T = P^{U_T^*\hat{Q}U_T}$.

## 3.5 The General Emergence of Operators

For Bohmian mechanics POVMs emerge naturally, not for discrete experiments, but for a general experiment (3.34). To see how this comes about consider the probability measure (3.35) giving the probability distribution of the result $Z = F(Q_T)$ of the experiment, where $Q_T$ is the final configuration of system and apparatus and $F$ is the calibration function expressing the numerical result, for example the orientation $\Theta$ of a pointer. Then the map

$$\psi \mapsto \mathsf{P}_\psi^Z = \mathsf{P}_{\Psi_T} \circ F^{-1}, \tag{3.79}$$

from the initial wave function of the system to the probability distribution of the result, is quadratic since it arises from the sequence of maps

$$\psi \mapsto \Psi = \psi \otimes \Phi_0 \mapsto \Psi_T = U(\psi \otimes \Phi_0) \mapsto \mathsf{P}_{\Psi_T}(dq) = \Psi_T^* \Psi_T dq \mapsto \mathsf{P}_\psi^Z$$
$$= \mathsf{P}_{\Psi_T} \circ F^{-1}, \tag{3.80}$$

where the middle map, to the quantum equilibrium distribution, is obviously quadratic, while all the other maps are linear, all but the second trivially so. Now, by (3.74), the notion of such a quadratic map (3.79) is completely equivalent to that of a POVM on the system Hilbert space $\mathscr{H}$. (The sesquilinear map $B$ associated with (3.80) is $B(\psi_1, \psi_2) = \Psi_{1T}^* \Psi_{2T} dq \circ F^{-1}$, where $\Psi_{iT} = U(\psi_i \otimes \Phi_0)$).

Thus the emergence and role of POVMs as generalized observables in Bohmian mechanics is merely an expression of the sesquilinearity of quantum equilibrium together with the linearity of the Schrödinger evolution. Thus the fact that with every experiment is associated a POVM, which forms a compact expression of the statistics for the possible results, is a near mathematical triviality. It is therefore rather dubious that the occurrence of POVMs—the simplest case of which is that of PVMs—as observables can be regarded as suggesting any deep truths about reality or about epistemology.

An explicit formula for the POVM defined by the quadratic map (3.79) follows immediately from (3.80):

$$\mathsf{P}_\psi^Z(d\lambda) = \langle \psi \otimes \Phi_0, U^* P^{\hat{Q}}(F^{-1}(d\lambda))U\psi \otimes \Phi_0\rangle$$
$$= \langle \psi \otimes \Phi_0, P_0 U^* P^{\hat{Q}}(F^{-1}(d\lambda))U P_0 \psi \otimes \Phi_0\rangle$$

where $P^{\hat{Q}}$ is the PVM for the position (configuration) operator (3.78) and $P_0$ is the projection onto $\mathscr{H} \otimes \Phi_0$, whence

3.5 The General Emergence of Operators

$$O(d\lambda) = 1_{\Phi_0}^{-1} P_0 U^* P^{\hat{Q}}(F^{-1}(d\lambda)) U P_0 1_{\Phi_0}, \quad (3.81)$$

where $1_{\Phi_0}\psi = \psi \otimes \Phi_0$ is the natural identification of $\mathcal{H}$ with $\mathcal{H} \otimes \Phi_0$. This is the obvious POVM reflecting the essential structure of the experiment.[23]

Note that the POVM (3.81) is unitarily equivalent to

$$P_0 P^{F(\hat{Q}_T)}(d\lambda) P_0 \quad (3.83)$$

where $\hat{Q}_T$ is the Heisenberg configuration of system and apparatus at time $T$. This POVM, acting on the subspace $\mathcal{H} \otimes \Phi_0$, is the projection to that subspace of a PVM, the spectral projections for $F(\hat{Q}_T)$. Naimark has shown (see, e.g., [4]) that every POVM is equivalent to one that arises in this way, as the orthogonal projection of a PVM to a subspace.[24]

We shall now illustrate the association of POVMs with experiments by considering some special cases of (3.80).

### 3.5.1 "No Interaction" Experiments

Let $U = U_S \otimes U_A$ in (3.80) (hereafter the indices "$S$" and "$A$" shall refer, respectively, to system and apparatus). Then for $F(x, y) = y$ the measure-valued quadratic map defined by (3.80) is

$$\psi \mapsto c(y)\|\psi\|^2 dy$$

where $c(y) = |U_A \Phi_0|^2(y)$, with POVM $O_1(dy) = c(y)dy\, I_S$, while for $F(q) = q = (x, y)$ the map is

$$\psi \mapsto c(y)|U_S\psi|^2(x)\, dq$$

---

[23] This POVM can also be written as

$$O(d\lambda) = \operatorname{tr}_A \left[ P_0 U^* P^{\hat{Q}}(F^{-1}(d\lambda)) U \right], \quad (3.82)$$

where $\operatorname{tr}_A$ is the partial trace over the apparatus variables. The partial trace is a map $\operatorname{tr}_A : W \mapsto \operatorname{tr}_A(W)$, from trace class operators on the Hilbert space $\mathcal{H}_S \otimes \mathcal{H}_A$ to trace class operators on $\mathcal{H}_S$, uniquely defined by $\operatorname{tr}_S(\operatorname{tr}_A(W)B) = \operatorname{tr}_{S+A}(WB \otimes I)$, where $\operatorname{tr}_{S+A}$ and $\operatorname{tr}_S$ are the usual (scalar-valued) traces of operators on $\mathcal{H}_S \otimes \mathcal{H}_A$ and $\mathcal{H}_S$, respectively. For a trace class operator $B$ on $L^2(dx) \otimes L^2(dy)$ with kernel $B(x, y, x', y')$ we have $\operatorname{tr}_A(B)(x, x') = \int B(x, y, x', y) dy$. In (3.82) $\operatorname{tr}_A$ is applied to operators that need not be trace class—nor need the operator on the left be trace class—since, e.g., $O(\Lambda) = I$. The formula nonetheless makes sense.

[24] If $O(d\lambda)$ is a POVM on $\Sigma$ acting on $\mathcal{H}$, then the Hilbert space on which the corresponding PVM acts is the natural Hilbert space associated with the data at hand, namely $L^2(\Sigma, \mathcal{H}, O(d\lambda))$, the space of $\mathcal{H}$-valued functions $\psi(\lambda)$ on $\Sigma$, with inner product given by $\int \langle \psi(\lambda), O(d\lambda)\phi(\lambda) \rangle$. (If this is not, in fact, positive definite, then the quotient with its kernel should be taken—$\psi(\lambda)$ should, in other words, be understood as the appropriate equivalence class.) Then $O(d\lambda)$ is equivalent to $PE(d\lambda)P$, where $E(\Delta) = \hat{1}_\Delta(\lambda)$, multiplication by $1_\Delta(\lambda)$, and $P$ is the orthogonal projection onto the subspace of constant $\mathcal{H}$-valued functions $\psi(\lambda) = \psi$.

with corresponding POVM $O_2(dq) = c(y) U_S^* P^{\hat{X}}(dx) U_S \, dy$. Neither $O_1$ nor $O_2$ is a PVM. However, if $F$ is independent of $y$, $F(x, y) = F(x)$, then the apparatus can be ignored in (3.80) or (3.81) and $O = U_S^* P^{\hat{X}} U_S \circ F^{-1}$, i.e.,

$$O(d\lambda) = U_S^* P^{\hat{X}}(F^{-1}(d\lambda)) U_S,$$

which is manifestly a PVM—in fact corresponding to $F(\hat{X}_T)$, where $\hat{X}_T$ is the Heisenberg configuration of the system at the end of the experiment.

This case is somewhat degenerate: with no interaction between system and apparatus it hardly seems anything like a measurement. However, it does illustrate that it is "true" POVMs (i.e., those that aren't PVMs) that typically get associated with experiments—i.e., unless some special conditions hold (here that $F = F(x)$).

### 3.5.2 "No X" Experiments

The map (3.80) is well defined even when the system (the $x$-system) has no translational degrees of freedom, so that there is no $x$ (or $X$). This will be the case, for example, when the system Hilbert space $\mathscr{H}_S$ corresponds to the spin degrees of freedom. Then $\mathscr{H}_S = \mathbb{C}^n$ is finite dimensional.

In such cases, the calibration $F$ of course is a function of $y$ alone, since there is no $x$. For $F = y$ the measure-valued quadratic map defined by (3.80) is

$$\psi \mapsto |[U(\psi \otimes \Phi_0)](y)|^2 dy, \tag{3.84}$$

where $|\cdots|$ denotes the norm in $\mathbb{C}^n$.

This case is physically more interesting than the previous one, though it might appear rather puzzling since until now our measured systems have always involved configurations. After all, without configurations there is no Bohmian mechanics! However, what is relevant from a Bohmian perspective is that the composite of system and apparatus be governed by Bohmian mechanics, and this may well be the case if the apparatus has configurational degrees of freedom, even if what is called the system doesn't. Moreover, this case provides the prototype of many real-world experiments, e.g., spin measurements.

For the measurement of a spin component of a spin–1/2 particle—recall the description of the Stern-Gerlach experiment given in Sect. 3.2.5—we let $\mathscr{H}_S = \mathbb{C}^2$, the spin space, with "apparatus" configuration $y = \mathbf{x}$, the position of the particle, and with suitable calibration $F(\mathbf{x})$. (For a real world experiment there would also have to be a genuine apparatus—a detector—that measures where the particle *actually is* at the end of the experiment, but this would not in any way affect our analysis. We shall elaborate upon this below.) The unitary $U$ of the experiment is the evolution operator up to time $T$ generated by the Pauli Hamiltonian (3.12), which under the

## 3.5 The General Emergence of Operators

assumption (3.14) becomes

$$H = -\frac{\hbar^2}{2m}\nabla^2 - (b + az)\sigma_z \tag{3.85}$$

Moreover, as in Sect. 3.2.5, we shall assume that the initial particle wave function has the form $\Phi_0(\mathbf{x}) = \Phi_0(z)\phi(x, y)$.[25] Then for $F(\mathbf{x}) = z$ the quadratic map (3.80) is

$$\psi \mapsto \left(|\langle \psi^+, \psi \rangle|^2 |\Phi_T^{(+)}(z)|^2 + |\langle \psi^-, \psi \rangle|^2 |\Phi_T^{(-)}(z)|^2\right) dz$$
$$= \left\langle \psi, \ |\psi^+\rangle\langle\psi^+| |\Phi_T^{(+)}(z)|^2 + |\psi^-\rangle\langle\psi^-| |\Phi_T^{(-)}(z)|^2 \ \psi \right\rangle dz$$

with POVM

$$O(dz) = \begin{pmatrix} |\Phi_T^{(+)}(z)|^2 & 0 \\ 0 & |\Phi_T^{(-)}(z)|^2 \end{pmatrix} dz, \tag{3.86}$$

where $\psi^\pm$ are the eigenvectors (3.13) of $\sigma_z$ and $\Phi_T^{(\pm)}$ are the solutions of (3.15) computed at $t = T$, for initial conditions $\Phi_0^{(\pm)} = \Phi_0(z)$.

Consider now the appropriate calibration for the Stern-Gerlach experiment, namely the function

$$F(\mathbf{x}) = \begin{cases} +1 & \text{if } z > 0, \\ -1 & \text{if } z < 0 \end{cases} \tag{3.87}$$

which assigns to the outcomes of the experiment the desired numerical results: if the particle goes up in the $z$- direction the spin is +1, while if the particle goes down the spin is -1. The corresponding POVM $O_T$ is defined by

$$O_T(+1) = \begin{pmatrix} p_T^+ & 0 \\ 0 & p_T^- \end{pmatrix} \qquad O_T(-1) = \begin{pmatrix} 1 - p_T^+ & 0 \\ 0 & 1 - p_T^- \end{pmatrix}$$

where

$$p_T^+ = \int_0^\infty |\Phi_T^{(+)}|^2(z) dz, \qquad p_T^- = \int_0^\infty |\Phi_T^{(-)}|^2(z) dz.$$

It should be noted that $O_T$ is not a PVM. However, as indicated in Sect. 3.2.5, as $T \to \infty$, $p_T^+ \to 1$ and $p_T^- \to 0$, and the POVM $O_T$ becomes the PVM of the operator $\sigma_z$, i.e., $O_T \to P^{\sigma_z}$, defined by

$$P(+1) = \begin{pmatrix} 1 & 0 \\ 0 & 0 \end{pmatrix} \qquad P(-1) = \begin{pmatrix} 0 & 0 \\ 0 & 1 \end{pmatrix} \tag{3.88}$$

and the experiment becomes a measurement of the operator $\sigma_z$.

---

[25] We abuse notation here in using the notation $y = \mathbf{x} = (x, y, z)$. The $y$ on the right should of course not be confused with the one on the left.

### 3.5.3 "No Y" Experiments

Suppose now that the "apparatus" involves no translational degrees of freedom, i.e., that there is no $y$ (or $Y$). For example, suppose the apparatus Hilbert space $\mathcal{H}_A$ corresponds to certain spin degrees of freedom, with $\mathcal{H}_A = \mathbb{C}^n$ finite dimensional. Then, of course, $F = F(x)$.

This case illustrates what measurements are not. If the apparatus has no configurational degrees of freedom, then neither in Bohmian mechanics nor in orthodox quantum mechanics is it a *bona fide* apparatus: Whatever virtues such an apparatus might otherwise have, it certainly can't generate any directly observable results (at least not when the system itself is microscopic). According to Bohr ([46], pp. 73 and 90): "Every atomic phenomenon is closed in the sense that its observation is based on registrations obtained by means of suitable amplification devices with irreversible functioning such as, for example, permanent marks on the photographic plate" and "the quantum-mechanical formalism permits well-defined applications only to such closed phenomena." To stress this point, discussing particle detection Bell has said [47]: "Let us suppose that a discharged counter pops up a flag sayings 'Yes' just to emphasize that it is a macroscopically different thing from an undischarged counter, in a very different region of configuration space."

Experiments based on certain micro-apparatuses, e.g., "one-bit detectors" [48], provide a nice example of "No Y" experiments. We may think of a one-bit detector as a spin-1/2-like system (e.g., a two-level atom), with "down" state $\Phi_0$ (the ready state) and "up" state $\Phi_1$ and which is such that its configurational degrees of freedom can be ignored. Suppose that this "spin-system," in its "down" state, is placed in a small spatial region $\Delta_1$ and consider a particle whose wave function has been prepared in such a way that at $t = 0$ it has the form $\psi = \psi_1 + \psi_2$, where $\psi_1$ is supported by $\Delta_1$ and $\psi_2$ by $\Delta_2$ disjoint from $\Delta_1$. Assume that the particle interacts locally with the spin-system, in the sense that were $\psi = \psi_1$ the "spin" would flip to the "up" state, while were $\psi = \psi_2$ it would remain in its "down" state, and that the interaction time is negligibly small, so that other contributions to the Hamiltonian can be ignored. Then the initial state $\psi \otimes \Phi_0$ undergoes the unitary transformation

$$U : \psi \otimes \Phi_0 \to \Psi = \psi_1 \otimes \Phi_1 + \psi_2 \otimes \Phi_0. \tag{3.89}$$

We may now ask whether $U$ defines an experiment genuinely measuring whether the particle is in $\Delta_1$ or $\Delta_2$. The answer of course is no (since in this experiment there is no apparatus property at all with which the position of the particle could be correlated) *unless* the experiment is (quickly) completed by a measurement of the "spin" by means of another (macroscopic) apparatus. In other words, we may conclude that the particle is in $\Delta_1$ only if the spin-system in effect pops up a flag saying "up".

### 3.5.4 "No Y no Φ" Experiments

Suppose there is no apparatus at all: no apparatus configuration $y$ nor Hilbert space $\mathscr{H}_A$, or, what amounts to the same thing, $\mathscr{H}_A = \mathbb{C}$. For calibration $F = x$ the measure-valued quadratic map defined by (3.80) is

$$\psi \mapsto |U\psi(x)|^2,$$

with POVM $U^* P^{\hat{X}} U$, while the POVM for general calibration $F(x)$ is

$$O(d\lambda) = U^* P^{\hat{X}}(F^{-1}(d\lambda))U. \tag{3.90}$$

$O$ is a PVM, as mentioned in Sect. 3.5.1, corresponding to the operator

$$U^* F(\hat{X}) U = F(\hat{X}_T),$$

where $\hat{X}_T$ is the Heisenberg position (configuration) operator at time $T$.

It is important to observe that even though these experiments suffer from the defect that no correlation is established between the system and an apparatus, this can easily be remedied—by adding a final *detection measurement* that measures the final actual configuration $X_T$—without in any way affecting the essential formal structure of the experiment. For these experiments the apparatus thus does not introduce any additional randomness, but merely reflects what was already present in $X_T$. All randomness in the final result

$$Z = F(X_T) \tag{3.91}$$

arises from randomness in the initial configuration of the system.[26]

For $F = x$ and $U = I$ the quadratic map is $\psi \mapsto |\psi(x)|^2$ with PVM $P^{\hat{X}}$, so that this (trivial) experiment corresponds to the simplest and most basic operator of quantum mechanics: the position operator. How other basic operators arise from experiments is what we are going to discuss next.

### 3.5.5 The Basic Operators of Quantum Mechanics

According to Bohmian mechanics, a particle whose wave function is real (up to a global phase), for example an electron in the ground state of an atom, has vanishing

---

[26] Though passive, the apparatus here plays an important role in recording the final configuration of the system. However, for experiments involving detections at different times, the apparatus plays an active role: Consider such an experiment, with detections at times $t_1, \ldots, t_n$, and final result $Z = F(X_{t_1}, \ldots, X_{t_n})$. Though the apparatus introduces no extra randomness, it plays an essential role by changing the wave function of the system at the times $t_1, \ldots, t_n$ and thus changing the evolution of its configuration. These changes are reflected in the POVM structure that governs the statistical distribution of $Z$ for such experiments (see Sect. 3.3.9).

velocity, even though the quantum formalism assigns a nontrivial probability distribution to its momentum. It might thus seem that we are faced with a conflict between the predictions of Bohmian mechanics and those of the quantum formalism. This, however, is not so. The quantum predictions about momentum concern the results of an experiment that happens to be called a momentum measurement and a conflict with Bohmian mechanics with regard to momentum must reflect disagreement about the results of such an experiment.

One may base such an experiment on free motion followed by a final measurement of position.[27] Consider a particle of mass $m$ whose wave function at $t = 0$ is $\psi = \psi(\mathbf{x})$. Suppose no forces are present, that is, that all the potentials acting on the particle are turned off, and let the particle evolve freely. Then we measure the position $\mathbf{X}_T$ that it has reached at the time $t = T$. It is natural to regard $\mathbf{V}_T = \mathbf{X}_T/T$ and $\mathbf{P}_T = m\mathbf{X}_T/T$ as providing, for large $T$, approximations to the asymptotic velocity and momentum of the particle. It turns out that the probability distribution of $\mathbf{P}_T$, in the limit $T \to \infty$, is exactly what quantum mechanics prescribes for the momentum, namely $|\tilde{\psi}(\mathbf{p})|^2$, where

$$\tilde{\psi}(\mathbf{p}) = (\mathscr{F}\psi)(\mathbf{p}) = \frac{1}{\sqrt{(2\pi\hbar)^3}} \int e^{-\frac{i}{\hbar}\mathbf{p}\cdot\mathbf{x}} \psi(\mathbf{x})\, d\mathbf{x}$$

is the Fourier transform of $\psi$.

This result can be easily understood: Observe that $|\psi_T(\mathbf{x})|^2\, d\mathbf{x}$, the probability distribution of $\mathbf{X}_T$, is the spectral measure $\mu_\psi^{\hat{\mathbf{X}}_T}(d\mathbf{x}) = \langle \psi, P^{\hat{\mathbf{X}}_T}(d\mathbf{x})\psi\rangle$ of $\hat{\mathbf{X}}_T = U_T^* \hat{\mathbf{X}} U_T$, the (Heisenberg) position operator at time $t = T$; here $U_t$ is the free evolution operator and $\hat{\mathbf{X}}$ is, as usual, the position operator at time $t = 0$. By elementary quantum mechanics (specifically, the Heisenberg equations of motion), $\hat{\mathbf{X}}_T = \frac{1}{m}\hat{\mathbf{P}} T + \hat{\mathbf{X}}$, where $\hat{\mathbf{P}} \equiv -i\hbar\nabla$ is the momentum operator. Thus as $T \to \infty$ the operator $m\hat{\mathbf{X}}_T/T$ converges to the momentum operator $\hat{\mathbf{P}}$, since $\hat{\mathbf{X}}/T$ is $O(1/T)$, and the distribution of the random variable $\mathbf{P}_T$ accordingly converges to the spectral measure of $\hat{\mathbf{P}}$, given by $|\tilde{\psi}(\mathbf{p})|^2$.[28]

---

[27] The emergence of the momentum operator in such so-called time-of-flight measurements was discussed by Bohm in his 1952 article [13]. A similar derivation of the momentum operator can be found in Feynman and Hibbs [49].

[28] This formal argument can be turned into a rigorous proof by considering the limit of the characteristic function of $\mathbf{P}_T$, namely of the function $f_T(\lambda) = \int e^{i\lambda\cdot\mathbf{p}} \rho_T(d\mathbf{p})$, where $\rho_T$ is the distribution of $m\mathbf{X}_T/T$: $f_T(\lambda) = \left\langle \psi, \exp\left(i\lambda\cdot m\hat{\mathbf{X}}_T/T\right)\psi\right\rangle$, and using the dominated convergence theorem [30] this converges as $T \to \infty$ to $f(\lambda) = \left\langle \psi, \exp\left(i\lambda\cdot\hat{\mathbf{P}}\right)\psi\right\rangle$, implying the desired result. The same result can also be obtained using the well known asymptotic formula (see, e.g., [50]) for the solution of the free Schrödinger equation with initial condition $\psi = \psi(\mathbf{x})$,

$$\psi_T(\mathbf{x}) \sim \left(\frac{m}{iT}\right)^{\frac{3}{2}} e^{i\frac{mx^2}{2\hbar T}} \tilde{\psi}\left(\frac{m\mathbf{x}}{T}\right) \quad \text{for} \quad T \to \infty.$$

## 3.5 The General Emergence of Operators

The momentum operator arises from a $(T \to \infty)$ limit of "no $Y$ no $\Phi$" single-particle experiments, each experiment being defined by the unitary operator $U_T$ (the free particle evolution operator up to time $T$) and calibration $F_T(\mathbf{x}) = m\mathbf{x}/T$. Other standard quantum-mechanical operators emerge in a similar manner, i.e., from a $T \to \infty$ limit of appropriate single-particle experiments.

This is the case, for example, for the spin operator $\sigma_z$. As in Sect. 3.5.2, consider the evolution operator $U_T$ generated by Hamiltonian (3.85), but instead of (3.87), consider the calibration $F_T(\mathbf{x}) = 2m\, z/a\, T^2$. This calibration is suggested by (3.16), as well as by the explicit form of the $z$-component of the position operator at time $t = T$,

$$\hat{Z}_T = U_T^* \hat{Z} U_T = \hat{Z} + \frac{\hat{P}_z}{m} T + \frac{a}{2m} \sigma_z T^2, \tag{3.92}$$

which follows from the Heisenberg equations

$$m \frac{d^2 \hat{Z}_t}{dt^2} = a\, \sigma_z, \quad m \frac{d \hat{Z}_t}{dt}\bigg|_{t=0} = \hat{P}_z \equiv -i\hbar \frac{\partial}{\partial z}, \quad \hat{Z}_0 = \hat{Z}.$$

Then, for initial state $\Psi = \psi \otimes \Phi_0$ with suitable $\Phi_0$, where $\psi = \alpha \psi^{(+)} + \beta \psi^{(-)}$, the distribution of the random variable

$$\Sigma_{zT} = F_T(\mathbf{X}_T) = \frac{2m Z_T}{a T^2}$$

converges as $T \to \infty$ to the spectral measure of $\sigma_z$, with values $+1$ and $-1$ occurring with probabilities $|\alpha|^2$ and $|\beta|^2$, respectively.[29] This is so, just as with the momentum, because as $T \to \infty$ the operator $\frac{2m \hat{Z}_T}{a T^2}$ converges to $\sigma_z$.

We remark that we've made use above of the fact that simple algebraic manipulations on the level of random variables correspond automatically to the same manipulations for the associated operators. More precisely, suppose that

$$Z \mapsto A \tag{3.93}$$

in the sense (of (3.37)) that the distribution of the random variable $Z$ is given by the spectral measure for the self-adjoint operator $A$. Then it follows from (3.54) that

$$f(Z) \to f(A) \tag{3.94}$$

for any (Borel) function $f$. For example, since $\mathbf{X}_T \mapsto \hat{\mathbf{X}}_T$, $m\mathbf{X}_T/T \mapsto m\hat{\mathbf{X}}_T/T$, and since $Z_T \to \hat{Z}_T$, $\frac{2m Z_T}{a T^2} \to \frac{2m \hat{Z}_T}{a T^2}$. Similarly, if a random variable $P \mapsto \hat{P}$, then $P^2/(2m) \mapsto H_0 = \hat{P}^2/(2m)$. This is rather trivial, but it is not as trivial as the failure even to distinguish $Z$ and $\hat{Z}$ would make it seem.

---

[29] For the Hamiltonian (3.85) no assumption on the initial state $\Psi$ is required here; however (3.85) will be a reasonably good approximation only when $\Psi$ has a suitable form, expressing in particular that the particle is appropriately moving towards the magnet.

## 3.5.6 From Positive-Operator-Valued Measures to Experiments

We wish here to point out that to a very considerable extent the association $\mathscr{E} \mapsto O(d\lambda)$ of experiments with POVMs is onto. It is more or less the case that every POVM arises from an experiment.

We have in mind two distinct remarks. First of all, it was pointed out in the first paragraph of Sect. 3.4.3 that every discrete POVM $O_\alpha$ (weak formal experiment) arises from some discrete experiment $\mathscr{E}$. Thus, for every POVM $O(d\lambda)$ there is a sequence $\mathscr{E}^{(n)}$ of discrete experiments for which the corresponding POVMs $O^{(n)}$ converge to $O$.

The second point we wish to make is that to the extent that every PVM arises from an experiment $\mathscr{E} = \{\Phi_0, U, F\}$, so too does every POVM. This is based on the fact, mentioned at the end of the introduction to Sect. 3.5, that every POVM $O(d\lambda)$ can be regarded as arising from the projection of a PVM $E(d\lambda)$, acting on $\mathscr{H}^{(1)}$, onto the subspace $\mathscr{H} \subset \mathscr{H}^{(1)}$. We may assume without loss of generality that both $\mathscr{H}$ and $\mathscr{H}^{(1)} \ominus \mathscr{H}$ are infinite dimensional (by some otherwise irrelevant enlargements if necessary). Thus we can identify $\mathscr{H}^{(1)}$ with $\mathscr{H} \otimes \mathscr{H}_{\text{apparatus}^{(1)}}$ and the subspace with $\mathscr{H} \otimes \Phi_0^{(1)}$, for any choice of $\Phi_0^{(1)}$. Suppose now that there is an experiment $\mathscr{E}^{(1)} = \{\Phi_0^{(2)}, U, F\}$ that measures the PVM $E$ (i.e., that measures the observable $A = \int \lambda E(d\lambda)$) where $\Phi_0^{(2)} \in \mathscr{H}_{\text{apparatus}^{(2)}}$, $U$ acts on $\mathscr{H} \otimes \mathscr{H}_{\text{apparatus}}$ where $\mathscr{H}_{\text{apparatus}} = \mathscr{H}_{\text{apparatus}^{(1)}} \otimes \mathscr{H}_{\text{apparatus}^{(2)}}$ and $F$ is a function of the configuration of the composite of the 3 systems: system, apparatus$^{(1)}$ and apparatus$^{(2)}$. Then, with $\Phi_0 = \Phi_0^{(1)} \otimes \Phi_0^{(2)}$, $\mathscr{E} = \{\Phi_0, U, F\}$ is associated with the POVM $O$.

## 3.5.7 Invariance Under Trivial Extension

Suppose we change an experiment $\mathscr{E}$ to $\mathscr{E}'$ by regarding its $x$-system as containing more of the universe that the $x$-system for $\mathscr{E}$, without in any way altering what is physically done in the experiment and how the result is specified. One would imagine that $\mathscr{E}'$ would be equivalent to $\mathscr{E}$. This would, in fact, be trivially the case classically, as it would if $\mathscr{E}$ were a genuine measurement, in which case $\mathscr{E}'$ would obviously measure the same thing as $\mathscr{E}$. This remains true for the more formal notion of measurement under consideration here. The only source of nontriviality in arriving at this conclusion is the fact that with $\mathscr{E}'$ we have to deal with a different, larger class of initial wave functions.

We will say that $\mathscr{E}'$ is a trivial extension of $\mathscr{E}$ if the only relevant difference between $\mathscr{E}$ and $\mathscr{E}'$ is that the $x$-system for $\mathscr{E}'$ has generic configuration $x' = (x, \hat{x})$, whereas the $x$-system for $\mathscr{E}$ has generic configuration $x$. In particular, the unitary operator $U'$ associated with $\mathscr{E}'$ has the form $U' = U \otimes \hat{U}$, where $U$ is the unitary associated with $\mathscr{E}$, implementing the interaction of the $x$-system and the apparatus, while $\hat{U}$ is a unitary operator describing the independent evolution of the $\hat{x}$-system, and the calibration $F$ for $\mathscr{E}'$ is the same as for $\mathscr{E}$. (Thus $F$ does not depend upon $\hat{x}$.)

3.5 The General Emergence of Operators

The association of experiments with (generalized) observables (POVMs) is *invariant under trivial extension*: if $\mathscr{E} \mapsto O$ in the sense of (3.66) and $\mathscr{E}'$ is a trivial extension of $\mathscr{E}$, then $\mathscr{E}' \mapsto O \otimes I$, where $I$ is the identity on the Hilbert space of the $\hat{x}$-system.

To see this note that if $\mathscr{E} \mapsto O$ then the sesquilinear map $B$ arising from (3.80) for $\mathscr{E}'$ is of the form

$$B(\psi_1 \otimes \hat{\psi}_1, \psi_2 \otimes \hat{\psi}_2) = \langle \psi_1, O\psi_2 \rangle \langle \hat{\psi}_1, \hat{\psi}_2 \rangle$$

on product wave functions $\psi' = \psi \otimes \hat{\psi}$, which easily follows from the form of $U'$ and the fact that $F$ doesn't depend upon $\hat{x}$, so that the $\hat{x}$-degrees of freedom can be integrated out. Thus the POVM $O'$ for $\mathscr{E}'$ agrees with $O \otimes I$ on product wave functions, and since such wave functions span the Hilbert space for the $(x, \hat{x})$-system, we have that $O' = O \otimes I$. Thus $\mathscr{E}' \mapsto O \otimes I$.

In other words, if $\mathscr{E}$ is a measurement of $O$, then $\mathscr{E}'$ is a measurement of $O \otimes I$. In particular, if $\mathscr{E}$ is a measurement the self-adjoint operator $A$, then $\mathscr{E}'$ is a measurement of $A \otimes I$. This result is not quite so trivial as it would be were it concerned with genuine measurements, rather than with the more formal notion under consideration here.

Now suppose that $\mathscr{E}'$ is a trivial extension of a discrete experiment $\mathscr{E}$, with state transformations given by $R_\alpha$. Then the state transformations for $\mathscr{E}'$ are given by $R'_\alpha = R_\alpha \otimes \hat{U}$. This is so because $R'_\alpha$ must agree with $R_\alpha \otimes \hat{U}$ on product wave functions $\psi' = \psi \otimes \hat{\psi}$, and these span the Hilbert space of the $(x, \hat{x})$-system.

### 3.5.8 POVMs and the Positions of Photons and Dirac Electrons

We have indicated how POVMs emerge naturally in association with Bohmian experiments. We wish here to indicate a somewhat different role for a POVM: to describe the probability distribution of the actual (as opposed to measured[30]) position. The probability distribution of the position of a Dirac electron in the state $\psi$ is $\psi^+\psi$. This is given by a PVM $E(d\mathbf{x})$ on the one-particle Hilbert space $\mathscr{H}$ spanned by positive and negative energy electron wave functions. However the physical one-particle Hilbert-space $\mathscr{H}_+$ consists solely of positive energy states, and this is not invariant under the projections $E$. Nonetheless the probability distribution of the position of the electron is given by the POVM $P_+ E(d\mathbf{x}) P_+$ acting on $\mathscr{H}_+$, where $P_+$ is the orthogonal projection onto $\mathscr{H}_+$. Similarly, constraints on the photon wave function require the use of POVMs for the localization of photons [51, 52].[31]

---

[30] The accurate measurement of the position of a Dirac electron is presumably impossible.

[31] For example, on the one-photon level, both the proposal $\Psi = \mathbf{E} + i\mathbf{B}$ (where $\mathbf{E}$ and $\mathbf{B}$ are the electric and the magnetic free fields) [53], and the proposal $\Psi = \mathbf{A}$ (where $\mathbf{A}$ is the vector potential in the Coulomb gauge) [52], require the constraint $\nabla \cdot \Psi = 0$.

## 3.6 Density Matrices

The notion of a density matrix, a positive (trace class) operator with unit trace on the Hilbert space of a system, is often regarded as providing the most general characterization of a quantum state of that system. According to the quantum formalism, when a system is described by the density matrix $W$, the expected value of an observable $A$ is given by $\operatorname{tr}(WA)$. If $A$ has PVM $O$, and more generally for any POVM $O$, the probability that the (generalized) observable $O$ has value in $\Delta$ is given by

$$\operatorname{Prob}(O \in \Delta) = \operatorname{tr}(W O(\Delta)). \tag{3.95}$$

A density matrix that is a one-dimensional projection, i.e., of the form $|\psi\rangle\langle\psi|$ where $\psi$ is a unit vector in the Hilbert space of the system, describes a *pure state* (namely, $\psi$), and a general density matrix can be decomposed into a *mixture* of pure states $\psi_k$,

$$W = \sum_k p_k |\psi_k\rangle\langle\psi_k| \quad \text{where} \quad \sum_k p_k = 1. \tag{3.96}$$

Naively, one might regard $p_k$ as the probability that the system *is* in the state $\psi_k$. This interpretation is, however, untenable, for a variety of reasons. First of all, the decomposition (3.96) is not unique. A density matrix $W$ that does not describe a pure state can be decomposed into pure states in a variety of different ways.

It is always possible to decompose a density matrix $W$ in such a way that its components $\psi_k$ are orthonormal. Such a decomposition will be unique except when $W$ is degenerate, i.e., when some $p_k$'s coincide. For example, if $p_1 = p_2$ we may replace $\psi_1$ and $\psi_2$ by any other orthonormal pair of vectors in the subspace spanned by $\psi_1$ and $\psi_2$. And even if $W$ were nondegenerate, it need not be the case that the system is in one of the states $\psi_k$ with probability $p_k$, because for any decomposition (3.96), regardless of whether the $\psi_k$ are orthogonal, if the wave function of the system were $\psi_k$ with probability $p_k$, this situation would be described by the density matrix $W$.

Thus a general density matrix carries no information—not even statistical information—about the actual wave function of the system. Moreover, a density matrix can describe a system that has no wave function at all! This happens when the system is a subsystem of a larger system whose wave function is entangled, i.e., does not properly factorize (in this case one usually speaks of the reduced density matrix of the subsystem).

This impossibility of interpreting density matrices as real mixtures of pure states has been regarded by many authors (e.g., von Neumann [3] and Landau [54]) as a further indication that quantum randomness is inexplicable within the realm of classical logic and probability. However, from the point of view of Bohmian mechanics, there is nothing mysterious about density matrices. Indeed, their role and status within the quantum formalism can be understood very easily in terms of the general framework of experiments of Sect. 3.5. (It can, we believe, be reasonably argued that even from the perspective of orthodox quantum theory, density matrices can be understood in a straightforward way.)

## 3.6.1 Density Matrices and Bohmian Experiments

Consider a general experiment $\mathscr{E} \mapsto O$ (see Eq. (3.66)) and suppose that the initial wave function of the system is random with probability distribution $p(d\psi)$ (on the set of unit vectors in $\mathscr{H}$). Then nothing will change in the general argument of Sect. 3.5 except that now $\mathsf{P}^Z_\psi$ in (3.66) and (3.80) should be interpreted as the conditional probability *given* $\psi$. It follows then from (3.95), using the fact that $\langle \psi, O(\Delta)\psi \rangle = \mathrm{tr}\,(|\psi\rangle\langle\psi|\,O(\Delta))$, that the probability that the result of $\mathscr{E}$ lies in $\Delta$ is given by

$$\int p(d\psi)\langle \psi, O(\Delta)\psi \rangle = \mathrm{tr}\left(\int p(d\psi)|\psi\rangle\langle\psi|\,O(\Delta)\right) = \mathrm{tr}\,(WO(\Delta)) \quad (3.97)$$

where[32]

$$W \equiv \int p(d\psi)\,|\psi\rangle\langle\psi| \quad (3.98)$$

is the *ensemble density matrix* arising from a random wave function with (ensemble) distribution $p$.

Now suppose that instead of having a random wave function, our system has no wave function at all because it is entangled with another system. Then there is still an object that can naturally be regarded as the state of our system, an object associated with the system itself in terms of which the results of experiments performed on our system can be simply expressed. This object is a density matrix $W$ and the results are governed by (3.95). $W$ is the *reduced density matrix* arising from the state of the larger system. This is more or less an immediate consequence of invariance under trivial extension, described in Sect. 3.5.7:

Consider a trivial extension $\mathscr{E}'$ of an experiment $\mathscr{E} \mapsto O$ on our system—precisely what we must consider if the larger system has a wave function $\psi'$ while our (smaller) system does not. The probability that the result of $\mathscr{E}'$ lies in $\Delta$ is given by

$$\langle \psi', O(\Delta)\otimes I\psi'\rangle = \mathrm{tr}'\left(|\psi'\rangle\langle\psi'|\,O(\Delta)\otimes I\right) = \mathrm{tr}\,(WO(\Delta)), \quad (3.99)$$

where $\mathrm{tr}'$ is the trace for the $x'$-system (the big system) and $\mathrm{tr}$ is the trace for the $x$-system. In agreement with standard quantum mechanics, the last equality of (3.99) defines $W$ as the reduced density matrix of the $x$-system, i.e,

$$W \equiv \widehat{\mathrm{tr}}\left(|\psi'\rangle\langle\psi'|\right) \quad (3.100)$$

where $\widehat{\mathrm{tr}}$ denotes the partial trace over the coordinates of the $\hat{x}$-system.

---

[32] Note that since $p$ is a probability measure on the unit sphere in $\mathscr{H}$, $W$ is a positive trace class operator with unit trace.

## 3.6.2 Strong Experiments and Density Matrices

A strong formal experiment $\mathscr{E} \equiv \{\lambda_\alpha, R_\alpha\}$ generates state transformations $\psi \to R_\alpha \psi$. This suggests the following action on an initial state described by a density matrix $W$: When the outcome is $\alpha$, we have the transformation

$$W \to \frac{\mathscr{R}_\alpha W}{\mathrm{tr}\,(\mathscr{R}_\alpha W)} \equiv \frac{R_\alpha W R_\alpha^*}{\mathrm{tr}\,(R_\alpha W R_\alpha^*)} \tag{3.101}$$

where

$$\mathscr{R}_\alpha W = R_\alpha W R_\alpha^*. \tag{3.102}$$

After all, (3.101) is a density matrix naturally associated with $R_\alpha$ and $W$, and it agrees with $\psi \to R_\alpha \psi$ for a pure state, $W = |\psi\rangle\langle\psi|$. In order to show that (3.101) is indeed correct, we must verify it for the two different ways in which our system might be assigned a density matrix $W$, i.e., for $W$ an ensemble density matrix and for $W$ a reduced density matrix.

Suppose the initial wave function is random, with distribution $p(d\psi)$. Then the initial state of our system is given by the density matrix (3.98). When the outcome $\alpha$ is obtained, two changes must be made in (3.98) to reflect this information: $|\psi\rangle\langle\psi|$ must be replaced by $(R_\alpha |\psi\rangle\langle\psi| R_\alpha^*)/\|R_\alpha \psi\|^2$, and $p(d\psi)$ must be replaced by $p(d\psi|\alpha)$, the conditional distribution of the initial wave function given that the outcome is $\alpha$. For the latter we have

$$p(d\psi|\alpha) = \frac{\|R_\alpha \psi\|^2}{\mathrm{tr}\,(R_\alpha W R_\alpha^*)} p(d\psi)$$

($\|R_\alpha \psi\|^2 p(d\psi)$ is the joint distribution of $\psi$ and $\alpha$ and the denominator is the probability of obtaining the outcome $\alpha$.) Therefore $W$ undergoes the transformation

$$W = \int p(d\psi)\, |\psi\rangle\langle\psi| \;\to\; \int p(d\psi|\alpha) \frac{R_\alpha |\psi\rangle\langle\psi| R_\alpha^*}{\|R_\alpha \psi\|^2}$$

$$= \int p(d\psi) \frac{R_\alpha |\psi\rangle\langle\psi| R_\alpha^*}{\mathrm{tr}\,(R_\alpha W R_\alpha^*)} = \frac{R_\alpha W R_\alpha^*}{\mathrm{tr}\,(R_\alpha W R_\alpha^*)}.$$

We wish to emphasize that this demonstrates in particular the nontrivial fact that the density matrix $\mathscr{R}_\alpha W / \mathrm{tr}\,(\mathscr{R}_\alpha W)$ produced by the experiment depends only upon the initial density matrix $W$. Though $W$ can arise in many different ways, corresponding to the multiplicity of different probability distributions $p(d\psi)$ yielding $W$ via (3.98), insofar as the final state is concerned, these differences don't matter.

This does not, however, establish (3.101) when $W$ arises not from a random wave function but as a reduced density matrix. To deal with this case we consider a trivial extension $\mathscr{E}'$ of a discrete experiment $\mathscr{E}$ with state transformations $R_\alpha$. Then $\mathscr{E}'$ has state transformations $R_\alpha \otimes \hat{U}$ (see Sect. 3.5.7). Thus, when the initial state of the $x'$-system is $\psi'$, the final state of the $x$-system is given by the partial trace

3.6 Density Matrices

$$\frac{\widehat{\mathrm{tr}}\left(R_\alpha \otimes \hat{U}|\psi'\rangle\langle\psi'|R_\alpha^* \otimes \hat{U}^*\right)}{\mathrm{tr}'\left(R_\alpha \otimes \hat{U}|\psi'\rangle\langle\psi'|R_\alpha^* \otimes \hat{U}^*\right)} = \frac{\widehat{\mathrm{tr}}\left(R_\alpha \otimes I|\psi'\rangle\langle\psi'|R_\alpha^* \otimes I\right)}{\mathrm{tr}'\left(R_\alpha \otimes I|\psi'\rangle\langle\psi'|R_\alpha^* \otimes I\right)}$$

$$= \frac{R_\alpha \widehat{\mathrm{tr}}(|\psi'\rangle\langle\psi'|)R_\alpha^*}{\mathrm{tr}\left(R_\alpha \widehat{\mathrm{tr}}(|\psi'\rangle\langle\psi'|)R_\alpha^*\right)} = \frac{R_\alpha W R_\alpha^*}{\mathrm{tr}\left(R_\alpha W R_\alpha^*\right)},$$

where the cyclicity of the trace has been used.

To sum up, when a strong experiment $\mathscr{E} \equiv \{\lambda_\alpha, R_\alpha\}$ is performed on a system described by the initial density matrix $W$ and the outcome $\alpha$ is obtained, the final density matrix is given by (3.101); moreover, from the results of the previous section it follows that the outcome $\alpha$ will occur with probability

$$p_\alpha = \mathrm{tr}\,(W O_\alpha) = \mathrm{tr}\,\left(W R_\alpha^* R_\alpha\right) = \mathrm{tr}\,(\mathscr{R}_\alpha W), \tag{3.103}$$

where the last equality follows from the cyclicity of the trace.

### 3.6.3 The Notion of Instrument

We shall briefly comment on the relationship between the notion of strong formal experiment and that of *instrument* (or *effect*) discussed by Davies [4].

Consider an experiment $\mathscr{E} \equiv \{\lambda_\alpha, R_\alpha\}$ on a system with initial density matrix $W$. Then a natural object associated with $\mathcal{E}$ is the set function

$$\mathscr{R}(\Delta)W \equiv \sum_{\lambda_\alpha \in \Delta} \mathscr{R}_\alpha W = \sum_{\lambda_\alpha \in \Delta} R_\alpha W R_\alpha^*. \tag{3.104}$$

The set function $\mathscr{R} : \Delta \mapsto \mathscr{R}(\Delta)$ compactly expresses both the statistics of $\mathcal{E}$ for a general initial system density matrix $W$ and the effect of $\mathcal{E}$ on $W$ *conditioned* on the occurrence of the event "the result of $\mathcal{E}$ is in $\Delta$".

To see this, note first that it follows from (3.103) that the probability that the result of the experiment lies in the set $\Delta$ is given by

$$p(\Delta) = \mathrm{tr}\,(\mathscr{R}(\Delta)W).$$

The conditional distribution $p(\alpha|\Delta)$ that the outcome is $\alpha$ given that the result $\lambda_\alpha \in \Delta$ is then $\mathrm{tr}\,(\mathscr{R}_\alpha W)/\mathrm{tr}\,(\mathscr{R}(\Delta)W)$. The density matrix that reflects the knowledge that the result is in $\alpha$, obtained by averaging (3.101) over $\Delta$ using $p(\alpha|\Delta)$, is thus $\mathscr{R}(\Delta)W/\mathrm{tr}\,(\mathscr{R}(\Delta)W)$.

It follows from (3.104) that $\mathscr{R}$ is a countably additive set function whose values are positive preserving linear transformations in the space of trace-class operators in $\mathscr{H}$. Any map with these properties, not necessarily of the special form (3.104), is called an *instrument*.

### 3.6.4 On the State Description Provided by Density Matrices

So far we have followed the standard terminology and have spoken of a density matrix as describing the *state* of a physical system. It is important to appreciate, however, that this is merely a frequently convenient way of speaking, for Bohmian mechanics as well as for orthodox quantum theory. Insofar as Bohmian mechanics is concerned, the significance of density matrices is neither more nor less than what is implied by their role in the quantum formalism as described in Sect. 3.6.1 and 3.6.2. While many aspects of the notion of (effective) wave function extend to density matrices, in particular with respect to weak and strong experiments, density matrices lack the dynamical implications of wave functions for the evolution of the configuration, a point that has been emphasized by Bell [47]:

> In the de Broglie-Bohm theory a fundamental significance is given to the wave function, and it cannot be transferred to the density matrix. .... Of course the density matrix retains all its usual practical utility in connection with quantum statistics.

That this is so should be reasonably clear, since it is the wave function that determines, in Bohmian mechanics, the evolution of the configuration, and the density matrix of a system does not determine its wave function, even statistically. To underline the point we shall recall the analysis of Bell [47]: Consider a particle described by a density matrix $W_t$ evolving autonomously, so that $W_t = U_t W_0 U_t^{-1}$, where $U_t$ is the unitary group generated by a Schrödinger Hamiltonian. Then $\rho^{W_t}(x) \equiv W_t(x, x) \equiv \langle x|W_t|x\rangle$ gives the probability distribution of the position of the particle. Note that $\rho^W$ satisfies the continuity equation

$$\frac{\partial \rho^W}{\partial t} + \text{div} J^W = 0 \quad \text{where}$$

$$J^W(x) = \frac{\hbar}{m} \text{Im} \left[\nabla_x W(x, x')\right]_{x'=x}.$$

This might suggest that the velocity of the particle should be given by $v = J^W/\rho^W$, which indeed agrees with the usual formula when $W$ is a pure state ($W(x, x') = \psi(x)\psi^*(x')$). However, this extension of the usual formula to arbitrary density matrices, though mathematically "natural," is not consistent with what Bohmian mechanics prescribes for the evolution of the configuration. Consider, for example, the situation in which the wave function of a particle is random, either $\psi_1$ or $\psi_2$, with equal probability. Then the density matrix is $W(x, x') = \frac{1}{2}\left(\psi_1(x)\psi_1^*(x') + \psi_2(x)\psi_2^*(x')\right)$. But the velocity of the particle will be always *either* $v_1$ or $v_2$ (according to whether the actual wave function is $\psi_1$ or $\psi_2$), and—unless $\psi_1$ and $\psi_2$ have disjoint supports—this does not agree with $J^W/\rho^W$, an average of $v_1$ and $v_2$.

What we have just said is correct, however, only when spin is ignored. For particles with spin a novel kind of density matrix emerges, a *conditional density matrix*, analogous to the conditional wave function (3.6) and with an analogous dynamical role: Even though no conditional wave function need exist for a system entangled with its environment when spin is taken into account, a conditional density matrix $W$

3.7 Genuine Measurements                                                                                    137

always exists, and is such that the velocity of the system is indeed given by $J^W/\rho^W$. See [17] for details.

A final remark: the statistical role of density matrices is basically different from that provided by statistical ensembles, e.g, by Gibbs states in classical statistical mechanics. This is because, as mentioned earlier, even when it describes a random wave function via (3.98), a density matrix $W$ does not determine the ensemble $p(d\psi)$ from which it emerges. The map defined by (3.98) from probability measures $p$ on the unit sphere in $\mathscr{H}$ to density matrices $W$ is many-to-one.[33] Consider, for example, the density matrix $\frac{1}{n}I$ where $I$ is the identity operator on an $n$-dimensional Hilbert space $\mathscr{H}$. Then a uniform distribution over the vectors of any given orthonormal basis of $\mathscr{H}$ leads to this density matrix, as well as does the continuous uniform measure on the sphere $\|\psi\| = 1$. However, since the statistical distribution of the results of any experiment depends on $p$ only through $W$, different $p$'s associated with the same $W$ are *empirically equivalent* in the sense that they can't be distinguished by experiments performed on a system prepared somehow in the state $W$.

## 3.7 Genuine Measurements

We have so far discussed various interactions between a system and an apparatus relevant to the quantum measurement formalism, from the very special ones formalized by "ideal measurements" to the general situation described in Sect. 3.6. It is important to recognize that nowhere in this discussion was there any implication that anything was actually being measured. The fact that an interaction with an apparatus leads to a pointer orientation that we call the result of the experiment or "measurement" in no way implies that this result reflects anything of significance concerning the system under investigation, let alone that it reveals some preexisting property of the system—and this is what is supposed to be meant by the word measurement. After all [29], "any old playing around with an indicating instrument in the vicinity of another body, whereby at any old time one then takes a reading, can hardly be called a measurement of this body," and the fact the experiment happens to be associated, say, with a self-adjoint operator in the manner we have described, so that the experiment is spoken of, in the quantum formalism, as a measurement of the corresponding observable, certainly offers little support for using language in this way.

We shall elaborate on this point later on. For now we wish to observe that the very generality of our analysis, particularly that of Sect. 3.6, covering as it does all

---

[33] This is relevant to the foundations of quantum statistical mechanics, for which the state of an isolated thermodynamic system is usually described by the microcanonical density matrix $\mathscr{Z}^{-1}\delta(H - E)$, where $\mathscr{Z} = \text{tr}\,\delta(H - E)$ is the partition function. Which ensemble of wave functions should be regarded as forming the thermodynamic ensemble? A natural choice is the uniform measure on the subspace $H = E$, which should be thought of as fattened in the usual way. Note that this choice is quite distinct from another one that people often have in mind: a uniform distribution over a basis of energy eigenstates of the appropriate energy. Depending upon the choice made, we obtain different notions of typical equilibrium wave function.

possible interactions between system and apparatus, covers as well those particular situations that in fact are genuine measurements. This allows us to make some definite statements about what can be measured in Bohmian mechanics.

For a physical quantity, describing an objective property of a system, to be measurable means that it is possible to perform an experiment on the system that measures the quantity, i.e., an experiment whose result conveys its value. In Bohmian mechanics a physical quantity $\xi$ is expressed by a function

$$\xi = f(X, \psi) \tag{3.105}$$

of the complete state $(X, \psi)$ of the system. An experiment $\mathscr{E}$ measuring $\xi$ is thus one whose result $Z = F(X_T, Y_T) \equiv Z(X, Y, \Psi)$ equals $\xi = f(X, \psi) \equiv \xi(X, \psi)$,

$$Z(X, Y, \Psi) = \xi(X, \psi), \tag{3.106}$$

where $X, Y, \psi$ and $\Psi$ refer, as in Sect. 3.5, to the initial state of system and apparatus, immediately prior to the measurement, and where the equality should be regarded as approximate, holding to any desired degree of accuracy.

The most basic quantities are, of course, the state components themselves, namely $X$ and $\psi$, as well as the velocities

$$\mathbf{v}_k = \frac{\hbar}{m_k} \mathrm{Im} \frac{\nabla_k \psi(X)}{\psi(X)} \tag{3.107}$$

of the particles. One might also consider quantities describing the future behavior of the system, such as the configuration of an isolated system at a later time, or the time of escape of a particle from a specified region, or the asymptotic velocity discussed in Sect. 3.5.5. (Because the dynamics is deterministic, all of these quantities are functions of the initial state of the system and are thus of the form (3.105).)

We wish to make a few remarks about the measurability of these quantities. In particular, we wish to mention, as an immediate consequence of the analysis at the beginning of Sect. 3.5, a condition that must be satisfied by any quantity if it is to be measurable.

### *3.7.1 A Necessary Condition for Measurability*

Consider any experiment $\mathscr{E}$ measuring a physical quantity $\xi$. We showed in Sect. 3.5 that the statistics of the result $Z$ of $\mathscr{E}$ must be governed by a POVM, i.e., that the probability distribution of $Z$ must be given by a measure-valued quadratic map on the system Hilbert space $\mathscr{H}$. Thus, by (3.106),

$\xi$ is measurable only if its probability distribution $\mu_\xi^\psi$ is a measure-valued quadratic map on $\mathscr{H}$. (3.108)

## 3.7 Genuine Measurements

As indicated earlier, the position **X** and the asymptotic velocity or momentum **P** have distributions quadratic in $\psi$, namely $\mu_\mathbf{X}^\psi(d\mathbf{x}) = |\psi(\mathbf{x})|^2$ and $\mu_\mathbf{P}^\psi(d\mathbf{p}) = |\tilde\psi(\mathbf{p})|^2$, respectively. Moreover, they are both measurable, basically because suitable local interactions exist to establish appropriate correlations with the relevant macroscopic variables. For example, in a bubble chamber a particle following a definite path triggers a chain of reactions that leads to the formation of (macroscopic) bubbles along the path.

The point we wish to make now, however, is simply this: the measurability of these quantities is not a consequence of the fact that these quantities obey this measurability condition. We emphasize that this condition is merely a necessary condition for measurability, and not a sufficient one. While it does follow that if $\xi$ satisfies this condition there exists a discrete experiment that is an approximate formal measurement of $\xi$ (in the sense that the distribution of the result of the experiment is approximately $\mu_\xi^\psi$), this experiment need not provide a genuine measurement of $\xi$ because the interactions required for its implementation need not exist and because, even if they did, the result $Z$ of the experiment might not be related to the quantity $\xi$ in the right way, i.e, via (3.106).

We now wish to illustrate the use of this condition, first transforming it into a weaker but more convenient form. Note that any quadratic map $\mu^\psi$ must satisfy

$$\mu^{\psi_1+\psi_2} + \mu^{\psi_1-\psi_2} = 2(\mu^{\psi_1} + \mu^{\psi_2})$$

and thus if $\mu^\psi$ is also positive we have the inequality

$$\mu^{\psi_1+\psi_2} \leq 2(\mu^{\psi_1} + \mu^{\psi_2}). \tag{3.109}$$

Thus it follows from (3.108) that a quantity[34]

> $\xi$ *must fail to be measurable if it has a possible value (one with nonvanishing probability or probability density) when the wave function of the system is $\psi_1 + \psi_2$ that is neither a possible value when the wave function is $\psi_1$ nor a possible value when the wave function is $\psi_2$.* (3.110)

(Here neither $\psi_1$ nor $\psi_2$ need be normalized.)

### 3.7.2 The Nonmeasurability of Velocity, Wave Function and Deterministic Quantities

It is an immediate consequence of (3.110) that neither the velocity nor the wave function is measurable, the latter because the value "$\psi_1 + \psi_2$" is neither "$\psi_1$" nor

---

[34] This conclusion is also a more or less direct consequence of the linearity of the Schrödinger evolution: If $\psi_i \otimes \Phi_0 \mapsto \Psi_i$ for all $i$, then $\sum \psi_i \otimes \Phi_0 \mapsto \sum \Psi_i$. But, again, our purpose here has been mainly to illustrate the use of the measurability condition itself.

"$\psi_2$," and the former because every wave function $\psi$ may be written as $\psi = \psi_1 + \psi_2$ where $\psi_1$ is the real part of $\psi$ and $\psi_2$ is $i$ times the imaginary part of $\psi$, for both of which the velocity (of whatever particle) is 0.

Note that this is a very strong and, in a sense, surprising conclusion, in that it establishes the *impossibility* of measuring what is, after all, a most basic dynamical variable for a *deterministic* mechanical theory of particles in motion. It should probably be regarded as even more surprising that the proof that the velocity—or wave function—is not measurable seems to rely almost on nothing, in effect just on the linearity of the evolution of the wave function. However, one should not overlook the crucial role of quantum equilibrium.

We observe that the nonmeasurability of the wave function is related to the *impossibility of copying* the wave function. (This question arises sometimes in the form, "Can one clone the wave function?" [55, 56, 57]). Copying would be accomplished, for example, by an interaction leading, for all $\psi$, from $\psi \otimes \phi_0 \otimes \Phi_0$ to $\psi \otimes \psi \otimes \Phi$, but this is clearly incompatible with unitarity. We wish here merely to remark that the impossibility of cloning can also be regarded as a consequence of the nonmeasurability of the wave function. In fact, were cloning possible one could—by making many copies—measure the wave function by performing suitable measurements on the various copies. After all, any wave function $\psi$ is determined by $\langle \psi, A\psi \rangle$ for sufficiently many observables $A$ and these expectation values can of course be computed using a sufficiently large ensemble.

By a deterministic quantity we mean any function $\xi = f(\psi)$ of the wave function alone (which thus does not inherit any irreducible randomness associated with the random configuration $X$). It follows easily from (3.110) that no (nontrivial) deterministic quantity is measurable.[35] In particular, the mean value $\langle \psi, A\psi \rangle$ of an observable $A$ (not a multiple of the identity) is not measurable—though it would be were it possible to copy the wave function, and it can of course be measured by a nonlinear experiment, see Sect. 3.7.4.

### 3.7.3 Initial Values and Final Values

Measurement is a tricky business. In particular, one may wonder how, if it is not measurable, we are ever able to know the wave function of a system—which in orthodox quantum theory often seems to be the only thing that we do know about it.

In this regard, it is important to appreciate that we were concerned in the previous section only with initial values, with the wave function and the velocity *prior* to the measurement. We shall now briefly comment upon the measurability of final values, produced by the experiment.

The nonmeasurability argument of Sect. 3.7.2 does not cover final values. This may be appreciated by noting that the crucial ingredient in the analysis involves a fundamental time-asymmetry: The probability distribution $\mu^\psi$ of the result of an

---

[35] Note also that $\mu_\xi^\psi(d\lambda) = \delta(\lambda - f(\psi))d\lambda$ seems manifestly nonquadratic in $\psi$ (unless $f$ is constant).

experiment is a quadratic functional of the *initial* wave function $\psi$, not the final one—of which it is not a functional at all. Moreover, the final velocity can indeed be measured, by a momentum measurement as described in Sect. 3.5.5. (That such a measurement yields also the final velocity follows from the formula in footnote 28 for the asymptotic wave function.) And the final wave function can be measured by an ideal measurement of any nondegenerate observable, and more generally by any strong formal measurement whose subspaces $\mathcal{H}_\alpha$ are one-dimensional, see Sect. 3.3.5: If the outcome is $\alpha$, the final wave function is $R_\alpha \psi = R_\alpha P_{\mathcal{H}_\alpha} \psi$, which is independent of the initial wave function $\psi$ (up to a scalar multiple).

We also wish to remark that this distinction between measurements of initial values and measurements of final values has no genuine significance for passive measurements, that merely reveal preexisting properties without in any way affecting the measured system. However, quantum measurements are usually active; for example, an ideal measurement transforms the wave function of the system into an eigenstate of the measured observable. But passive or active, a measurement, by its very meaning, is concerned strictly speaking with properties of a system just before its performance, i.e., with initial values. At the same time, to the extent that any property of a system is conveyed by a typical quantum "measurement," it is a property defined by a final value.

For example, according to orthodox quantum theory a position measurement on a particle with a spread-out wave function, to the extent that it measures anything at all, measures the final position of the particle, created by the measurement, rather than the initial position, which is generally regarded as not existing prior to the measurement. And even in Bohmian mechanics, in which such a measurement may indeed reveal the initial position, which—if the measurement is suitably performed—will agree with the final position, this measurement will still be active since the wave function of the system must be transformed by the measurement into one that is compatible with the sharper knowledge of the position that it provides, see Sect. 3.2.2.

### *3.7.4 Nonlinear Measurements and the Role of Prior Information*

The basic idea of measurement is predicated on initial ignorance. We think of a measurement of a property of a system as conveying that property by a procedure that does not seriously depend upon the state of the system,[36] any details of which must after all be unknown prior to at least some engagement with the system. Be that as it may, the notion of measurement as codified by the quantum formalism is indeed rooted in a standpoint of ignorance: the experimental procedures involved in the

---

[36] This statement must be taken with a grain of salt. Some things must be known about the system prior to measurement, for example, that it is in the vicinity the measurement apparatus, or that an atom whose angular momentum we wish to measure is moving towards the relevant Stern Gerlach magnets, as well as a host of similar, often unnoticed, pieces of information. This sort of thing does not much matter for our purposes in this chapter and can be safely ignored. Taking them into account would introduce pointless complications without affecting the analysis in an essential way.

measurement do not depend upon the state of the measured system. And our entire discussion of measurement up to now has been based upon that very assumption, that $\mathcal{E}$ itself does not depend on $\psi$ (and certainly not on $X$).

If, however, some prior information on the initial system wave function $\psi$ were available, we could exploit this information to measure quantities that would otherwise fail to be measurable. For example, for a single-particle system, if we somehow knew its initial wave function $\psi$ then a measurement of the initial position of the particle would convey its initial velocity as well, via (3.107)—even though, as we have shown, this quantity isn't measurable without such prior information.

By a nonlinear measurement or experiment $\mathcal{E} = \mathcal{E}^\psi$ we mean one in which, unlike those considered so far, one or more of the defining characteristics of the experiment depends upon $\psi$. For example, in the measurement of the initial velocity described in the previous paragraph, the calibration function $F = F^\psi$ depends upon $\psi$.[37] More generally we might have that $U = U^\psi$ or $\Phi_0 = \Phi_0^\psi$.

The wave function can of course be measured by a nonlinear measurement—just let $F^\psi \equiv \psi$. Somewhat less trivially, the initial wave function can be measured, at least formally, if it is known to be a member of a given orthonormal basis, by measuring any nondegenerate observable whose eigenvectors form that basis. The proposals of Aharonov, Anandan and Vaidman [58] for measuring the wave function, though very interesting, are of this character—they involve nonlinear measurements that depend upon a choice of basis containing $\psi$—and thus remain controversial.[38]

### 3.7.5 A Position Measurement that Does not Measure Position

We began this section by observing that what is spoken of as a measurement in quantum theory need not really measure anything. We mentioned, however, that in Bohmian mechanics the position can be measured, and the experiment that accomplishes this would of course be a measurement of the position operator. We wish here to point out, by means of a very simple example, that the converse is not true, i.e., that a measurement of the position operator need not be a measurement of the position.

Consider the harmonic oscillator in 2 dimensions with Hamiltonian

$$H = -\frac{\hbar^2}{2m}\left(\frac{\partial^2}{\partial x^2} + \frac{\partial^2}{\partial y^2}\right) + \frac{\omega^2 m}{2}(x^2 + y^2).$$

Except for an irrelevant time-dependent phase factor, the evolution $\psi_t$ is periodic, with period $\tau = 2\pi/\omega$. The Bohm motion of the particle, however, need not have

---

[37] Suppose that $Z_1 = F_1(Q_T) = X$ is the result of the measurement of the initial position. Then $F^\psi = G^\psi \circ F_1$ where $G^\psi(\,\cdot\,) = \frac{\hbar}{m}\text{Im}\frac{\nabla\psi}{\psi}(\,\cdot\,)$.

[38] In one of their proposals the wave function is "protected" by a procedure that depends upon the basis; in another, involving adiabatic interactions, $\psi$ must be a nondegenerate eigenstate of the Hamiltonian $H$ of the system, but it is not necessary that the latter be known.

## 3.7 Genuine Measurements

period $\tau$. For example, the $(n = 1, m = 1)$-state, which in polar coordinates is of the form

$$\psi_t(r,\phi) = \frac{m\omega}{\hbar\sqrt{\pi}} r e^{-\frac{m\omega}{2\hbar}r^2} e^{i\phi} e^{-i\frac{3}{2}\omega t}, \qquad (3.111)$$

generates a circular motion of the particle around the origin with angular velocity $\hbar/(mr^2)$, and hence with periodicity depending upon the initial position of the particle—the closer to the origin, the faster the rotation. Thus, in general,

$$\mathbf{X}_\tau \neq \mathbf{X}_0.$$

Nonetheless, $\mathbf{X}_\tau$ and $\mathbf{X}_0$ are identically distributed random variables, since $|\psi_\tau|^2 = |\psi_0|^2 \equiv |\psi|^2$.

We may now focus on two different experiments: Let $\mathscr{E}$ be a measurement of the actual position $\mathbf{X}_0$, the *initial* position, and hence of the position operator, and let $\mathscr{E}'$ be an experiment beginning at the same time as $\mathscr{E}$ but in which it is the position $\mathbf{X}_\tau$ at time $\tau$ that is actually measured. Since for all $\psi$ the result of $\mathscr{E}'$ has the same distribution as the result of $\mathscr{E}$, $\mathscr{E}'$ is also a measurement of the position operator. But $\mathscr{E}'$ is not a measurement of the initial position since the position at time $\tau$ does not in general agree with the initial position: A measurement of the position at time $\tau$ is not a measurement of the position at time 0. Thus, while a measurement of position is always a measurement of the position operator,

*A measurement of the position operator is not necessarily a genuine measurement of position!*

### *3.7.6 Theory Dependence of Measurement*

The harmonic oscillator example provides a simple illustration of an elementary point that is often ignored: in discussions of measurement it is well to keep in mind the theory under consideration. The theory we have been considering here has been Bohmian mechanics. If, instead, we were to analyze the harmonic oscillator experiments described above using different theories our conclusions about results of measurements would in general be rather different, even if the different theories were empirically equivalent. So we shall analyze the above experiment $\mathscr{E}'$ in terms of various other formulations or interpretations of quantum theory.

In strict orthodox quantum theory there is no such thing as a genuine particle, and thus there is no such thing as the genuine position of a particle. There is, however, a kind of operational definition of position, in the sense of an experimental setup, where a measurement device yields results the statistics of which are given by the position operator.

In naive orthodox quantum theory one does speak loosely about a particle and its position, which is thought of—in a somewhat uncritical way—as being revealed by measuring the position operator. Any experiment that yields statistics given by the position operator is considered a genuine measurement of the particle's position.[39] Thus $\mathscr{E}'$ would be considered as a measurement of the position of the particle at time zero.

The decoherent (or consistent) histories formulation of quantum mechanics [18, 19, 23] is concerned with the probabilities of certain coarse-grained histories, given by the specification of finite sequences of events, associated with projection operators, together with their times of occurrence. These probabilities are regarded as governing the occurrence of the histories, regardless of whether any of the events are measured or observed, but when they are observed, the probabilities of the observed histories are the same as those of the unobserved histories. The experiments $\mathscr{E}$ and $\mathscr{E}'$ are measurements of single-event histories corresponding to the position of the particle at time 0 and at time $\tau$, respectively. Since the Heisenberg position operators $\hat{\mathbf{X}}_\tau = \hat{\mathbf{X}}_0$ for the harmonic oscillator, it happens to be the case, according to the decoherent histories formulation of quantum mechanics, that for this system the position of the particle at time $\tau$ is the same as its position at time 0 when the positions are unobserved, and that $\mathscr{E}'$ in fact measures the position of the particle at time 0 (as well as the position at time $\tau$).

The spontaneous localization or dynamical reduction models [59, 60] are versions of quantum theory in which there are no genuine particles; in these theories reality is represented by the wave function alone (or, more accurately, by entities entirely determined by the wave function). In these models Schrödinger's equation is modified by the addition of a stochastic term that causes the wave function to collapse during measurement in a manner more or less consistent with the quantum formalism. In particular, the performance of $\mathscr{E}$ or $\mathscr{E}'$ would lead to a random collapse of the oscillator wave function onto a narrow spatial region, which might be spoken of as the position of the particle at the relevant time. But $\mathscr{E}'$ could not be regarded in any sense as measuring the position at time 0, because the localization does not occur for $\mathscr{E}'$ until time $\tau$.

Finally we mention stochastic mechanics [15], a theory ontologically very similar to Bohmian mechanics in that the basic entities with which it is concerned are particles described by their positions. Unlike Bohmian mechanics, however, the positions evolve randomly, according to a diffusion process. Just as with Bohmian mechanics, for stochastic mechanics the experiment $\mathscr{E}'$ is not a measurement of the position at time zero, but in contrast to the situation in Bohmian mechanics, where the result of the position measurement at time $\tau$ determines, given the wave function, the position at time zero (via the Bohmian equation of motion), this is not so in stochastic mechanics because of the randomness of the motion.

## 3.8 Hidden Variables

The issue of hidden variables concerns the question of whether quantum randomness arises in a completely ordinary manner, merely from the fact that in orthodox quantum theory we deal with an incomplete description of a quantum system. According to

---

[39] This, and the failure to appreciate the theory dependence of measurements, has been a source of unfounded criticisms of Bohmian mechanics (see [61, 62, 63]).

3.8 Hidden Variables 145

the hidden-variables hypothesis, if we had at our disposal a sufficiently complete description of the system, provided by supplementary parameters traditionally called hidden variables, the totality of which is usually denoted by $\lambda$, the behavior of the system would thereby be determined, as a function of $\lambda$ (and the wave function). In such a hidden-variables theory, the randomness in results of measurements would arise solely from randomness in the unknown variables $\lambda$. On the basis of a variety of "impossibility theorems," the hidden-variables hypothesis has been widely regarded as having been discredited.

Note that Bohmian mechanics is just such a hidden-variables theory, with the hidden variables $\lambda$ given by the configuration $Q$ of the total system. We have seen in particular that in a Bohmian experiment, the result $Z$ is determined by the initial configuration $Q = (X, Y)$ of the system and apparatus. Nonetheless, there remains much confusion about the relationship between Bohmian mechanics and the various theorems supposedly establishing the impossibility of hidden variables. In this section we wish to make several comments on this matter.

### 3.8.1 Experiments and Random Variables

In Bohmian mechanics we understand very naturally how random variables arise in association with experiments: the initial complete state $(Q, \Psi)$ of system and apparatus evolves deterministically and uniquely determines the outcome of the experiment; however, as the initial configuration $Q$ is in quantum equilibrium, the outcome of the experiment is random.

A general experiment $\mathscr{E}$ is then *always* associated a random variable (RV) $Z$ describing its result. In other words, according to Bohmian mechanics, there is a natural association

$$\mathscr{E} \mapsto Z, \tag{3.112}$$

between experiments and RVs. Moreover, whenever the statistics of the result of $\mathscr{E}$ is governed by a self-adjoint operator $A$ on the Hilbert space of the system, with the spectral measure of $A$ determining the distribution of $Z$, for which we shall write $Z \mapsto A$ (see (3.37), Bohmian mechanics establishes thereby a natural association between $\mathscr{E}$ and $A$

$$\mathscr{E} \mapsto A. \tag{3.113}$$

While for Bohmian mechanics the result $Z$ depends in general on both $X$ and $Y$, the initial configurations of the system and of the apparatus, for many real-world experiments $Z$ depends only on $X$ and the randomness in the result of the experiment is thus due solely to randomness in the initial configuration of the system alone. This is most obvious in the case of genuine position measurements (for which $Z(X, Y) = X$). That in fact the apparatus need not introduce any extra randomness for many other real-world experiments as well follows then from the observation that the role of the apparatus in many real-world experiments is to provide suitable background fields, which introduce no randomness, as well as a final detection, a measurement of the actual positions of the particles of the system. In particular, this is the case for those experiments most relevant to the issue of hidden variables, such as Stern-Gerlach measurements of spin, as well as for momentum measurements and more generally scattering experiments, which are completed by a final detection of position.

The result of these experiments is then given by a random variable

$$Z = F(X_T) = G(X),$$

where $T$ is the final time of the experiment,[40] on the probability space $\{\Omega, \mathbb{P}\}$, where $\Omega = \{X\}$ is the set of initial configurations of the system and $\mathbb{P}(dx) = |\psi|^2 dx$ is the quantum equilibrium distribution associated with the initial wave function $\psi$ of the system. For these experiments (see Sect. 3.5.4) the distribution of $Z$ is always governed by a PVM, corresponding to some self-adjoint operator $A$, $Z \mapsto A$, and thus Bohmian mechanics provides in these cases a natural map $\mathscr{E} \mapsto A$.

### 3.8.2 Random Variables, Operators, and the Impossibility Theorems

We would like to briefly review the status of the so-called impossibility theorems for hidden variables, the most famous of which are due to von Neumann [3], Gleason [64], Kochen and Specker [65], and Bell [66]. Since Bohmian mechanics exists, these theorems can't possibly establish the impossibility of hidden variables, the widespread belief to the contrary notwithstanding. What these theorems do establish, in great generality, is that there is no "*good*" map from self-adjoint operators on a Hilbert space $\mathscr{H}$ to random variables on a common probability space,

$$A \mapsto Z \equiv Z_A, \qquad (3.114)$$

where $Z_A = Z_A(\lambda)$ should be thought of as the result of "measuring $A$" when the hidden variables, that complete the quantum description and restore determinism, have value $\lambda$. Different senses of "good" correspond to different impossibility theorems.

For any particular choice of $\lambda$, say $\lambda_0$, the map (3.114) is transformed to a *value* map

$$A \mapsto v(A) \qquad (3.115)$$

---

[40] Concerning the most common of all real-world quantum experiments, scattering experiments, although they are completed by a final detection of position, this detection usually occurs, not at a definite time $T$, but at a random time, for example when a particle enters a localized detector. Nonetheless, for computational purposes the final detection can be regarded as taking place at a definite time $T$. This is a consequence of the flux-across-surfaces theorem [67, 68, 69], which establishes an asymptotic equivalence between flux across surfaces (detection at a random time) and scattering into cones (detection at a definite time).

## 3.8 Hidden Variables

from self-adjoint operators to real numbers (with $v(A) = Z_A(\lambda_0)$). The stronger impossibility theorems establish the impossibility of a good value map, again with different senses of "good" corresponding to different theorems.

Note that such theorems are not very surprising. One would not expect there to be a "good" map from a noncommutative algebra to a commutative one.

One of von Neumann's assumptions was, in effect, that the map (3.114) be linear. While mathematically natural, this assumption is physically rather unreasonable and in any case is entirely unnecessary. In order to establish that there is no good map (3.114), it is sufficient to require that the map be good in the minimal sense that the following *agreement condition* is satisfied:

> *Whenever the quantum mechanical joint distribution of a set of self-adjoint operators $(A_1, \ldots, A_m)$ exists, i.e., when they form a commuting family, the joint distribution of the corresponding set of random variables, i.e., of $(Z_{A_1}, \ldots, Z_{A_m})$, agrees with the quantum mechanical joint distribution.*

The agreement condition implies that all deterministic relationships among commuting observables must be obeyed by the corresponding random variables. For example, if $A$, $B$ and $C$ form a commuting family and $C = AB$, then we must have that $Z_C = Z_A Z_B$ since the joint distribution of $Z_A$, $Z_B$ and $Z_C$ must assign probability 0 to the set $\{(a, b, c) \in \mathbb{R}^3 | c \neq ab\}$. This leads to a minimal condition for a good value map $A \mapsto v(A)$, namely that it preserve functional relationships among commuting observables: For any commuting family $A_1, \ldots, A_m$, whenever $f(A_1, \ldots, A_m) = 0$ (where $f : \mathbb{R}^m \to \mathbb{R}$ represents a linear, multiplicative, or any other relationship among the $A_i$'s), the corresponding values must satisfy the same relationship, $f(v(A_1), \ldots, v(A_m)) = 0$.

The various impossibility theorems correctly demonstrate that there are no maps, from self-adjoint operators to random variables or to values, that are good, merely in the minimal senses described above.[41]

We note that while the original proofs of the impossibility of a good value map, in particular that of the Kochen-Specker theorem, were quite involved, in more recent years drastically simpler proofs have been found (for example, by Peres [70], by Greenberger, Horne, and Zeilinger [71], and by Mermin [72]).

In essence, one establishes the impossibility of a good map $A \mapsto Z_A$ or $A \mapsto v(A)$ by showing that the $v(A)$'s, or $Z_A$'s, would have to satisfy impossible relationships. These impossible relationships are very much like the following: $Z_A = Z_B = Z_C \neq Z_A$. However no impossible relationship can arise for only three quantum observables, since they would have to form a commuting family, for which quantum mechanics would supply a joint probability distribution. Thus the quantum relationships can't possibly lead to an inconsistency for the values of the random variables in this case.

---

[41] Another natural sense of good map $A \mapsto v(A)$ is given by the requirement that $v(\mathbf{A}) \in \text{sp}(\mathbf{A})$, where $\mathbf{A} = (A_1, \ldots, A_m)$ is a commuting family, $v(\mathbf{A}) = (v(A_1), \ldots, v(A_m)) \in \mathbb{R}^m$ and $\text{sp}(\mathbf{A})$ is the joint spectrum of the family. That a map good in this sense is impossible follows from the fact that if $\alpha = (\alpha_1, \ldots \alpha_m) \in \text{sp}(\mathbf{A})$, then $\alpha_1, \ldots \alpha_m$ must obey all functional relationships for $A_1, \ldots, A_m$.

With four observables $A, B, C$, and $D$ it may easily happen that $[A, B] = 0$, $[B, C] = 0$, $[C, D] = 0$, and $[D, A] = 0$ even though they don't form a commuting family (because, say, $[A, C] \neq 0$). It turns out, in fact, that four observables suffice for the derivation of impossible quantum relationships. Perhaps the simplest example of this sort is due to Hardy [73], who showed that for almost every quantum state for two spin 1/2 particles there are four observables $A, B, C$, and $D$ (two of which happen to be spin components for one of the particles while the other two are spin components for the other particle) whose quantum mechanical pair-wise distributions for commuting pairs are such that a good map to random variables must yield random variables $Z_A, Z_B, Z_C$, and $Z_D$ obeying the following relationships:

(1) The event $\{Z_A = 1 \text{ and } Z_B = 1\}$ has positive probability (with an optimal choice of the quantum state, about .09).
(2) If $\{Z_A = 1\}$ then $\{Z_D = 1\}$.
(3) If $\{Z_B = 1\}$ then $\{Z_C = 1\}$.
(4) The event $\{Z_D = 1 \text{ and } Z_C = 1\}$ has probability 0.

Clearly, there exist no such random variables.

The point we wish to emphasize here, however, is that although they are correct and although their hypotheses may seem minimal, these theorems are nonetheless far less relevant to the possibility of a deterministic completion of quantum theory than one might imagine. In the next subsection we will elaborate on how that can be so. We shall explain why we believe such theorems have little physical significance for the issues of determinism and hidden variables. We will separately comment later in this section on Bell's related nonlocality analysis [66], which does have profound physical implications.

## 3.8.3 Contextuality

It is a simple fact there can be no map $A \mapsto Z_A$, from self-adjoint operators on $\mathcal{H}$ (with $\dim(\mathcal{H}) \geq 3$) to random variables on a common probability space, that is good in the minimal sense that the joint probability distributions for the random variables agree with the corresponding quantum mechanical distributions, whenever the latter ones are defined. But does not Bohmian mechanics yield precisely such a map? After all, have we not emphasized how Bohmian mechanics naturally associates with any experiment a random variable $Z$ giving its result, in a manner that is in complete agreement with the quantum mechanical predictions for the result of the experiment? Given a quantum observable $A$, let $Z_A$ be then the result of a measurement of $A$. What gives?

Before presenting what we believe to be the correct response, we mention some possible responses that are off-target. It might be objected that measurements of different observables will involve different apparatuses and hence different probability spaces. However, one can simultaneously embed all the relevant probability spaces into a huge common probability space. It might also be objected that not all

## 3.8 Hidden Variables

self-adjoint operators can be realistically be measured. But to arrive at inconsistency one need consider, as mentioned in the last subsection, only 4 observables, each of which are spin components and are thus certainly measurable, via Stern-Gerlach experiments. Thus, in fact, no enlargement of probability spaces need be considered to arrive at a contradiction, since as we emphasized at the end of Sect. 3.8.1, the random variables giving the results of Stern-Gerlach experiments are functions of initial particle positions, so that for joint measurements of pairs of spin components for 2-particles the corresponding results are random variables on the common probability space of initial configurations of the 2 particles, equipped with the quantum equilibrium distribution determined by the initial wave function.

There must be a mistake. But where could it be? The mistake occurs, in fact, so early that it is difficult to notice it. It occurs at square one. The difficulty lies not so much in any conditions on the map $A \mapsto Z_A$, but in the conclusion that Bohmian mechanics supplies such a map at all.

What Bohmian mechanics naturally supplies is a map $\mathscr{E} \mapsto Z_\mathscr{E}$, from experiments to random variables. When $Z_\mathscr{E} \mapsto A$, so that we speak of $\mathscr{E}$ as a measurement of $A$ ($\mathscr{E} \mapsto A$), this very language suggests that insofar as the random variable is concerned all that matters is that $\mathscr{E}$ measures $A$, and the map $\mathscr{E} \mapsto Z_\mathscr{E}$ becomes a map $A \mapsto Z_A$. After all, if $\mathscr{E}$ were a genuine measurement of $A$, revealing, that is, the preexisting (i.e., prior to the experiment) value of the observable $A$, then $Z$ would have to agree with that value and hence would be an unambiguous random variable depending only on $A$.

But this sort of argument makes sense only if we take the quantum talk of operators as observables too seriously. We have emphasized in this chapter that operators do naturally arise in association with quantum experiments. But there is little if anything in this association, beyond the unfortunate language that is usually used to describe it, that supports the notion that the operator $A$ associated with an experiment $\mathscr{E}$ is in any meaningful way genuinely measured by the experiment. From the nature of the association itself, it is difficult to imagine what this could possibly mean. And for those who think they imagine some meaning in this talk, the impossibility theorems show they are mistaken.

The bottom line is this: in Bohmian mechanics the random variables $Z_\mathscr{E}$ giving the results of experiments $\mathscr{E}$ depend, of course, on the experiment, and there is no reason that this should not be the case when the experiments under consideration happen to be associated with the same operator. Thus with any self-adjoint operator $A$, Bohmian mechanics naturally may associate many different random variables $Z_\mathscr{E}$, one for each different experiment $\mathscr{E} \mapsto A$ associated with $A$. A crucial point here is that the map $\mathscr{E} \mapsto A$ is many-to-one.[42]

---

[42] We wish to remark that, quite aside from this many-to-oneness, the random variables $Z_\mathscr{E}$ cannot generally be regarded as corresponding to any sort of natural property of the "measured" system. $Z_\mathscr{E}$, in general a function of the initial configuration of the system-apparatus composite, may fail to be a function of the configuration of the system alone. And even when, as is often the case, $Z_\mathscr{E}$ does depend only on the initial configuration of the system, owing to chaotic dynamics this dependence could have an extremely complex character.

Suppose we define a map $A \mapsto Z_A$ by selecting, for each $A$, one of the experiments, call it $\mathscr{E}_A$, with which $A$ is associated, and define $Z_A$ to be $Z_{\mathscr{E}_A}$. Then the map so defined can't be good, because of the impossibility theorems; moreover there is no reason to have expected the map to be good. Suppose, for example, that $[A, B] = 0$. Should we expect that the joint distribution of $Z_A$ and $Z_B$ will agree with the joint quantum mechanical distribution of $A$ and $B$? Only if the experiments $\mathscr{E}_A$ and $\mathscr{E}_B$ used to define $Z_A$ and $Z_B$ both involved a common experiment that "simultaneously measures $A$ and $B$," i.e., an experiment that is associated with the commuting family $(A, B)$. If we consider now a third operator $C$ such that $[A, C] = 0$, but $[B, C] \neq 0$, then there is no choice of experiment $\mathscr{E}$ that would permit the definition of a random variable $Z_A$ relevant both to a "simultaneous measurement of $A$ and $B$" and a "simultaneous measurement of $A$ and $C$" since no experiment is a "simultaneous measurement of $A$, $B$, and $C$." In the situation just described we must consider at least two random variables associated with $A$, $Z_{A,B}$ and $Z_{A,C}$, depending upon whether we are considering an experiment "measuring $A$ and $B$" or an experiment "measuring $A$ and $C$." It should be clear that when the random variables are assigned to experiments in this way, the possibility of conflict with the predictions of orthodox quantum theory is eliminated. It should also be clear, in view of what we have repeatedly stressed, that quite aside from the impossibility theorems, this way of associating random variables with experiments is precisely what emerges in Bohmian mechanics.

The dependence of the result of a "measurement of the observable $A$" upon the other observables, if any, that are "measured simultaneously together with $A$"—e.g., that $Z_{A,B}$ and $Z_{A,C}$ may be different—is called *contextuality*: the result of an experiment depends not just on "what observable the experiment measures" but on more detailed information that conveys the "context" of the experiment. The essential idea, however, if we avoid misleading language, is rather trivial: that the result of an experiment depends on the experiment.

To underline this triviality we remark that for two experiments, $\mathscr{E}$ and $\mathscr{E}'$, that "measure $A$ and only $A$" and involve no simultaneous "measurement of another observable," the results $Z_\mathscr{E}$ and $Z_{\mathscr{E}'}$ may disagree. For example in Sect. 3.7.5 we described experiments $\mathscr{E}$ and $\mathscr{E}'$ both of which "measured the position operator" but only one of which measured the actual initial position of the relevant particle, so that for these experiments in general $Z_\mathscr{E} \neq Z_{\mathscr{E}'}$.

One might feel, however, that in the example just described the experiment that does not measure the actual position is somewhat disreputable—even though it is in fact a "measurement of the position operator." We shall therefore give another example, due to D. Albert [74], in which the experiments are as simple and canonical as possible and are entirely on the same footing. Let $\mathscr{E}_\uparrow$ and $\mathscr{E}_\downarrow$ be Stern-Gerlach measurements of $A = \sigma_z$, with $\mathscr{E}_\downarrow$ differing from $\mathscr{E}_\uparrow$ only in that the polarity of the Stern-Gerlach magnet for $\mathscr{E}_\downarrow$ is the reverse of that for $\mathscr{E}_\uparrow$. (In particular, the geometry of the magnets for $\mathscr{E}_\uparrow$ and $\mathscr{E}_\downarrow$ is the same.) If the initial wave function $\psi_{\text{symm}}$ and the magnetic field $\pm B$ have sufficient reflection symmetry with respect to a plane between the poles of the Stern-Gerlach magnets, the particle whose spin component is being "measured" cannot cross this plane of symmetry, so that if the particle

3.8 Hidden Variables                                                                 151

is initially above, respectively below, the symmetry plane, it will remain above, respectively below, that plane. But because their magnets have opposite polarity, $\mathcal{E}_\uparrow$ and $\mathcal{E}_\downarrow$ involve opposite calibrations: $F_\uparrow = -F_\downarrow$. It follows that

$$Z_{\mathcal{E}_\uparrow}^{\psi_{\text{symm}}} = -Z_{\mathcal{E}_\downarrow}^{\psi_{\text{symm}}}$$

and the two experiments completely disagree about the "value of $\sigma_z$" in this case.

The essential point illustrated by the previous example is that instead of having in Bohmian mechanics a natural association $\sigma_z \mapsto Z_{\sigma_z}$, we have a rather different pattern of relationships, given in the example by

$$\begin{array}{l} \mathcal{E}_\uparrow \to Z_{\mathcal{E}_\uparrow} \\ \mathcal{E}_\downarrow \to Z_{\mathcal{E}_\downarrow} \end{array} \searrow \!\!\!\!\!\nearrow \sigma_z,$$

### 3.8.4 Against "Contextuality"

The impossibility theorems require the assumption of noncontextuality, that the random variable $Z$ giving the result of a "measurement of quantum observable $A$" should depend on $A$ alone, further experimental details being irrelevant. How big a deal is contextuality, the violation of this assumption? Here are two ways of describing the situation:

1. In quantum mechanics (or quantum mechanics supplemented with hidden variables), observables and properties have a novel, highly nonclassical aspect: they (or the result of measuring them) depend upon which other compatible properties, if any, are measured together with them.
   In this spirit, Bohm and Hiley [75] write that (p. 109)

   > the quantum properties imply ... that measured properties are not intrinsic but are inseparably related to the apparatus. It follows that the customary language that attributes the results of measurements ... to the observed system alone can cause confusion, unless it is understood that these properties are actually dependent on the total relevant context.

   They later add that (p. 122)

   > The context dependence of results of measurements is a further indication of how our interpretation does not imply a simple return to the basic principles of classical physics. It also embodies, in a certain sense, Bohr's notion of the indivisibility of the combined system of observing apparatus and observed object.

2. The result of an experiment depends upon the experiment. Or, as expressed by Bell [76] (p. 166),

   > A final moral concerns terminology. Why did such serious people take so seriously axioms which now seem so arbitrary? I suspect that they were misled by the pernicious misuse of the word 'measurement' in contemporary theory. This word very strongly

suggests the ascertaining of some preexisting property of some thing, any instrument involved playing a purely passive role. Quantum experiments are just not like that, as we learned especially from Bohr. The results have to be regarded as the joint product of 'system' and 'apparatus,' the complete experimental set-up. But the misuse of the word 'measurement' makes it easy to forget this and then to expect that the 'results of measurements' should obey some simple logic in which the apparatus is not mentioned. The resulting difficulties soon show that any such logic is not ordinary logic. It is my impression that the whole vast subject of 'Quantum Logic' has arisen in this way from the misuse of a word. I am convinced that the word 'measurement' has now been so abused that the field would be significantly advanced by banning its use altogether, in favour for example of the word 'experiment.'

With one caveat, we entirely agree with Bell's observation. The caveat is this: We do not believe that the difference between quantum mechanics and classical mechanics is quite as crucial for Bell's moral as his language suggests it is. For any experiment, quantum or classical, it would be a mistake to regard any instrument involved as playing a purely passive role, unless the experiment is a genuine measurement of a property of a system, in which case the result is determined by the initial conditions of the system alone. However, a relevant difference between classical and quantum theory remains: Classically it is usually taken for granted that it is in principle possible to measure any observable without seriously affecting the observed system, which is clearly false in quantum mechanics (or Bohmian mechanics).[43]

Mermin has raised a similar question [72] (p.811):

Is noncontextuality, as Bell seemed to suggest, as silly a condition as von Neumann's ... ?

To this he answers:

I would not characterize the assumption of noncontextuality as a silly constraint on a hidden-variables theory. It is surely an important fact that the impossibility of embedding quantum mechanics in a noncontextual hidden-variables theory rests not only on Bohr's doctrine of the inseparability of the objects and the measuring instruments, but also on a straightforward contradiction, independent of one's philosophic point of view, between some quantitative consequences of noncontextuality and the quantitative predictions of quantum mechanics.

This is a somewhat strange answer. First of all, it applies to von Neumann's assumption (linearity), which Mermin seems to agree is silly, as well as to the assumption of noncontextuality. And the statement has a rather question-begging flavor, since the importance of the fact to which Mermin refers would seem to depend on the nonsilliness of the assumption which the fact concerns.

Be that as it may, Mermin immediately supplies his real argument for the nonsilliness of noncontextuality. Concerning two experiments for "measuring observable $A$," he writes that

it is ... an elementary theorem of quantum mechanics that the joint distribution ... for the first experiment yields precisely the same marginal distribution (for $A$) as does the joint distribution ... for the second, in spite of the different experimental arrangements. ... The obvious way to account for this, particularly when entertaining the possibility of a hidden-variables theory, is to propose that both experiments reveal a set of values for $A$ in the

---

[43] The assumption could (and probably should) also be questioned classically.

## 3.8 Hidden Variables

individual systems that is the same, regardless of which experiment we choose to extract them from.... A *contextual* hidden-variables account of this fact would be as mysteriously silent as the quantum theory on the question of why nature should conspire to arrange for the marginal distributions to be the same for the two different experimental arrangements.

A bit later, Mermin refers to the "striking insensitivity of the distribution to changes in the experimental arrangement."

For Mermin there is a mystery, something that demands an explanation. It seems to us, however, that the mystery here is very much in the eye of the beholder. It is first of all somewhat odd that Mermin speaks of the mysterious silence of quantum theory concerning a question whose answer, in fact, emerges as an "elementary theorem of quantum mechanics." What better way is there to answer questions about nature than to appeal to our best physical theories?

More importantly, the "two different experimental arrangements," say $\mathscr{E}_1$ and $\mathscr{E}_2$, considered by Mermin are not merely any two randomly chosen experimental arrangements. They obviously must have something in common. This is that they are both associated with the same self-adjoint operator $A$ in the manner we have described: $\mathscr{E}_1 \mapsto A$ and $\mathscr{E}_2 \mapsto A$. It is quite standard to say in this situation that both $\mathscr{E}_1$ and $\mathscr{E}_2$ measure the observable $A$, but both for Bohmian mechanics and for orthodox quantum theory the very meaning of the association with the operator $A$ is merely that the distribution of the result of the experiment is given by the spectral measures for $A$. Thus there is no mystery in the fact that $\mathscr{E}_1$ and $\mathscr{E}_2$ have results governed by the same distribution, since, when all is said and done, it is on this basis, and this basis alone, that we are comparing them.

(One might wonder how it could be possible that there are two different experiments that are related in this way. This is a somewhat technical question, rather different from Mermin's, and it is one that Bohmian mechanics and quantum mechanics readily answer, as we have explained in this chapter. In this regard it would probably be good to reflect further on the simplest example of such experiments, the Stern-Gerlach experiments $\mathscr{E}_\uparrow$ and $\mathscr{E}_\downarrow$ discussed in the previous subsection.)

It is also difficult to see how Mermin's proposed resolution of the mystery, "that both experiments reveal a set of values for $A$ ... that is the same, regardless of which experiment we choose to extract them from," could do much good. He is faced with a certain pattern of results in two experiments that would be explained if the experiments did in fact genuinely measure the same thing. The experiments, however, as far as any detailed quantum mechanical analysis of them is concerned, don't appear to be genuine measurements of anything at all. He then suggests that the mystery would be resolved if, indeed, the experiments did measure the same thing, the analysis to the contrary notwithstanding. But this proposal merely replaces the original mystery with a bigger one, namely, of how the experiments could in fact be understood as measuring the same thing, or anything at all for that matter. It is like explaining the mystery of a talking cat by saying that the cat is in fact a human being, appearances to the contrary notwithstanding.

A final complaint about contextuality: the terminology is misleading. It fails to convey with sufficient force the rather definitive character of what it entails: *"Properties" that are merely contextual are not properties at all; they do not exist, and their failure to do so is in the strongest sense possible!*

## 3.8.5 Nonlocality, Contextuality and Hidden Variables

There is, however, a situation where contextuality is physically relevant. Consider the EPRB experiment, outlined at the end of Sect. 3.3.6. In this case the dependence of the result of a measurement of the spin component $\sigma_1 \cdot \mathbf{a}$ of a particle upon which spin component of a distant particle is measured together with it—the difference between $Z_{\sigma_1 \cdot \mathbf{a}, \sigma_2 \cdot \mathbf{b}}$ and $Z_{\sigma_1 \cdot \mathbf{a}, \sigma_2 \cdot \mathbf{c}}$ (using the notation described in the seventh paragraph of Sect. 3.8.3)—is an expression of *nonlocality*, of, in Einstein words, a "spooky action at distance." More generally, whenever the relevant context is distant, contextuality implies nonlocality.

Nonlocality is an essential feature of Bohmian mechanics: the velocity, as expressed in the guiding Eq. (3.2), of any one of the particles of a many-particle system will typically depend upon the positions of the other, possibly distant, particles whenever the wave function of the system is entangled, i.e., not a product of single-particle wave functions. In particular, this is true for the EPRB experiment under examination. Consider the extension of the single particle Hamiltonian (3.12) to the two-particle case, namely

$$H = -\frac{\hbar^2}{2m_1}\nabla_1^2 - \frac{\hbar^2}{2m_2}\nabla_2^2 - \mu_1 \sigma_1 \cdot B(\mathbf{x}_1) - \mu_2 \sigma_2 \cdot B(\mathbf{x}_2).$$

Then for initial singlet state, and spin measurements as described in Sects. 3.2.5 and 3.5.2, it easily follows from the laws of motion of Bohmian mechanics that

$$Z_{\sigma_1 \cdot \mathbf{a}, \sigma_2 \cdot \mathbf{b}} \neq Z_{\sigma_1 \cdot \mathbf{a}, \sigma_2 \cdot \mathbf{c}}.$$

This was observed long ago by Bell [12]. In fact, Bell's examination of Bohmian mechanics led him to his celebrated nonlocality analysis. In the course of his investigation of Bohmian mechanics he observed that ([76], p. 11)

> in this theory an explicit causal mechanism exists whereby the disposition of one piece of apparatus affects the results obtained with a distant piece.
> Bohm of course was well aware of these features of his scheme, and has given them much attention. However, it must be stressed that, to the present writer's knowledge, there is no *proof* that *any* hidden variable account of quantum mechanics *must* have this extraordinary character. It would therefore be interesting, perhaps, to pursue some further "impossibility proofs," replacing the arbitrary axioms objected to above by some condition of locality, or of separability of distant systems.

In a footnote, Bell added that "Since the completion of this paper such a proof has been found." This proof was published in his 1964 paper [66], "On the Einstein-Podolsky-Rosen Paradox," in which he derives Bell's inequality, the basis of his conclusion of quantum nonlocality.

We find it worthwhile to reproduce here the analysis of Bell, deriving a simple inequality equivalent to Bell's, in order to highlight the conceptual significance of Bell's analysis and, at the same time, its mathematical triviality. The analysis involves two parts. The first part, the Einstein-Podolsky-Rosen argument applied to the EPRB experiment, amounts to the observation that for the singlet state the assumption of

## 3.8 Hidden Variables

locality implies the existence of noncontextual hidden variables. More precisely, it implies, for the singlet state, the existence of random variables $Z_\alpha^i = Z_{\alpha \cdot \sigma_i}$, $i = 1, 2$, corresponding to all possible spin components of the two particles, that obey the agreement condition described in Sect. 3.8.2. In particular, focusing on components in only 3 directions **a**, **b** and **c** for each particle, locality implies the existence of 6 random variables

$$Z_\alpha^i \qquad i = 1, 2 \quad \alpha = \mathbf{a, b, c}$$

such that

$$Z_\alpha^i = \pm 1 \tag{3.116}$$
$$Z_\alpha^1 = -Z_\alpha^2 \tag{3.117}$$

and, more generally,

$$\text{Prob}(Z_\alpha^1 \neq Z_\beta^2) = q_{\alpha\beta}, \tag{3.118}$$

the corresponding quantum mechanical probabilities. This conclusion amounts to the idea that measurements of the spin components reveal preexisting values (the $Z_\alpha^i$), which, assuming locality, is implied by the perfect quantum mechanical anticorrelations [66]:

> Now we make the hypothesis, and it seems one at least worth considering, that if the two measurements are made at places remote from one another the orientation of one magnet does not influence the result obtained with the other. Since we can predict in advance the result of measuring any chosen component of $\sigma_2$, by previously measuring the same component of $\sigma_1$, it follows that the result of any such measurement must actually be predetermined.

People very often fail to appreciate that the existence of such variables, given locality, is not an assumption but a consequence of Bell's analysis. Bell repeatedly stressed this point (by determinism Bell here means the existence of hidden variables):

> It is important to note that to the limited degree to which *determinism* plays a role in the EPR argument, it is not assumed but *inferred*. What is held sacred is the principle of 'local causality' – or 'no action at a distance'. ...
> It is remarkably difficult to get this point across, that determinism is not a *presupposition* of the analysis. ([76], p. 143)
> Despite my insistence that the determinism was inferred rather than assumed, you might still suspect somehow that it is a preoccupation with determinism that creates the problem. Note well then that the following argument makes no mention whatever of determinism. ... Finally you might suspect that the very notion of particle, and particle orbit ... has somehow led us astray. ... So the following argument will not mention particles, nor indeed fields, nor any other particular picture of what goes on at the microscopic level. Nor will it involve any use of the words 'quantum mechanical system', which can have an unfortunate effect on the discussion. The difficulty is not created by any such picture or any such terminology. It is created by the predictions about the correlations in the visible outputs of certain conceivable experimental set-ups. ([76], p. 150)

The second part of the analysis, which unfolds the "difficulty ... created by the ... correlations", involves only very elementary mathematics. Clearly,

$$\text{Prob}\left(\{Z_a^1 = Z_b^1\} \cup \{Z_b^1 = Z_c^1\} \cup \{Z_c^1 = Z_a^1\}\right) = 1\ .$$

since at least two of the three (2-valued) variables $Z_\alpha^1$ must have the same value. Hence, by elementary probability theory,

$$\text{Prob}\left(Z_a^1 = Z_b^1\right) + \text{Prob}\left(Z_b^1 = Z_c^1\right) + \text{Prob}\left(Z_c^1 = Z_a^1\right) \geq 1,$$

and using the perfect anticorrelations (3.117) we have that

$$\text{Prob}\left(Z_a^1 = -Z_b^2\right) + \text{Prob}\left(Z_b^1 = -Z_c^2\right) + \text{Prob}\left(Z_c^1 = -Z_a^2\right) \geq 1, \quad (3.119)$$

which is equivalent to Bell's inequality and in conflict with (3.118). For example, when the angles between **a**, **b** and **c** are 120° the 3 relevant quantum correlations $q_{\alpha\beta}$ are all $1/4$.

To summarize the argument, let H be the hypothesis of the existence of the noncontextual hidden variables we have described above. Then the logic of the argument is:

| | | | |
|---|---|---|---|
| Part 1: | quantum mechanics + locality | $\Rightarrow$ H | (3.120) |
| Part 2: | quantum mechanics | $\Rightarrow$ not H | (3.121) |
| Conclusion: | quantum mechanics | $\Rightarrow$ not locality | (3.122) |

To fully grasp the argument it is important to appreciate that the identity of H—the existence of the noncontextual hidden variables—is of little substantive importance. What is important is not so much the identity of H as the fact that H is incompatible with the predictions of quantum theory. The identity of H is, however, of great historical significance: It is responsible for the misconception that Bell proved that hidden variables are impossible, a belief until recently almost universally shared by physicists.

Such a misconception has not been the only reaction to Bell's analysis. Roughly speaking, we may group the different reactions into three main categories, summarized by the following statements:

1. Hidden variables are impossible.
2. Hidden variables are possible, but they must be contextual.
3. Hidden variables are possible, but they must be nonlocal.

Statement 1 is plainly wrong. Statement 2 is correct but not terribly significant. Statement 3 is correct, significant, but nonetheless rather misleading. It follow from (3.120) and (3.121) that *any* account of quantum phenomena must be nonlocal, not just any hidden variables account. Bell's argument shows that nonlocality is implied by the predictions of standard quantum theory itself. Thus if nature is governed by these predictions, then *nature is nonlocal*. (That nature is so governed, even in the crucial EPR-correlation experiments, has by now been established by a great many experiments, the most conclusive of which is perhaps that of Aspect [77].)

## 3.9 Against Naive Realism About Operators

Traditional naive realism is the view that the world *is* pretty much the way it *seems*, populated by objects which force themselves upon our attention as, and which in fact are, the locus of sensual qualities. A naive realist regards these "secondary qualities," for example color, as objective, as out there in the world, much as perceived. A decisive difficulty with this view is that once we understand, say, how our perception of what we call color arises, in terms of the interaction of light with matter, and the processing of the light by the eye, and so on, we realize that the presence out there of color per se would play no role whatsoever in these processes, that is, in our understanding what is relevant to our perception of "color." At the same time, we may also come to realize that there is, in the description of an object provided by the scientific world-view, as represented say by classical physics, nothing which is genuinely "color-like."

A basic problem with quantum theory, more fundamental than the measurement problem and all the rest, is a naive realism about operators, a fallacy which we believe is far more serious than traditional naive realism: With the latter we are deluded partly by language but in the main by our senses, in a manner which can scarcely be avoided without a good deal of scientific or philosophical sophistication; with the former we are seduced by language alone, to accept a view which can scarcely be taken seriously without a large measure of (what often passes for) sophistication.

Not many physicists—or for that matter philosophers—have focused on the issue of naive realism about operators, but Schrödinger and Bell have expressed similar or related concerns:

> ... the new theory [quantum theory] ... considers the [classical] model suitable for guiding us as to just which measurements can in principle be made on the relevant natural object.... Would it not be pre-established harmony of a peculiar sort if the classical-epoch researchers, those who, as we hear today, had no idea of what *measuring* truly is, had unwittingly gone on to give us as legacy a guidance scheme revealing just what is fundamentally measurable for instance about a hydrogen atom!? [29]

> Here are some words which, however legitimate and necessary in application, have no place in a *formulation* with any pretension to physical precision: *system; apparatus; environment; microscopic, macroscopic; reversible, irreversible; observable; information; measurement.* ... The notions of "microscopic" and "macroscopic" defy precise definition. ... Einstein said that it is theory which decides what is "observable." I think he was right. ... "observation" is a complicated and theory-laden business. Then that notion should not appear in the *formulation* of fundamental theory. ...
> On this list of bad words from good books, the worst of all is "measurement". It must have a section to itself. [24]

We agree almost entirely with Bell here. We insist, however, that "observable" is just as bad as "measurement," maybe even a little worse. Be that as it may, after listing Dirac's measurement postulates Bell continues:

> It would seem that the theory is exclusively concerned about "results of measurement", and has nothing to say about anything else. What exactly qualifies some physical systems to play the role of "measurer"? Was the wave function of the world waiting to jump for thousands

of millions of years until a single-celled living creature appeared? Or did it have to wait a little longer, for some better qualified system ... with a Ph.D.? If the theory is to apply to anything but highly idealized laboratory operations, are we not obliged to admit that more or less "measurement-like" processes are going on more or less all the time, more or less everywhere. Do we not have jumping then all the time?

The first charge against "measurement", in the fundamental axioms of quantum mechanics, is that it anchors the shifty split of the world into "system" and "apparatus". A second charge is that the word comes loaded with meaning from everyday life, meaning which is entirely inappropriate in the quantum context. When it is said that something is "measured" it is difficult not to think of the result as referring to some *preexisting property* of the object in question. This is to disregard Bohr's insistence that in quantum phenomena the apparatus as well as the system is essentially involved. If it were not so, how could we understand, for example, that "measurement" of a component of "angular momentum" ... *in an arbitrarily chosen direction* ... yields one of a discrete set of values? When one forgets the role of the apparatus, as the word "measurement" makes all too likely, one despairs of ordinary logic ... hence "quantum logic". When one remembers the role of the apparatus, ordinary logic is just fine.

In other contexts, physicists have been able to take words from ordinary language and use them as technical terms with no great harm done. Take for example the "strangeness", "charm", and "beauty" of elementary particle physics. No one is taken in by this "baby talk". ... Would that it were so with "measurement". But in fact the word has had such a damaging effect on the discussion, that I think it should now be banned altogether in quantum mechanics. (*Ibid.*)

While Bell focuses directly here on the misuse of the word "measurement" rather than on that of "observable," it is worth noting that the abuse of "measurement" is in a sense inseparable from that of "observable," i.e., from naive realism about operators. After all, one would not be very likely to speak of measurement unless one thought that something, some "observable" that is, was somehow there to be measured.

Operationalism, so often used without a full appreciation of its consequences, may lead many physicists to beliefs which are the opposite of what one might expect. Namely, by believing somehow that a physical property *is* and *must be* defined by an operational definition, many physicists come to regard properties such as spin and polarization, which can easily be operationally defined, as intrinsic properties of the system itself, the electron or photon, despite all the difficulties that this entails. If operational definitions were banished, and "real definitions" were required, there would be far less reason to regard these "properties" as intrinsic, since they are not defined in any sort of intrinsic way; in short, we have no idea what they really mean, and there is no reason to think they mean anything beyond the behavior exhibited by the system in interaction with an apparatus.

There are two primary sources of confusion, mystery and incoherence in the foundations of quantum mechanics: the insistence on the completeness of the description provided by the wave function, despite the dramatic difficulties entailed by this dogma, as illustrated most famously by the measurement problem; and naive realism about operators. While the second seems to point in the opposite direction from the first, the dogma of completeness is in fact nourished by naive realism about operators. This is because naive realism about operators tends to produce the belief that a more complete description is impossible because such a description should involve preexisting values of the quantum observables, values that are revealed by measurement. And this is impossible. But without naive realism about operators—without being misled by all the quantum talk of the measurement of observables—most of what is shown to be impossible by the impossibility theorems would never have been expected to begin with.

# References

1. J. S. Bell. Quantum Mechanics for Cosmologists. In C. Isham, R. Penrose, and D. Sciama, editors, *Quantum Gravity 2*, pages 611–637. Oxford University Press, New York, 1981. Reprinted in [26].
2. J. S. Bell. On the Impossible Pilot wave. *Foundations of Physics*, 12:989–999, 1982. Reprinted in [26].
3. J. von Neumann. *Mathematische Grundlagen der Quantenmechanik*. Springer Verlag, New York-Heidelberg-Berlin, 1932. English translation by R. T. Beyer, *Mathematical Foundations of Quantum Mechanics*. Princeton University Press, Princeton, N.J., 1955.
4. E. B. Davies. *Quantum Theory of Open Systems*. Academic Press, London-New York-San Francisco, 1976.
5. A. S. Holevo. *Probabilistic and Statistical Aspects of Quantum Theory*, volume 1 of *North-Holland Series in Statistics and Probability*. North-Holland, Amsterdam-New York-Oxford, 1982.
6. K. Kraus. States, Effects, and Operations. *Lectures Notes in Physics*, 190, 1983.
7. W. Heisenberg. *Physics and Philosophy*. Harper and Row, New York, 1958.
8. D. Dürr, S. Goldstein, and N. Zanghì. Quantum Equilibrium and the Origin of Absolute Uncertainty. *Journal of Statistical Physics*, **67**:843–907, 1992.
9. D. Dürr, S. Goldstein, R. Tumulka, and N. Zanghì. Trajectories and Particle Creation and Annihilation in Quantum Field Theory. *Journal of Physics A*, 36:4143–4150, 2003.
10. D. Dürr, S. Goldstein, R. Tumulka, and N. Zanghì. Quantum Hamiltonians and Stochastic Jumps. *Communications in Mathematical Physics*, 254:129–166, 2005.
11. D. Dürr, S. Goldstein, R. Tumulka, and N. Zanghì. Bell-Type Quantum Field Theories. *Journal of Physics A*, 38:R1–R43, 2005.
12. J. S. Bell. On the Problem of Hidden Variables in Quantum Mechanics. *Reviews of Modern Physics*, 38:447–452, 1966. Reprinted in [211] and in [26].
13. D. Bohm. A Suggested Interpretation of the Quantum Theory in Terms of "Hidden" Variables: Part I. *Physical Review*, 85:166–179, 1952. Reprinted in [211].
14. D. Bohm. A Suggested Interpretation of the Quantum Theory in Terms of "Hidden" Variables: Part II. *Physical Review*, 85:180–193, 1952. Reprinted in [211].
15. E. Nelson. *Quantum Fluctuations*. Princeton University Press, Princeton, N.J., 1985.
16. S. Goldstein. Stochastic Mechanics and Quantum Theory. *Journal of Statistical Physics*, 47:645–667, 1987.
17. D. Dürr, S. Goldstein, R. Tumulka, and N. Zanghì. On the Role of Density Matrices in Bohmian Mechanics. *Foundations of Physics*, 35:449–467, 2005.
18. R. B. Griffiths. Consistent Histories and the Interpretation of Quantum Mechanics. *Journal of Statistical Physics*, 36:219–272, 1984.
19. R. Omnes. Logical Reformulation of Quantum Mechanics. *Journal of Statistical Physics*, 53:893–932, 1988.
20. A. J. Leggett. Macroscopic Quantum Systems and the Quantum Theory of Measurement. *Supplement of the Progress of Theoretical Physics*, 69:80–100, 1980.
21. W. H. Zurek. Environment-Induced Superselection Rules. *Physical Review D*, 26:1862–1880, 1982.
22. E. Joos and H. D. Zeh. The Emergence of Classical Properties Through Interaction with the Environment. *Zeitschrift für Physik B*, 59:223–243, 1985.

23. M. Gell-Mann and J. B. Hartle. Quantum Mechanics in the Light of Quantum Cosmology. In W. Zurek, editor, *Complexity, Entropy, and the Physics of Information*, pages 425–458. Addison-Wesley, Reading, 1990. Also in [130].
24. J. S. Bell. Against "Measurement". *Physics World*, 3:33–40, 1990. Also in [157].
25. D. Bohm. *Quantum Theory*. Prentice-Hall, Englewood Cliffs, N.J., 1951.
26. E. P. Wigner. The Problem of Measurement. *American Journal of Physics*, 31:6–15, 1963. Reprinted in [214], and in [211].
27. E. P. Wigner. Interpretation of Quantum Mechanics. In [211], 1976.
28. A. Böhm. *Quantum Mechanics*. Springer-Verlag, New York-Heidelberg-Berlin, 1979.
29. E. Schrödinger. Die Gegenwärtige Situation in der Quantenmechanik. *Naturwissenschaften*, 23:807–812, 1935. English translation by J. D. Trimmer, *The Present Situation in Quantum Mechanics: A Translation of Schrödinger's "Cat Paradox" Paper*, Proceedings of the American Philosophical Society, 124:323–338, 1980. Reprinted in [211].
30. M. Reed and B. Simon. *Methods of Modern Mathematical Physics I. Functional Analysis, revised and enlarged edition*. Academic Press, New York, 1980.
31. P. A. M. Dirac. *The Principles of Quantum Mechanics*. Oxford University Press, Oxford, 1930.
32. W. Pauli. In S. Flügge, editor, *Encyclopedia of Physics*, volume 60. Springer, New York-Heidelberg-Berlin, 1958.
33. V. B. Braginsky, Y. I. Vorontsov, and K. S. Thorne. Quantum Nondemolition Measurements. *Science*, 209:547–57, 1980. Reprinted in [211].
34. F. Riesz and B. Sz-Nagy. *Functional Analysis*. F. Ungar, New York, 1955.
35. E. Prugovecki. *Quantum Mechanics in Hilbert Space*. Academic Press, New York and London, 1971.
36. A. Einstein, B. Podolsky, and N. Rosen. Can Quantum-Mechanical Description of Physical Reality Be Considered Complete? *Physical Review*, 47:777–780, 1935.
37. S. Goldstein and D. N. Page. Linearly Positive Histories: Probabilities for a Robust Family of Sequences of Quantum Events. *Physical Review Letters*, 74:3715–3719, 1995.
38. S. Goldstein. Quantum Theory Without Observers. Part One. *Physics Today*, 51(3):42–46, 1998.
39. S. Goldstein. Quantum Theory Without Observers. Part Two. *Physics Today*, 51(4):38–42, 1988.
40. G. Lüders. Über die Zustundsänderung Durch des Messprozess. *Annalen der Physik*, 8:322–328, 1951.
41. M. Daumer, D. Dürr, S. Goldstein, and N. Zanghì. On the Quantum Probability Flux Through Surfaces. *Journal of Statistical Physics*, 88:967–977, 1997.
42. C. R. Leavens. Transmission, Reflection and Dwell Times Within Bohm's Causal Interpretation of Quantum Mechanics. *Solid State Communications*, 74:923–928, 1990.
43. W.R. McKinnon and C.R. Leavens. Distributions of Delay Times and Transmission Times in Bohm's Causal Interpretation of Quantum Mechanics. *Physical Review A*, 51:2748–2757, 1995.
44. C.R. Leavens. Time of Arrival in Quantum and Bohmian Mechanics. *Physical Review A*, 58:840–847, 1998.
45. G. Grübl and K. Rheinberger. Time of Arrival from Bohmian Flow. *Journal of Physics A*, 35:2907–2924, 2002.
46. N. Bohr. *Atomic Physics and Human Knowledge*. Wiley, New York, 1958.
47. J. S. Bell. De Broglie-Bohm, Delayed-Choice Double-Slit Experiment, and Density Matrix. *International Journal of Quantum Chemistry: A Symposium*, 14:155–159, 1980. Reprinted in [26].
48. M. O. Scully, B. G. Englert, and H. Walther. Quantum Optical Tests of Complementarity. *Nature*, 351:111–116, 1991.
49. R. P. Feynman and A. R. Hibbs. *Quantum Mechanics and Path Integrals*. McGraw-Hill, New York, 1965.
50. M. Reed and B. Simon. *Methods of Modern Mathematical Physics II*. Academic Press, New York, 1975.

# References

51. K. Kraus. Position Observables of the Photon. In W.C. Price and S.S. Chissick, editors, *The Uncertainty Principle and Foundations of Quantum Mechanics*, pages 293–320. Wiley, New York, 1977.
52. S.T. Ali and G. G. Emch. Fuzzy Observables in Quantum Mechanics. *Journal of Mathematical Physics*, 15:176–182, 1974.
53. I. Bialynicki-Birula. On the Wave Function of the Photon. *Acta Physica Polonica*, 86:97–116, 1994.
54. L. D. Landau and E. M. Lifshitz. *Quantum Mechanics: Non-relativistic Theory*. Pergamon Press, Oxford and New York, 1958. Translated from the Russian by J. B. Sykes and J. S. Bell.
55. G. C. Ghirardi. Private communication, 1981.
56. W. K. Wooters and W. H. Zurek. A Single Quantum Cannot Be Cloned. *Nature*, 299:802–803, 299.
57. G. C. Ghirardi and T. Weber. Quantum Mechanics and Faster-Than-Light Communication: Methodological Considerations. *II Nuovo Cimento B*, 78:9–20, 1983.
58. Y. Aharonov, J. Anandan, and L. Vaidman. Meaning of the Wave Function. *Physical Review A*, 47:4616–4626, 1993.
59. G. C. Ghirardi, A. Rimini, and T. Weber. Unified Dynamics for Microscopic and Macroscopic Systems. *Physical Review D*, 34:470–491, 1986.
60. G. C. Ghirardi, P. Pearle, and A. Rimini. Markov Processes in Hilbert Space and Continous Spontaneous Localization of System of Identical Particles. *Physical Review A*, 42:78–89, 1990.
61. E. Englert, M. O. Scully, G. Süssman, and H. Walther. Surrealistic Bohm Trajectories. *Z. f. Naturforsch.*, 47a:1175, 1992.
62. D. Dürr, W. Fusseder, S. Goldstein, and N. Zanghì. Comment on: Surrealistic Bohm Trajectories. *Z. f. Naturforsh.*, 48a:1261–1262, 1993.
63. C. Dewdney, L. Hardy, and E. J. Squires. How Late Meaurements of Quantum Trajectories Can Fool a Detector. *Physics Letters A*, 184:6–11, 1993.
64. A. Gleason. Measures on the Closed Subspaces of a Hilbert Space. *J. Math. Mech.*, 6:885–893, 1957.
65. S. Kochen and E. P. Specker. The Problem of Hidden Variables in Quantum Mechanics. *Journal of Mathematics and Mechanics*, 17:59–87, 1967.
66. J. S. Bell. On the Einstein Podolsky Rosen Paradox. *Physics*, 1:195–200, 1964. Reprinted in [211], and in [26].
67. M. Daumer, D. Dürr, S. Goldstein, and N. Zanghì. On the Flux-Across-Surfaces Theorem. *Letters in Mathematical Physics*, 38(103–116), 1996.
68. D. Dürr, K. Münch-Berndl, and S. Teufel. The Flux Across Surfaces Theorem for Short Range Potentials Without Energy Cutoffs. *Journal of Mathematical Physics*, 40:1901–1922, 1999.
69. D. Dürr, S. Goldstein, S. Teufel, and N. Zanghì. Scattering Theory From Microscopic First Principles. *Physica A*, 279:416–431, 2000.
70. A. Peres. Two Simple Proofs of the Kochen-Specker Theorem. *Journal of Physics*, A24:175–178, 1991.
71. D. M. Greenberg, M. Horne, S. Shimony, and A. Zeilinger. Bell's Theorem Without Inequalities. *American Journal of Physics*, 58:1131–1143, 1990.
72. D. N. Mermin. Hidden Variables and the Two Theorems of John Bell. *Review of Modern Physics*, 65:803–815, 1993.
73. L. Hardy. Nonlocality for Two Particles Without Inequalities for Almost All Entangled States. *Physical Review Letters*, 71:1665–1668, 1993.
74. D. Z. Albert. *Quantum Mechanics and Experience*. Harvard University Press, Cambridge, MA, 1992.
75. D. Bohm and B. J. Hiley. *The Undivided Universe: An Ontological Intepretation of Quantum Theory*. Routledge & Kegan Paul, London, 1993.
76. J. S. Bell. *Speakable and Unspeakable in Quantum Mechanics*. Cambridge University Press, Cambridge, 1987.
77. A. Aspect, J. Dalibard, and G. Roger. Experimental Test of Bell's Inequalities Using Time-Varying Analyzers. *Physical Review Letters*, 49:1804–1807, 1982.

# Chapter 4
# Quantum Philosophy: The Flight from Reason in Science

We want to pause and comment on a rather delicate matter concerning a notoriously difficult subject, the foundations of quantum mechanics, a subject that has inspired a great many peculiar proclamations. Some examples:

> ... the idea of an objective real world whose smallest parts exist objectively in the same sense as stones or trees exist, independently of whether or not we observe them ... is impossible ... (Heisenberg, in [1] p. 129)

and

> We can no longer speak of the behavior of the particle independently of the process of observation. As a final consequence, the natural laws formulated mathematically in quantum theory no longer deal with the elementary particles themselves but with our knowledge of them. Nor is it any longer possible to ask whether or not these particles exist in space and time objectively ...
>
> ... Science no longer confronts nature as an objective observer, but sees itself as an actor in this interplay between man and nature. The scientific method of analysing, explaining, and classifying has become conscious of its limitations ... method and object can no longer be separated. (Heisenberg, in [2], pp. 15 and 29)

and

> A complete elucidation of one and the same object may require diverse points of view which defy a unique description. Indeed, strictly speaking, the conscious analysis of any concept stands in a relation of exclusion to its immediate application. (Bohr, quoted in [3], p. 102)

This last quotation is an expression of what has traditionally been called complementarity—but what might nowadays be called multiphysicalism.

For our purposes here, what is most relevant about these sentiments is that they were expressed, not by lay popularizers of modern science, nor by its postmodern critics, but by Werner Heisenberg and Niels Bohr, the two physicists most responsible, with the possible exception of Erwin Schrödinger, for the creation of quantum theory. It does not require great imagination to suggest that there is little in these sentiments with which a postmodernist would be inclined to disagree and much that he or she would be happy to regard as compelling support for the postmodern enterprise (see, for example, [4, 5, 6]).

The "quantum philosophy" expressed by such statements is part of the Copenhagen interpretation of quantum theory, which, in addition to the vagueness and subjectivity suggested by the preceding quotes, also incorporated as a central ingredient the notion that in the microscopic quantum domain the laws of nature involve irreducible randomness. The Copenhagen interpretation was widely, we would say at one time almost universally, accepted within the physics community, though there were some notable exceptions, such as Einstein and Schrödinger. Here is Schrödinger in 1926 (letter to Wien, quoted in [7], p. 228):

> Bohr's ... approach to atomic problems ... is really remarkable. He is completely convinced that any understanding in the usual sense of the word is impossible. Therefore the conversation is almost immediately driven into philosophical questions, and soon you no longer know whether you really take the position he is attacking, or whether you really must attack the position he is defending.

and Schrödinger in 1959 (letter to Wien, quoted in [7], p. 472):

> With very few exceptions (such as Einstein and Laue) all the rest of the theoretical physicists were unadulterated asses and I was the only sane person left. ... The one great dilemma that ails us ... day and night is the wave-particle dilemma. In the last decade I have written quite a lot about it and have almost tired of doing so: just in my case the effect is null ... because most of my friendly (truly friendly) nearer colleagues ( ... theoretical physicists) ... have formed the opinion that I am—naturally enough—in love with 'my' great success in life (viz., wave mechanics) reaped at the time I still had all my wits at my command and therefore, so they say, I insist upon the view that 'all is waves'. Old-age dotage closes my eyes towards the marvelous discovery of 'complementarity'. So unable is the good average theoretical physicist to believe that any sound person could refuse to accept the Kopenhagen oracle ...

Einstein in 1949 [8] (pp. 666 and 672) offered a somewhat more constructive response:

> I am, in fact, rather firmly convinced that the essentially statistical character of contemporary quantum theory is solely to be ascribed to the fact that this (theory) operates with an incomplete description of physical systems ...
>
> [In] a complete physical description, the statistical quantum theory would ... take an approximately analogous position to the statistical mechanics within the framework of classical mechanics ...

Part of what Einstein is saying here is that (much of) the apparent peculiarity of quantum theory, and in particular its randomness, arises from mistaking an incomplete description for a complete one.

In view of the radical character of quantum philosophy, the arguments offered in support of it have been surprisingly weak. More remarkable still is the fact that it is not at all unusual, when it comes to quantum philosophy, to find the very best physicists and mathematicians making sharp emphatic claims, almost of a mathematical character, that are trivially false and profoundly ignorant. For example, John von Neumann, one of the greatest mathematicians of this century, claimed to have mathematically proven that Einstein's dream, of a deterministic completion or reinterpretation of quantum theory, was mathematically impossible. He concluded that [9] (p. 325)

… 4 Quantum Philosophy: The Flight from Reason in Science

> It is therefore not, as is often assumed, a question of a re-interpretation of quantum mechanics—the present system of quantum mechanics would have to be objectively false, in order that another description of the elementary processes than the statistical one be possible.

This claim of von Neumann was, of course, just about universally accepted. For example, Max Born, who formulated the statistical interpretation of the wave function, assures us that [10] (p. 109)

> No concealed parameters can be introduced with the help of which the indeterministic description could be transformed into a deterministic one. Hence if a future theory should be deterministic, it cannot be a modification of the present one but must be essentially different. (Born 1949)

However, in 1952 David Bohm, through a refinement of de Broglie's pilot wave model of 1927, found just such a reformulation of quantum theory[11, 12]. Bohm's theory, Bohmian mechanics, was precise, objective, and deterministic—not at all congenial to quantum philosophy and a counterexample to the claims of von Neumann. Nonetheless, we still find, more than a quarter of a century after the discovery of Bohmian mechanics, statements such as these:

> The proof he [von Neumann] published … , though it was made much more convincing later on by Kochen and Specker, still uses assumptions which, in my opinion, can quite reasonably be questioned. … In my opinion, the most convincing argument against the theory of hidden variables was presented by Bell (1964). (Eugene Wigner 1976 [13])

and

> This [hidden variables] is an interesting idea and even though few of us were ready to accept it, it must be admitted that the truly telling argument against it was produced as late as 1965, by J. S. Bell. … This appears to give a convincing argument against the hidden variables theory. (Wigner 1983 [14], p. 53)

Now there are many more statements of a similar character that we could have cited; we chose these partly because Wigner was not only one of the leading physicists of his generation, but, unlike most of his contemporaries, he was also profoundly concerned with the conceptual foundations of quantum mechanics and wrote on the subject with great clarity and insight.

There was, however, one physicist who wrote on this subject with even greater clarity and insight than Wigner himself, namely the very J. S. Bell whom Wigner praises for demonstrating the impossibility of a deterministic completion of quantum theory such as Bohmian mechanics. So let's see how Bell himself reacted to Bohm's discovery:

> But in 1952 I saw the impossible done. It was in papers by David Bohm. Bohm showed explicitly how parameters could indeed be introduced, into nonrelativistic wave mechanics, with the help of which the indeterministic description could be transformed into a deterministic one. More importantly, in my opinion, the subjectivity of the orthodox version, the necessary reference to the 'observer,' could be eliminated. [15] (p. 160)

and Bell again

> Bohm's 1952 papers on quantum mechanics were for me a revelation. The elimination of indeterminism was very striking. But more important, it seemed to me, was the elimination

> of any need for a vague division of the world into "system" on the one hand, and "apparatus" or "observer" on the other. I have always felt since that people who have not grasped the ideas of those papers ... and unfortunately they remain the majority ... are handicapped in any discussion of the meaning of quantum mechanics. [15] (p. 173)

Wigner to the contrary notwithstanding, Bell did not establish the impossibility of a deterministic reformulation of quantum theory, nor did he ever claim to have done so. On the contrary, over the course of the past several decades, until his untimely death several years ago, Bell was the prime proponent, for a good part of this period almost the sole proponent, of the very theory, Bohmian mechanics, that he is supposed to have demolished. What Bell did demonstrate is the remarkable conclusion that nature, if governed by the predictions of quantum theory, must be nonlocal, exhibiting surprising connections between distant events. And unlike the claims of quantum philosophy, this nonlocality *is* well founded, and, with the experiments of Aspect [16], rather firmly established. Nonetheless, *it* is far from universally accepted by the physics community. Here is Bell, expressing his frustration at the obtuseness of his critics, and insisting that his argument for nonlocality involves no unwarranted assumptions:

> Despite my insistence that the determinism was inferred rather than assumed, you might still suspect somehow that it is a preoccupation with determinism that creates the problem. Note well then that the following argument makes no mention whatever of determinism. ... Finally you might suspect that the very notion of particle, and particle orbit ... has somehow led us astray. ... So the following argument will not mention particles ... nor any other picture of what goes on at the microscopic level. Nor will it involve any use of the words 'quantum mechanical system', which can have an unfortunate effect on the discussion. The difficulty is not created by any such picture or any such terminology. It is created by the predictions about the correlations in the visible outputs of certain conceivable experimental set-ups. [15] (p. 150)

So what is the relevance of what we have described to the theme of this conference?[1] Well, there's some bad news and some good news. The bad news, nothing you didn't already know anyway, is that objectivity is difficult to maintain and that physicists, even in their capacity as scientists, are only human. Nothing new. We must say, however, that the complacency of the physics establishment with regard to the foundations of quantum mechanics has been, it seems to me, somewhat astonishing, though we must admit to lacking sufficient historical perspective to have genuine confidence that what has occurred is at all out of the ordinary. But let us once again quote Bell:

> But why then had Born not told me of this 'pilot wave'? If only to point out what was wrong with it? Why did von Neumann not consider it? ... Why is the pilot wave picture ignored in text books? Should it not be taught, not as the only way, but as an antidote to the prevailing complacency? To show us that vagueness, subjectivity, and indeterminism, are not forced on us by experimental facts, but by deliberate theoretical choice? [15] (p. 160)

---

[1] This chapter is the contribution of one of us (S. G.) to a conference entitled *The Flight from Science and Reason*, which was sponsored by the New York Academy of Sciences and held in New York, NY on May 31–June 2, 1995.

The last quoted sentence refers, of course, to the good news: that when we consider, not the behavior of physicists but the physics itself, we find, in the stark contrast between the claims of quantum philosophy and the actual facts of quantum physics, compelling support for the objectivity and rationality of nature herself.

Here is one more bit of information somewhat relevant in this regard. You may well be wondering how, in fact, Bohm managed to accomplish what was so widely regarded as impossible, and what his completion of quantum theory involves. But you probably imagine that what eluded so many great minds could not be conveyed in but a few minutes, even were this an audience of experts. However, the situation is quite otherwise. In order to arrive at Bohmian mechanics from standard quantum theory one need do almost nothing! One need only avoid quantum philosophy and complete the usual quantum description in what is really the most obvious way: by simply including the positions of the particles of a quantum system as part of the state description of that system, allowing these positions to evolve in the most natural way (see Chap. 2). The entire quantum formalism, including the uncertainty principle and quantum randomness, emerges from an analysis of this evolution (see Chap. 2 and [17]). This can be expressed succinctly—though in fact not succinctly enough—by declaring that the essential innovation of Bohmian mechanics is the insight that *particles move*! Bell, referring to the double-slit interference experiment, put the matter this way:

> Is it not clear from the smallness of the scintillation on the screen that we have to do with a particle? And is it not clear, from the diffraction and interference patterns, that the motion of the particle is directed by a wave? De Broglie showed in detail how the motion of a particle, passing through just one of two holes in screen, could be influenced by waves propagating through both holes. And so influenced that the particle does not go where the waves cancel out, but is attracted to where they cooperate. This idea seems to me so natural and simple, to resolve the wave-particle dilemma in such a clear and ordinary way, that it is a great mystery to me that it was so generally ignored. [15] (p. 191)

We think this should be a bit of a mystery for all of us!

# References

1. W. Heisenberg. *Physics and Philosophy*. Harper and Row, New York, 1958.
2. W. Heisenberg. *The Physicist's Conception of Nature*. Harcourt Brace, 1958. Trans. Arnold J. Pomerans.
3. M. Jammer. *The Philosophy of Quantum Mechanics*. Wiley, New York, 1974.
4. A. Plotnitsky. *Complementarity: Anti-Epistemology after Bohr and Derrida*. Duke University Press, Durham, N.C., 1994.
5. S. Aronowitz. *Science as Power: Discourse and Ideology in Modern Society*. University of Minnesota Press, Minneapolis, 1988.
6. E. Fox Keller. Cognitive Repression in Contemporary Physics. *American Journal of Physics*, 47:718–721, 1979.
7. W. Moore. *Schrödinger*. Cambridge University Press, New York, 1989.
8. P. A. Schilpp, editor. *Albert Einstein, Philosopher-Scientist*. Library of Living Philosophers, Evanston, Ill., 1949.

9. J. von Neumann. *Mathematische Grundlagen der Quantenmechanik*. Springer Verlag, New York-Heidelberg-Berlin, 1932. English translation by R. T. Beyer, Mathematical Foundations of Quantum Mechanics. Princeton University Press, Princeton, N.J., 1955.
10. M. Born. *Natural Philosophy of Cause and Chance*. Oxford University Press, Oxford, 1949.
11. D. Bohm. A Suggested Interpretation of the Quantum Theory in Terms of "Hidden" Variables: Part I. *Physical Review*, 85:166–179, 1952. Reprinted in [?].
12. D. Bohm. A Suggested Interpretation of the Quantum Theory in Terms of "Hidden" Variables: Part II. *Physical Review*, 85:180–193, 1952. Reprinted in [?].
13. E. P. Wigner. Interpretation of Quantum Mechanics. In [?], 1976
14. E. P. Wigner. Review of the Quantum Mechanical Measurement Problem. In P. Meystre and M. O. Scully, editors, Quantum Optics, *Experimental Gravity and Measurement Theory*, pages 43–63. Plenum, New York, 1983.
15. J. S. Bell. *Speakable and Unspeakable in Quantum Mechanics*. Cambridge University Press, Cambridge, 1987.
16. A. Aspect, J. Dalibard, and G. Roger. Experimental Test of Bell's Inequalities Using Time-Varying Analyzers. *Physical Review Letters*, 49:1804–1807, 1982.
17. K. Berndl, M. Daumer, D. Dürr, S. Goldstein, and N. Zanghì. A Survey of Bohmian Mechanics. Il *Nuovo Cimento*, 110B:737–750, 1995.

# Part II
# Quantum Motion

# Chapter 5
# Seven Steps Towards the Classical World

## 5.1 Introduction

The classical world, say the world of objects of familiar experience that obey Newtonian laws, seems far removed from the "wavy" world of quantum mechanics. In this chapter we shall sketch what we believe are the basic steps to be taken in going from the quantum world to the classical world.

1. The first step is the crucial one: As is well known, and as Bell has emphasized [1], standard quantum mechanics is not a precise microscopic theory because the division between the microscopic and the macroscopic world, which is essential to the very formulation of that theory, is not made precise by the theory [2, 3]. In fact, the following conclusion seems inevitable: quantum mechanics does not contain the means for describing the classical world in any approximate sense and one needs to go beyond quantum mechanics in order to do so. There are two natural possibilities for amending ordinary quantum mechanics: either the wave function is not all there is, or Schrödinger's equation is wrong. In this chapter we'll formulate the problem of the classical limit within the framework of Bohmian mechanics, a theory which follows the first path.
2. To get a handle on a problem, one should first simplify it as much as possible. The complex motion of a macroscopic body can be drastically simplified by making some rather standard approximations and reducing the problem to that of a "particle" moving in an external potential. This is what we shall do in Sect. 5.2.
3. Good textbooks on quantum mechanics contain enlightening ideas. One of these ideas is the so called Ehrenfest theorem that we shall use in Sect. 5.4 in order to obtain a necessary condition for the classicality for wave packets.
4. The structure of Bohmian mechanics contains the means for extending the condition for classicality to more general wave functions, namely wave functions which locally look like plane waves, as we shall see in Sect. 5.5. In Sect. 5.6 we shall then show how the problem of classical limit for general wave functions can be reduced to that for local plane waves.

5. Simplicity is good, but it has its limitations: the reduction of the motion of the center of mass to a one body problem doesn't explain the robustness and stability of classical behavior. This however can be explained by making the model a little more realistic, say by including in an effective way the external as well as the internal environment. We shall briefly touch this point in Sect. 5.7.
6. This step is the crucial one from a mathematical point of view. In Sect. 5.8 we shall put forward a mathematical conjecture on the emergence of classical behavior. Unfortunately we cannot provide any rigorous mathematical justification for it. Mathematical work on it would be valuable since this conjecture goes beyond the standard mathematical work of semiclassical analysis (see, e.g., [4], [5]) or, in more modern terms, microlocal analysis (see, e.g., [6]).
7. This is the last step in what we believe is the main structure of the classical limit:

$$(\psi, Q) \to (P, Q),$$

where on the two sides of the arrow are represented the complete state description of Bohmian mechanics, in terms of wave function and position, and of classical mechanics, in terms of momentum and position.

## 5.2 Motion in an External Potential

Our goal is to study the classical behavior of a macroscopic body composed of $N$ particles with $N \gg 1$ (one may think of an apple falling from a tree or a planet moving around the sun). It is rather clear that one expects classical behavior only for appropriate macroscopic functions of the particle configuration $Q = (\mathbf{Q}_1, \ldots, \mathbf{Q}_N)$. The relevant macroscopic variable, whose classical behavior we wish to investigate here, is the center of mass of the body

$$\mathbf{X} = \frac{\sum_i m_i \mathbf{Q}_i}{m},$$

where $m_1, \ldots, m_N$ are the masses of the particles composing the body and $m = \sum_i m_i$ is the total mass of the body.

We shall assume that the particles interact through internal forces as well as being subjected to an external potential, so that the potential energy is of the form

$$U(q) = \sum_{i<j} U(\mathbf{q}_i, \mathbf{q}_j) + \sum_i V_i(\mathbf{q}_i).$$

Thus, for non relativistic spinless particles, the Bohmian state $(\Psi, Q)$ evolves according to the equations

## 5.3 The Classical Limit in Bohmian Mechanics

$$\frac{d\mathbf{Q}_k}{dt} = \frac{\hbar}{m_k} \text{Im} \frac{\nabla_k \Psi(Q)}{\Psi(Q)} \tag{5.1}$$

$$i\hbar \frac{\partial \Psi}{\partial t} = -\sum_{k=1}^{N} \frac{\hbar^2}{2m_k} \Delta_k \Psi + U(q)\Psi. \tag{5.2}$$

Let $y = (\mathbf{y}_1, \ldots, \mathbf{y}_{N-1})$ be a suitable set of coordinates[1] relative to the center of mass $\mathbf{x} = \sum_i m_i \mathbf{q}_i / m$. Then under the change of variables $q = (\mathbf{x}, y)$ Schrödinger's equation (5.2) assumes the form

$$i\hbar \frac{\partial \Psi}{\partial t} = \left( H^{\mathbf{x}} + H^y + H^{(\mathbf{x},y)} \right) \Psi \tag{5.3}$$

where

$$H^{\mathbf{x}} = -\frac{\hbar^2}{2m} \Delta + V(\mathbf{x}), \qquad V(\mathbf{x}) \equiv \sum_i V_i(\mathbf{x}),$$

and $\Delta$ is the Laplacian with respect to the x-variables. $H^y$ in (5.3) is the free Hamiltonian associated with the relative coordinates $y$ and the operator $H^{(\mathbf{x},y)}$ describes the interaction between the center of mass and the relative coordinates. If $V_i$ are slowly varying on the size of the body, $H^{(\mathbf{x},y)}$ can be treated as a small perturbation, and, in first approximation, neglected. Thus, if $\Psi = \psi(\mathbf{x})\phi(y)$ at some time, the time evolution of the center of mass decouples from that of the relative coordinates and we end up with a very simple one particle problem: the wave function $\psi$ of the center of mass evolves according to one-particle Schrödinger's equation

$$i\hbar \frac{\partial \psi}{\partial t} = -\frac{\hbar^2}{2m} \Delta \psi + V(\mathbf{x})\psi \tag{5.4}$$

and its position $\mathbf{X}$ evolves according to

$$\frac{d\mathbf{X}}{dt} = \frac{\hbar}{m} \text{Im} \frac{\nabla \psi}{\psi}(\mathbf{X}), \tag{5.5}$$

where $\nabla$ is the gradient with respect to the x-variables. From now on, whenever no ambiguity will arise, we shall treat the center of mass as a "particle" and we shall refer to $\mathbf{X}$ and $\psi$ as the position and the wave function of such a particle.

## 5.3 The Classical Limit in Bohmian Mechanics

In order to investigate the conditions under which $\mathbf{X}$ evolves classically it is useful to write the wave function $\psi = \psi(\mathbf{x})$ in the polar form

---

[1] For sake of concreteness one may think, e.g., of the so called Jacobi coordinates.

$$\psi(\mathbf{x}) = R(\mathbf{x})e^{\frac{i}{\hbar}S(\mathbf{x})}, \tag{5.6}$$

From Schrödinger's equation (5.4) one obtains, following Bohm [7], the continuity equation for $R^2$,

$$\frac{\partial R^2}{\partial t} + \operatorname{div}\left[\left(\frac{\boldsymbol{\nabla} S}{m}\right)R^2\right] = 0, \tag{5.7}$$

and the modified Hamilton-Jacobi equation for $S$

$$\frac{\partial S}{\partial t} + \frac{(\boldsymbol{\nabla} S)^2}{2m} + V - \frac{\hbar^2}{2m}\frac{\Delta R}{R} = 0. \tag{5.8}$$

Note that Eq. (5.8) is the usual classical Hamilton-Jacobi equation with an additional term

$$V_Q \equiv -\frac{\hbar^2}{2m}\frac{\Delta R}{R}, \tag{5.9}$$

called the quantum potential. Since $\frac{\boldsymbol{\nabla} S}{m}$ is the right hand side of (5.5), one then sees that the (size of the) quantum potential provides a rough measure of the deviation of Bohmian evolution from its classical approximation.

Analogously, consider the modified Newton equation associated with (5.8), and obtained by differentiating both sides of Eq. (5.5) with respect to time,

$$m\frac{d^2\mathbf{X}}{dt^2} = \mathbf{F} + \mathbf{F}_Q, \tag{5.10}$$

where $\mathbf{F} = -\boldsymbol{\nabla} V(\mathbf{X})$ and $\mathbf{F}_Q = -\boldsymbol{\nabla} V_Q(\mathbf{X})$ are respectively the *classical* force and the *"quantum"* force. Eq. (5.10) shows that all the *deviations* from classicality are embodied in the quantum force $\mathbf{F}_Q$.

Thus, the *formulation* of the classical limit in Bohmian mechanics turns out to be rather simple: classical behavior emerges whenever the particle trajectory $\mathbf{X} = \mathbf{X}(t)$, satisfying (5.10), approximately satisfies the classical Newton equation, i.e.,

$$m\frac{d^2\mathbf{X}}{dt^2} \simeq \mathbf{F}. \tag{5.11}$$

The problem is to determine the physical conditions ensuring (5.11). Usually, physicists consider classical behavior as ensured by the limit $\hbar \to 0$, meaning by this

$$\hbar \ll A_0, \tag{5.12}$$

where $A_0$ is *some* characteristic action of the corresponding classical motion (see, e.g., [4, 8, 9]). Condition (5.12) is often regarded as equivalent to another standard condition of classicality which involves the length scales of the motion (see, e.g., [10]): if the de Broglie wave length $\lambda$ is small with respect to the characteristic dimension $L$ determined by the scale of variation of the potential $V$, the behavior of the system should be close to the classical behavior in the same potential $V$. This is very reminiscent of how geometrical optics can be deduced from wave optics. We regard this condition, i.e.,

$$\lambda \ll L, \tag{5.13}$$

as the most natural condition of classicality since it relates in a completely transparent way a property of the state, namely its de Broglie wave length $\lambda$, and a property of the dynamics, namely the scale of variation of the potential $L$. In the remainder of this chapter we shall argue that (5.13) is indeed a necessary and sufficient condition for (5.11).

## 5.4 Wave Packets

To explain the physical content of (5.13) and its implications we shall consider first the case for which the wave function has a well-defined de Broglie wave length: we shall assume that $\psi$ is a wave packet with diameter $\sigma$, with mean wave vector $\mathbf{k}$ and associated wave length $\lambda = 2\pi/|\mathbf{k}|$.

As we shall see, the analysis of this situation will allow us to find a precise characterization of the scale $L$ of variation of the potential. Our analysis will be rather standard—it is basically the *Ehrenfest's Theorem*—and can be found in good textbooks (see, e.g., [11]). We reproduce it here both for the sake of completeness and because we believe that it attains, within the Bohmian framework, a deeper and much more general significance than within standard formulations of quantum mechanics.

From the equivariance of $\rho(q) = |\psi(q)|^2$, we have that the mean particle position at time $t$ is given by

$$\langle \mathbf{X} \rangle = \int \mathbf{x} |\psi_t(\mathbf{x})|^2 d^3\mathbf{x}.$$

From (5.4) it follows that

$$m \frac{d^2}{dt^2} \langle \mathbf{X} \rangle = -\int \nabla V(\mathbf{x}) |\psi_t(\mathbf{x})|^2 d^3\mathbf{x}.$$

By expanding $\mathbf{F}(\mathbf{x}) = -\nabla V(\mathbf{x})$ in Taylor series around $\langle \mathbf{X} \rangle$ one obtains

$$m \frac{d^2}{dt^2} \langle \mathbf{X} \rangle = \mathbf{F}(\langle \mathbf{X} \rangle) + \frac{1}{2} \sum_{j,k} \Delta_{j,k} \frac{\partial^2 \mathbf{F}}{\partial x_j \partial x_k}(\langle \mathbf{X} \rangle) + ..., \tag{5.14}$$

where

$$\Delta_{j,k} = \langle X_j X_k \rangle - \langle X_j \rangle \langle X_k \rangle$$

is of order $\sigma^2$, where $\sigma$ is the diameter of the packet. Therefore, the mean particle position should satisfy the classical Newton equation whenever

$$\sigma^2 \left| \frac{\partial^3 V}{\partial x_i \partial x_j \partial x_k} \right| \ll \left| \frac{\partial V}{\partial x_i} \right|, \tag{5.15}$$

i.e.,

$$\sigma \ll \sqrt{\left| \frac{V'}{V'''} \right|} \tag{5.16}$$

where $V'$ and $V'''$ denote respectively suitable estimates of the first and third derivatives (e.g., by taking a sup over the partial derivatives).

The minimum value of the diameter of the packet $\sigma$ is of order $\lambda$. Hence (5.16) becomes

$$\lambda \ll \sqrt{\left| \frac{V'}{V'''} \right|} \tag{5.17}$$

This last equation gives a *necessary* condition for the classicality of the particle motion and, by comparing it with (5.13), a precise definition of the notion of scale of variation of the potential, namely,

$$L = L(V) = \sqrt{\left| \frac{V'}{V'''} \right|}. \tag{5.18}$$

In the following we shall argue that (5.13), with $L$ given by (5.18), is indeed also *sufficient* for classical behavior of Bohmian trajectories. For wave packets this follows easily from the equivariance of $|\psi|^2$: over the lapse of time for which the spreading of the packet can be neglected, the overwhelming majority[2] of trajectories $\mathbf{X} = \mathbf{X}(t)$ will stick around their mean value $\langle \mathbf{X} \rangle$ and follow its classical time evolution. Thus we expect (5.11) to hold for the overwhelming majority of trajectories.

## 5.5 Local Plane Waves

Suppose now that $\psi$ is not a packet but a wave function that locally looks like a packet. By this we mean, referring to the polar representation (5.6), that the amplitude $R(\mathbf{x})$ and the *local* wave vector

$$\mathbf{k} = \mathbf{k}(\mathbf{x}) \equiv \nabla S(\mathbf{x})/\hbar \tag{5.19}$$

are slowly varying over distances of order $\lambda(\mathbf{x}) \equiv h/|\nabla S(\mathbf{x})|$, the *local* de Broglie wave length. We may call such a $\psi$ a "local plane wave".

---

[2] With respect to the equivariant measure $|\psi|^2$.

At any given time the local plane wave can be thought as composed of a sum of wave packets: Consider a partition of physical space into a union of disjoint sets $\Delta_i$ chosen in such a way that the local wave vector $\mathbf{k}(\mathbf{x})$ doesn't vary appreciably inside of each of them and denote by $\mathbf{k}_i$ the almost constant value $\mathbf{k}(\mathbf{x})$ for $\mathbf{x} \in \Delta_i$. Let $\chi_{\Delta_i}$ be the characteristic function of the set $\Delta_i$ ($\chi_{\Delta_i}(\mathbf{x}) = 1$ if $\mathbf{x} \in \Delta_i$ and 0 otherwise). Since $\sum_i \chi_{\Delta_i} = 1$, we have

$$\psi(\mathbf{x}) = \sum_i \chi_{\Delta_i}(\mathbf{x})\psi(\mathbf{x}) = \sum_i \psi_i(\mathbf{x}). \tag{5.20}$$

Note that this decomposition is somewhat arbitrary: provided that $\mathbf{k}(\mathbf{x})$ is almost constant in $\Delta_i$, the extent of these sets can be of the order of many wave lengths down to a minimal size $\sigma_i \simeq |\Delta_i|^{1/3}$ of the same order of $\lambda_i$.[3]

At any time, the position $\mathbf{X}$ of the particle will be in the support of one of the packets forming the decomposition (5.20), say in the support of $\psi_i$. If the condition (5.16) holds for $\sigma_i$, we may then proceed as in the previous section: the minimal size of the packet $\psi_i$ can be taken of order $\sigma_i = \lambda(\mathbf{x})$ and the condition of classicality is again (5.17) for $\lambda = \lambda(\mathbf{x})$.

Note that this straightforward reduction of the classical limit for local plane waves to that for wave packets is possible only within Bohmian mechanics: since the particle has at any time a well-defined position $\mathbf{X}$ and the different components of the local plane wave (5.20) don't interfere, we may "collapse" $\psi$ to the wave packet $\psi_i$ relevant to the dynamics of $\mathbf{X}$.

## 5.6 General Wave Functions

We wish now to investigate the physical content of (5.13) and its implications for a general wave function. The first issue to address is what notion of wave length should be appropriate for this case. A rough estimate of $\lambda$ could be given in terms of mean kinetic energy associated with $\psi$,

$$E_{\text{kin}}(\psi) = \langle \psi, -\frac{\hbar^2}{2m}\Delta\psi \rangle, \tag{5.21}$$

with associated wave length

$$\lambda = \lambda(\psi) = \frac{h}{\sqrt{2m E_{\text{kin}}(\psi)}}. \tag{5.22}$$

Suppose now that (5.17), with $\lambda$ given by (5.22), is satisfied. We claim that in this case the Schrödinger evolution should "quickly" produce a local plane wave, that can be effectively regarded as built of pieces that are wave packets satisfying (5.17) for

---

[3] The use of the characteristic function may introduce an undesirable lack of smoothness, but this can be easily taken care by replacing the $\chi_{\Delta_i}$ with functions $\theta_i$ forming a smooth partition of unity.

$\lambda = \lambda(\mathbf{x})$ and hence themselves evolving classically as we have seen in the previous section.

In fact, if $\lambda \ll L$ the kinetic energy dominates the potential energy and the free Schrödinger evolution provides a rough approximation of the dynamics up to the time needed for the potential to affect the evolution significantly. During this time, the Schrödinger evolution produces a spatial separation of the different wave vectors contained in $\psi$, more or less as Newton's prism separates white light into the different colors of the rainbow. In other words, the formation of a local plane wave originates in the "*dispersive*" character of free Schrödinger evolution.

So, in order to gain some appreciation of this phenomenon consider the free Schrödinger evolution

$$\psi_t(\mathbf{x}) = \frac{1}{(2\pi)^{3/2}} \int e^{it\left[\mathbf{k}\cdot\frac{\mathbf{x}}{t} - \frac{\hbar k^2}{2m}\right]} \hat{\psi}(\mathbf{k}) d^3\mathbf{k}, \tag{5.23}$$

where $\hat{\psi}$ is the Fourier transform of the initial wave function $\psi$. The stationary phase method yields straightforwardly the long time asymptotics of $\psi_t$,

$$\psi_t(\mathbf{x}) \sim \left(\frac{im}{\hbar t}\right)^{3/2} e^{i\frac{m}{2\hbar}\frac{x^2}{t}} \hat{\psi}(\mathbf{k}), \quad \text{where} \quad \mathbf{k} = \frac{m}{\hbar}\frac{\mathbf{x}}{t}, \tag{5.24}$$

which is indeed a local plane wave with local wave vector $\mathbf{k} = m\mathbf{x}/(\hbar t)$.

We said above that the local plane wave is "quickly" produced. But how quickly? In order to estimate such a time, consider the simple example of an initial wave function $\psi$ composed of two overlapping wave packets with the same position spread $\Delta x$ and with opposite momenta $p$ and $-p$. The time $\tau$ of formation of a local plane wave should be of the order of the time for separation of the packets, which is basically the time needed to cover a space equal to $\Delta x$. From $\Delta x \Delta p \sim \hbar$ and $\Delta p \sim p$ we obtain

$$\tau \sim \frac{\Delta x}{p/m} \sim \frac{\hbar}{p^2/m} \sim \frac{\hbar}{\langle E \rangle}, \tag{5.25}$$

where $\langle E \rangle$ is the mean kinetic energy of the particle. It is reasonable to suggest that (5.25), with $\langle E \rangle$ given by (5.21), could give a very rough estimate of the time of formation of a local plane wave for a general wave function $\psi$. Note that the time needed for the potential to produce significant effects on the evolution is of order

$$T = \frac{L}{v}, \quad \text{where} \quad v = \frac{h}{m\lambda}. \tag{5.26}$$

Thus, if $\lambda \ll L$ we have that $\tau \ll T$, which means that the local plane wave gets formed on a time scale much shorter than the time scale over which the potential affects the dynamics.

We arrive in this way at a sharp (or, at least, sharper than usually encountered) mathematical formulation of the classical limit for a general wave function $\psi$. First of all, consider the dimensionless parameter

$$\varepsilon = \frac{\lambda(\psi)}{L(V)}. \tag{5.27}$$

Secondly, consider the Bohm motion **X** on the "macroscopic" length and time scales defined by $\psi$ and $V$. By this we mean $\mathbf{X}' = \mathbf{X}'(t')$, where

$$\mathbf{X}' = \mathbf{X}/L \quad \text{and} \quad t' = t/T \tag{5.28}$$

with $T$ given by (5.26). Finally, consider $\mathbf{F}_Q/m$, the "quantum" contribution to the total acceleration in (5.10), on the macroscopic scales (5.28), namely

$$\mathbf{D} = \frac{T^2}{L}\mathbf{F}_Q(\mathbf{X}'L, t'T) \tag{5.29}$$

Then the Bohm motion on the macroscopic length and time scales will be approximately classical, with deviation from classicality **D** tending to 0 as $\varepsilon \to 0$.

We'd like to point out that the use of macroscopic coordinates (5.28) for the formulation of the classical limit is rather natural from a physical point of view. First of all, the scales $L$ and $T$ are the fundamental units of measure for the motion: $L$ is the scale on which the potential varies and $T$ provides an estimate of the time necessary for the particle to see its effects. More importantly, in the limit $\varepsilon \to 0$ the nonclassical behavior–occurring during the time $\tau$ of formation of the local plane wave—disappears, since, as we have argued above, in this limit $\tau \ll T$. In other words, on the macroscopic scales on which we expect classical behavior the local plane wave has been formed.

## 5.7 Limitations of the Model: Interference and the Role of the Environment

Before commenting on the mathematics of the limit $\varepsilon \to 0$ we should stress a physical caveat. For motion in unbounded space, the expanding character of the Schrödinger evolution makes the set of local plane waves an "attractor" for the dynamics—so that the local plane wave form is in this sense "typical". However, for motion in a bounded region (with wave functions which are superpositions of bound states) the "typical" wave function is composed by a sum of local plane waves, this being due to interference between the waves reflected by the "edges" of the confining potential. Consider for example an infinite potential well of size $L$ in one dimension and initial wave function $\psi$, well localized in the center of the well which is the superposition of two packets with opposite momenta $p$ and $-p$. Suppose that $\lambda(\psi) \ll L$. Then the two packets move classically and at a certain time, say $t_r$, are reflected from the walls of the potential. At the time $t_c = 2t_r$, they interfere in the middle of the well. $t_c$ is the "*first caustic time*," the time at which the classical action $S_{cl}(\mathbf{x}, t)$ becomes multivalued. In general, we should not expect classical behavior for times larger than the first caustic time $t_c$.

What is going on? The emergence of classical behavior should be robust and stable, which would not be the case if it were restricted to times smaller than $t_c$. However, if one remembers that the model we are investigating is a strong idealization, the problem evaporates. We are in fact dealing with the one-body problem defined by (5.4) and (5.5), an approximation to the complete dynamics defined by (5.3) in which the term $H^{(x,y)}$, describing the interaction between the center of mass and the relative coordinates, is neglected. Note than even (5.3) is an idealization since it does not include the unavoidable interaction of the body with its *external* environment: in a more realistic model $H^{(x,y)}$ would take into account both the internal and external environment of the center of mass (with $y$ now including both the relative coordinates and the degrees of freedom of the external environment). These interactions—even for very small interaction energy—should produce *entanglement* between the center of mass $\mathbf{x}$ of the system and the other degrees of freedom $y$, so that their effective role is that of "measuring" the position $\mathbf{X}$ and suppressing superpositions of spatially separated wave functions. (Taking these interactions into account is what people nowadays call decoherence, see, e.g., [12] and the references therein). Referring to the above example, the effect of the environment should be to select (as relevant to the dynamics of $\mathbf{X}$, see Chap. 2) one of the two packets on a time scale much shorter than the first caustic time $t_c$.

## 5.8 Towards a Mathematical Conjecture

The mathematical content of Sect. 5.6 and 5.7 is summarized by the following (not yet sharply formulated) conjecture:

**Conjecture.** *Let $\epsilon$ be the dimensionless parameter defined by (5.27) and $\mathbf{D}$ be the quantity given by (5.29). Then there are environmental interactions such that $\mathbf{D} \rightsquigarrow 0$ as $\epsilon \to 0$, uniformly in $\psi$ and $V$.*

Concerning this conjecture, we'd like to make here just a few remarks.

1. "$\mathbf{D} \rightsquigarrow 0$" means convergence to 0 in a "suitable" probabilistic sense since $\mathbf{D}$ is a random variable. $\mathbf{D}$ is a function of $\mathbf{X}$, and $\mathbf{X}$ is random with probability distribution given by $|\psi|^2$. To require almost sure convergence is probably too strong a demand. Convergence in probability, or $L^2$ convergence, would seem more appropriate. Moreover, $\lambda$ is defined in (5.22) in terms of the *average* kinetic energy. This average could be large even when there is a significant probability for a very small kinetic energy. Thus it is probably necessary to regard $\lambda$ as random (with randomness inherited from the kinetic energy) and to understand $\varepsilon \to 0$ also in a probabilistic sense.
2. Uniformity of the limit in $\psi$ and $V$ could be expressed as follows: let $(V_n, \psi_n)$ be any sequence for which $\varepsilon_n = \frac{\lambda_n}{L_n} \to 0$, with $\lambda_n = \lambda(\psi_n)$ given by (5.22), and $L_n = L(V_n)$. Then $\mathbf{D} \rightsquigarrow 0$ as $n \to +\infty$. Understanding $\mathbf{D} \rightsquigarrow 0$ as convergence in probability, we could also express uniformity in the following way: for any

## 5.8 Towards a Mathematical Conjecture

$\eta > 0$ and for any $\delta > 0$, there exists an $\epsilon_0 > 0$ such that $\mathbb{P}(\mathbf{D} > \delta)$ is smaller than $\eta$ whenever $\varepsilon < \varepsilon_0$. Here $\mathbb{P}$ is the probability measure defined by $|\Psi|^2$, i.e., $\mathbb{P}(dq) = |\Psi(q)|^2 dq$, which includes randomness arising from the environment.

3. For quadratic potentials (including free motion and motion in a uniform force field) $L = \infty$ so that $\varepsilon = 0$. In this case the conjecture should be modified as follows: let $L_o$ be *any* length scale and $T_o$ the corresponding time scale $T_o = \frac{L_o}{v}$. Then for the Bohm motion on the scales given by $L_o$ and $T_o$, $\mathbf{D} \rightsquigarrow 0$ uniformly in $\psi$ and $L_o$ whenever $\tilde{\varepsilon} \equiv \frac{\lambda(\psi)}{L_o} \to 0$.

4. There is an enormous amount of mathematical work, called semiclassical analysis or, in more modern terms, microlocal analysis, in which the limit $\hbar \to 0$ of Schrödinger evolutions is rigorously studied. It should be stressed that the limit $\varepsilon \to 0$ is *much more general* than the limit $\hbar \to 0$. In fact $\varepsilon = \lambda/L = h/mvL$. So keeping $L$ and the momentum $mv$ fixed, the limit $\hbar \to 0$ implies $\epsilon \to 0$. But there are many ways in which $\varepsilon$ could go to zero. The classical limit, as expressed by the above conjecture, is (at the very least) a two-parameters limit, involving $\lambda$ and $L$, and $\hbar \to 0$ is just a very special case. Moreover, these two parameters themselves live on infinite dimensional spaces since $\lambda = \lambda(\psi)$, with $\psi$ varying in the Hilbert space of the system's wave functions, and $L = L(V)$, with $V$ varying in the class of admissible one particle potentials (that is, potentials leading to a self-adjoint Hamiltonian).

5. Exactly for the reason expressed in the previous remark, the conjecture is really *very* hard to prove: it requires a lot of uniformity both in the wave function $\psi$ and in the potential $V$. Just to have an idea of the difficulties, one may think of the analogous problem in statistical mechanics, namely the problem of studying the deviations from thermodynamic behavior of a large but finite system about which not so much is known.

6. While the conjecture is difficult to prove, it is still not completely satisfactory from a physical point of view. The conjecture states only that $\mathbf{D}$ depends on $\varepsilon$ in such a way that $\mathbf{D} \rightsquigarrow 0$ as $\varepsilon \to 0$, uniformly in $\psi$ and $V$. A physically more relevant result would be to estimate how rapidly $\mathbf{D}$ is tending to 0 (e.g., like $\varepsilon$, or $\varepsilon^2$ or whatever). Note that only this last kind of result can be of practical value: given $V$ and $\psi$, it provides an estimate for the deviation from classicality, while any other results do not quite do this.

7. With the conjecture, and even with the refinement proposed in the previous remark, there is a further difficulty to consider: even if $H^{(x,y)}$ is treated as a small perturbation in (5.3), the suggestion of Sect. sec:env might not be too realistic. In fact, the autonomous Schrödinger evolution, even of a very narrow wave function $\psi = \psi(\mathbf{x})$, could be destroyed in very short times. This is a serious difficulty; one resolution might be found in the notion of *conditional wave function* of the **x**-system, $\psi(\mathbf{x}) = \Psi(\mathbf{x}, Y)$, where $Y$ is the actual configuration of the environment (this notion has been introduced and analyzed in Chap. 2). We regard the extension of the conjecture to this more realistic framework as the most interesting open problem on the classical limit, which we leave for future work.

## 5.9 The Classical Limit in a Nutshell

The key ingredient in our analysis of the emergence of the classical world, is that as soon as the local plane wave has formed, each configuration **X** is attached to a guiding wave packet with a definite wave vector $\mathbf{k}(\mathbf{x},t)$ that locally determines the particle dynamics according to the local de Broglie relation

$$\mathbf{p}(\mathbf{x},t) = \hbar \mathbf{k}(\mathbf{x},t),$$

which, for $\lambda \ll L$, evolves according to classical laws. This means that the classical limit can be symbolically expressed as

$$(\psi, \mathbf{X}) \rightarrow (\mathbf{P}, \mathbf{X}),$$

where $(\psi, \mathbf{X})$ is the complete quantum state description in terms of wave function and position, while $(\mathbf{P}, \mathbf{X})$ is the complete classical state description in terms of momentum and position. All the relevant macroscopic information contained in the pair $(\psi, \mathbf{X})$ is, in the classical limit, embodied in the pair $(\mathbf{P}, \mathbf{X})$—the only robust, stable quantity. In other words, as far as the macroscopic dynamics of **X** is concerned, only the information carried by **P** is relevant.

## References

1. J. S. Bell. *Speakable and Unspeakable in Quantum Mechanics*. Cambridge University Press, Cambridge, 1987.
2. S. Goldstein. Quantum Theory Without Observers. Part One. *Physics Today*, 51(3):42–46, 1998.
3. S. Goldstein. Quantum Theory Without Observers. Part Two. *Physics Today*, 51(4):38–42, 1988.
4. V. P. Maslov and M. V. Fedoriuk. *Semi–Classical Approximation in Quantum Mechanics*. D. Reidel Publ. Co., Dordrecht, Holland, 1981.
5. D. Robert. *Autour de l'Approximation Semi-Classique*. Birkhauser, Boston, 1987.
6. A. Martinez. *An Introduction to Semiclassical and Microlocal Analysis*. Springer, New York-Heidelberg-Berlin, 2001.
7. D. Bohm. A Suggested Interpretation of the Quantum Theory in Terms of "Hidden" Variables: Part I. *Physical Review*, 85:166–179, 1952. Reprinted in [211].
8. L. I. Schiff. *Quantum Mechanics*. McGraw–Hill, New York, 1949.
9. M. Berry. Chaos and the Semiclassical Limit of Quantum Mechanics (Is the Moon There When Somebody Looks?). In R. J. Russell, P. Clayton, K. Wegter-McNelly, and J. Polkinghorne, editors, *Quantum Mechanics: Scientfic Perpectives on Divine Action*, pages 41–54. Vatican Observatory – CTNS Publications, 2001.
10. L. D. Landau and E. M. Lifshitz. *Quantum Mechanics: Non-relativistic Theory*. Pergamon Press, Oxford and New York, 1958. Translated from the Russian by J. B. Sykes and J. S. Bell.
11. K. Gottfried. *Quantum Mechanics*: Vol. I. W. A. Benjamin Inc, New York, 1976.
12. D. Giulini, E. Joos, C. Kiefer, J. Kumpsch, I. O. Stamatescu, and H. D. Zeh. *Decoherence and the Appearance of a Classical World in Quantum Theory*. Springer, New York-Heidelberg-Berlin, 1996.

# Chapter 6
# On the Quantum Probability Flux Through Surfaces

## 6.1 Introduction

In Born's interpretation of the wave function $\psi_t$ at time $t$ of a single particle of mass $m$, $\rho_t(\mathbf{x}) = |\psi_t(\mathbf{x})|^2$ is the probability density for finding the particle at $\mathbf{x}$ at that time. The consistency of this interpretation is ensured by the continuity equation

$$\frac{\partial \rho_t}{\partial t} + \operatorname{div} \cdot \mathbf{j}^{\psi_t} = 0,$$

where $\mathbf{j}^{\psi_t} = \frac{1}{m} \operatorname{Im} \psi_t^* \nabla \psi_t$ is the quantum current ($\hbar = 1$).

The quantum current is usually not considered to be of any operational significance (see however [1]). It is not related to any standard quantum mechanical measurement in the way, for example, that the density $\rho$, as the spectral measure of the position operator, gives the statistics for a position measurement. Nonetheless, it is hard to resist the suggestion that the quantum current integrated over a surface gives the probability that the particle crosses that surface, i.e., that

$$\mathbf{j}^{\psi_t} \cdot d\mathbf{S} dt \tag{6.1}$$

is the probability that a particle crosses the surface element $d\mathbf{S}$ in the time $dt$. However, this suggestion must be taken "cum grano salis" since $\mathbf{j}^{\psi_t} \cdot d\mathbf{S} dt$ may be somewhere negative, in which case it cannot be a probability. But before discussing the situations where $\mathbf{j}^{\psi_t} \cdot d\mathbf{S} dt$ can be negative we want to consider first a regime for which we can expect this quantity to be positive, so that its meaning could in fact be the crossing probability, namely, the regime described by scattering theory.

## 6.2 Standard Scattering Theory

In textbooks on quantum mechanics the principal objects of interest for scattering phenomena are nonnormalized stationary solutions of the Schrödinger equation with the asymptotic behavior

$$\psi(\mathbf{x}) \overset{x \to \infty}{\sim} e^{i\mathbf{p}_{in}\cdot\mathbf{x}} + f(\theta,\phi)\frac{e^{ipx}}{x},$$

where $e^{i\mathbf{p}_{in}\cdot\mathbf{x}}$ represents an incoming wave, $p = |\mathbf{p}_{in}|$, and $f(\theta,\phi)\frac{e^{ipx}}{x}$ is the scattered wave with angular dependent amplitude. $f(\theta,\phi)$ gives the probability for deflection of the particle in the direction specified by $\theta,\phi$ by the well-known formula for the differential cross section

$$d\sigma = |f(\theta,\phi)|^2 \sin\theta d\theta d\phi \tag{6.2}$$

This representation of a scattering process is, however, not entirely convincing since Born's rule is not directly applicable to non-normalizable wave functions. More important, this picture is entirely time-independent whereas the physical scattering event is certainly a process in space and time. Indeed, according to some experts, the arguments leading to the formula (6.2) for the cross section "wouldn't convince an educated first grader" ([2], p. 97).

It is widely accepted that the stationary treatment is justified by an analysis of wave packets evolving with time. Using a normalized wave packet $\psi_t(\mathbf{x}) = e^{-iHt}\psi(\mathbf{x})$ one immediately obtains by Born's rule the probabilities for position measurements. But what are the relevant probabilities in a scattering experiment? In mathematical physics (e.g. [3], p. 356, and [4]) an answer to this is provided by Dollard's scattering-into-cones theorem [5]:

$$\lim_{t \to \infty} \int_C d^3x |\psi_t(\mathbf{x})|^2 = \int_C d^3p |\widehat{\Omega_-^* \psi}(\mathbf{p})|^2.$$

This connects the asymptotic probability of finding the particle in some cone $C$ with the probability of finding its asymptotic momentum $\mathbf{p}$ in that cone, where $\Omega_- =$ s-$\lim_{t\to\infty} e^{iHt}e^{-iH_0 t}$ is the wave operator ("s-lim" denotes the strong limit), $H = H_0 + V$, with $H_0 = -\frac{1}{2m}\nabla^2$, and $\widehat{\phantom{x}}$ denotes the Fourier transform. It is generally believed that the left hand side of the scattering-into-cones theorem is exactly what the scattering experiment measures, as if the fundamental cross section associated with the solid angle $\Sigma$ (to be identified with a subset of the unit sphere) were

$$\sigma_{\text{cone}}(\Sigma) := \lim_{t\to\infty} \int_{C_\Sigma} d^3x |\psi_t(\mathbf{x})|^2,$$

where $C_\Sigma$ is the cone with apex at the origin subtended by $\Sigma$ (see Fig. 6.1). To connect this with (6.2), which is independent of the details of the initial wave function, one may invoke the right hand side of the scattering-into-cones theorem to recover the

6.2 Standard Scattering Theory

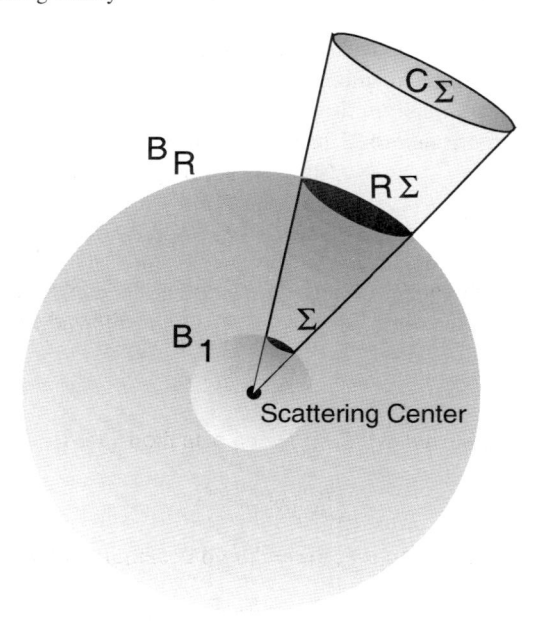

**Fig. 6.1** The geometry of the scattering-into-cones and the flux-across-surfaces theorems

usual formula with additional assumptions on the initial wave packet (see [3] p. 356 for a discussion of this.)

So far the mathematics. But back to physics. The left hand side of the scattering-into-cones theorem is the probability that at some large but fixed time, when the position of the particle is measured, the particle is found in the cone $C$. But does one actually measure in a scattering experiment in what cone the particle happens to be found at some large but *fixed* time? Is it not rather the case that one of a collection of distant detectors surrounding the scattering center fires at some *random* time, a time that is not chosen by the experimenter? And isn't that random time simply the time at which, roughly speaking, the particle crosses the detector surface subtended by the cone?

This suggests that the relevant quantity for the scattering experiment should be the quantum current. If the detectors are sufficiently distant from the scattering center the current will typically be outgoing and (6.1) will be positive. We obtain as the probability that the particle has crossed some distant surface during some time interval the integral of (6.1) over that time interval and that surface. The integrated current thus provides us with a physical definition (see also [6], p. 164) of the cross section:

$$\sigma_{\text{flux}}(\Sigma) := \lim_{R \to \infty} \int_0^\infty dt \int_{R\Sigma} \mathbf{j}^{\psi_t} \cdot d\mathbf{S}, \quad (6.3)$$

where $R\Sigma$ is the intersection of the cone $C_\Sigma$ with the sphere of radius $R$ (see Fig. 6.1). As before, one would like to connect this with the usual formulas and hence we need the counterpart of the scattering-into-cones theorem—the flux-across-surfaces theorem—which provides us with a formula for $\sigma_{\text{flux}}$:

$$\lim_{R\to\infty} \int_0^\infty dt \int_{R\Sigma} \mathbf{j}^{\psi_t} \cdot d\mathbf{S} = \int_{C_\Sigma} d^3 p |\widehat{\Omega_-^* \psi}(\mathbf{p})|^2. \tag{6.4}$$

The fundamental importance of the flux-across-surfaces theorem was first recognized by Combes, Newton and Shtokhamer [7]. To our knowledge there exists no rigorous proof of this theorem, although the heuristic argument for it is straightforward. Let us consider first the "free flux-across-surfaces theorem," where $\psi_t := e^{-iH_0 t}\psi$:

$$\lim_{R\to\infty} \int_0^\infty dt \int_{R\Sigma} \mathbf{j}^{\psi_t} \cdot d\mathbf{S} = \int_{C_\Sigma} d^3 p |\hat{\psi}(\mathbf{p})|^2 \tag{6.5}$$

(This free theorem, by the way, should be physically sufficient, since the scattered wave packet should in any case move almost freely after the scattering has essentially been completed (see also [7]).)

Now the current should contribute to the integral in (6.5) only for large times, because the packet must travel a long time before it reaches the distant sphere at radius $R$. Thus we may use the long-time asymptotics of the free evolution. We split $\psi_t(\mathbf{x}) = (e^{-iH_0 t}\psi)(\mathbf{x})$ into

$$\psi_t(\mathbf{x}) = \left(\frac{m}{2\pi i t}\right)^{3/2} \int d^3 y\, e^{im\frac{|\mathbf{x}-\mathbf{y}|^2}{2t}} \psi(\mathbf{y})$$

$$= \left(\frac{m}{it}\right)^{3/2} e^{im\frac{\mathbf{x}^2}{2t}} \hat{\psi}\left(\frac{m\mathbf{x}}{t}\right)$$

$$+ \left(\frac{m}{it}\right)^{3/2} e^{im\frac{\mathbf{x}^2}{2t}} \int \frac{d^3 y}{(2\pi)^{3/2}} e^{-im\frac{\mathbf{x}\cdot\mathbf{y}}{t}} (e^{im\frac{\mathbf{y}^2}{2t}} - 1)\psi(\mathbf{y}).$$

Since $(e^{im\frac{\mathbf{y}^2}{2t}} - 1) \to 0$ as $t \to \infty$, we may neglect the second term, so that as $t\to\infty$ we have that

$$\psi_t(\mathbf{x}) \sim \left(\frac{m}{it}\right)^{3/2} e^{im\frac{\mathbf{x}^2}{2t}} \hat{\psi}\left(\frac{m\mathbf{x}}{t}\right). \tag{6.6}$$

(This asymptotics has long been recognized as important for scattering theory, e.g. [5, 8].) From (6.6) we now find that

$$\mathbf{j}^{\psi_t}(\mathbf{x}) = \frac{1}{m}\mathrm{Im}\,\psi_t^*(\mathbf{x})\nabla\psi_t(\mathbf{x}) \approx \frac{\mathbf{x}}{t}\left(\frac{m}{t}\right)^3 |\hat{\psi}\left(\frac{m\mathbf{x}}{t}\right)|^2. \tag{6.7}$$

(Note that by (6.7) the current is strictly radial for large times, so that $\mathbf{j}^{\psi_t} \cdot d\mathbf{S}$ is indeed positive.)

Using now the approximation (6.7) and substituting $\mathbf{p} := m\frac{\mathbf{x}}{t}$ we readily arrive at

## 6.3 Near Field Scattering

$$\int_0^\infty dt \int_{R\Sigma} \mathbf{j}^{\psi_t} \cdot d\mathbf{S} \approx \int_0^\infty dt \int_{R\Sigma} (\frac{m}{t})^3 |\hat{\psi}(\frac{m\mathbf{x}}{t})|^2 \frac{\mathbf{x}}{t} \cdot d\mathbf{S}$$
$$= \int_0^\infty dp p^2 \int_\Sigma d\sigma |\hat{\psi}(\mathbf{p})|^2 = \int_{C_\Sigma} d^3p |\hat{\psi}(\mathbf{p})|^2.$$

This heuristic argument for the free flux-across-surfaces theorem (6.5) is so simple and intuitive that one may wonder why it does not appear in any primer on scattering theory. (For a rigorous proof see [9]).

To arrive at the general result (6.4) one may use the fact that the long time behavior of $\psi_t(\mathbf{x}) := e^{-iHt}\psi(\mathbf{x})$ is governed by $e^{-iH_0 t} \Omega_-^* \psi$ (see, e.g., [5]) so that the asymptotic current is simply

$$\mathbf{j}^{\psi_t}(\mathbf{x}) = \mathrm{Im}\psi_t^*(\mathbf{x}) \nabla \psi_t(\mathbf{x}) \approx \frac{\mathbf{x}}{t}(\frac{m}{t})^3 |\widehat{\Omega_-^* \psi}(\frac{m\mathbf{x}}{t})|^2,$$

yielding (6.4).

## 6.3 Near Field Scattering

We turn now to a much more subtle question (see also [10]): What happens if we place the detectors *not* too distant from the scattering center and prepare the wave function near the scattering center, i.e., what happens if we do not take the limit $R \to \infty$ so central to scattering theory? The detectors will of course again fire at some random time and position, but what now of the statistics? This question is not quite as innocent as it sounds; it concerns in fact one of the most debated problems in quantum theory: what we are considering here is the problem of time measurement, specifically the problem of escape time (and position at such time) of a particle from a region $G$. It is well known that there is no self-adjoint time observable of any sort and there is a huge and controversial literature on this and on what to do about it. (See [11]–[22] and references therein.) Note also that since the exit position is the position of the particle at a random time, it cannot be expressed as a Heisenberg position operator in any obvious way.

The obvious answer (see [22, 23] for a one-dimensional version) is, of course, provided by (6.1), provided that the boundary of $G$ is crossed at most once by the particle (whatever this is supposed to mean for a quantum particle), so that every crossing of the boundary of $G$ is a first crossing, and provided of course that (6.1) is nonnegative.[1] Notice that the preceding provisos might well be expected to be intimately connected. We thus propose that (6.1) indeed gives the first exit statistics whenever the following current positivity condition (a condition on both the wave function and on the surface)

---

[1] The wave function $\psi_t$ in (6.1) should of course be understood as referring to the Schrödinger evolution with no detectors present.

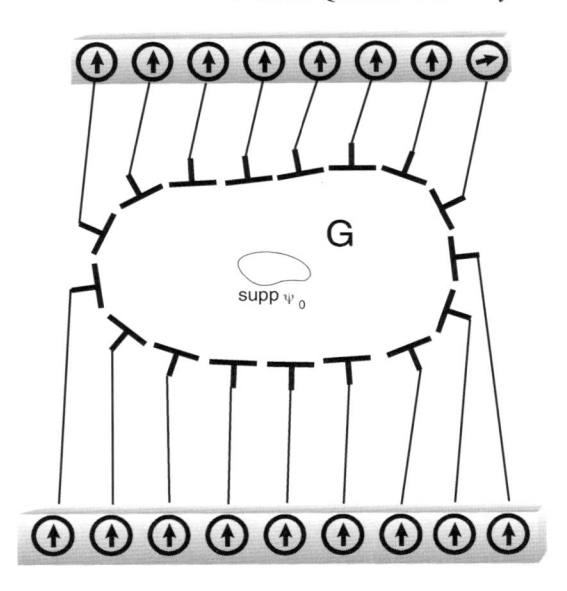

**Fig. 6.2** Escape experiment: A region $G$ is defined by an array of detectors, which surround a smaller region, supp $\psi_0$, in which a particle's wave function is initially localized. The detectors record the time at which they fire. Typically only one of the detectors will fire, and the position of this detector yields the measured exit position

$$\text{CPC}: \quad \forall t > 0 \quad \text{and} \quad \forall \mathbf{x} \in \text{boundary of the region } G$$
$$\mathbf{j}^{\psi_t}(\mathbf{x}, t) \cdot d\mathbf{S} > 0$$

is satisfied.

We predict that the statistics given by (6.1) will (approximately) be obtained in an experiment on an ensemble of particles prepared with (approximately) CPC wave function $\psi$ which is initially well localized in some region $G$ whenever the detectors around the boundary of $G$ (see Fig. 6.2) are *sufficiently passive*, a condition that needs to be more carefully delineated but which should widely be satisfied. As to how widely the CPC is satisfied, this is not easy to say. We do note, however, that since whether or not it is satisfied depends upon the region $G$ upon which we focus and around which we place our detectors, it may often be possible to suitably adjust the region $G$ so that the CPC becomes satisfied, at least approximately, even if the CPC fails to be satisfied for our original choice of $G$.

A simple example of a situation where the CPC does hold and where one may easily compute the exit-time statistics is the following. A spherically symmetric Gaussian wave packet, with initial width $\sigma$, which is initially located at the center of $G$, a sphere with radius $R$, evolves freely. One readily finds for the exit time probability density $\rho(t) := \int \mathbf{j}^{\psi_t} \cdot d\mathbf{S}$ that

$$\rho(t) = \frac{1}{\sqrt{8\pi}} \frac{R^3}{m^2 \sigma^7} \left[1 + \left(\frac{t}{2m\sigma^2}\right)^2\right]^{-\frac{5}{2}} e^{-\frac{R^2}{2\sigma^2}\left[1+\left(\frac{t}{2m\sigma^2}\right)^2\right]}.$$

Of course, some important questions remain: The expression (6.1) is not a probability for all wave functions—so what if anything does it physically represent in general? And what in the general case is the formula for the first exit statistics?

We stress again that the prediction (6.1) for the exit statistics is not of the standard form, as given by the quantum formalism, since it is not concerned with the measurement of an operator as observable.[2] However, no claim is made that the expression (6.1) and its interpretation cannot also be arrived at from standard quantum mechanics—it presumably can—e.g., by including the measurement devices in the quantum mechanical analysis. (See however [11]–[13].) After all, though there is no standard quantum observable (i.e., self-adjoint operator) to directly describe the escape time, the "pointer variable" for the detectors *is* a standard quantum observable, whose probability distribution after the experiment can in principle be computed in the standard way.

In the next section we shall explain how the current as the central object for escape and scattering phenomena arises naturally within Bohmian mechanics, where the physical meaning of (6.1) turns out to be the measure for the expected number of *signed* crossings, which of course can be negative.

## 6.4 Bohmian Mechanics

In Bohmian mechanics a particle moves along a trajectory $\mathbf{X}(t)$ determined by

$$\frac{d}{dt}\mathbf{X}(t) = \mathbf{v}^{\psi_t}(\mathbf{X}(t)) = \frac{1}{m}\text{Im}\frac{\nabla \psi_t}{\psi_t}(\mathbf{X}(t)), \tag{6.8}$$

where $\psi_t$ is the particle's wave function, evolving according to Schrödinger's equation. Moreover, if an ensemble of particles with wave function $\psi$ is prepared, the positions $\mathbf{X}$ of the particles are distributed according to the quantum equilibrium measure $\mathbb{P}$ with density $\rho = |\psi|^2$ ($\psi$ normalized) [24].

In particular, the continuity equation for the probability shows that the probability flux $(|\psi_t|^2, |\psi_t|^2 \mathbf{v}^{\psi_t})$ is conserved, since $|\psi_t|^2 \mathbf{v}^{\psi_t} = \mathbf{j}^{\psi_t}$.

Hence, given $\psi_t$, the solutions $\mathbf{X}(\mathbf{X}_0, t)$ of Eq. (6.8) are random trajectories, where the randomness comes from the $\mathbb{P}$-distributed random initial position $\mathbf{X}_0$, $\psi$ being the initial wave function.

Consider now, at time $t = 0$, a particle with wave function $\psi$ localized in some region $G \subset \mathbb{R}^3$ with smooth boundary. Consider the number $N(dS, dt)$ of crossings by $\mathbf{X}(t)$ of the surface element $dS$ of the boundary of $G$ in the time $dt$ (see Fig. 6.3).

---

[2] Nor are they given by a positive-operator-valued measure (POV), which has been proposed as a generalized quantum observable, see [25].

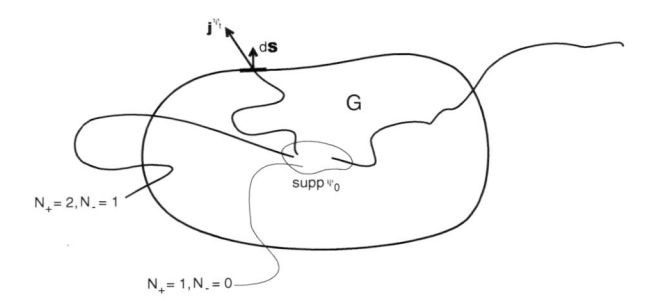

**Fig. 6.3** In Bohmian mechanics the flow lines of the current represent the possible trajectories of the Bohmian particle. Some Bohmian trajectories leaving $G$ are drawn (for the Schrödinger evolution without detectors, see footnote 1)

Splitting $N(dS, dt) =: N_+(dS, dt) + N_-(dS, dt)$, where $N_+(dS, dt)$ denotes the number of outward crossings and $N_-(dS, dt)$ the number of backward crossings of $dS$ in time $dt$, we define the number of signed crossings by $N_s(dS, dt) := N_+(dS, dt) - N_-(dS, dt)$.

We can now compute the expectation values with respect to the probability $\mathbb{P}$ of these numbers in the usual statistical mechanics manner. Note that for a crossing of $dS$ in the time interval $(t, t + dt)$ to occur, the particle has to be in a cylinder of size $|\mathbf{v}^{\psi_t} dt \cdot d\mathbf{S}|$ at time $t$. Thus we obtain for the expectation value

$$\mathbb{E}(N(dS, dt)) = |\psi_t|^2 |\mathbf{v}^{\psi_t} dt \cdot d\mathbf{S}| = |\mathbf{j}^{\psi_t} \cdot d\mathbf{S}| dt,$$

and similarly $\mathbb{E}(N_s(dS, dt)) = \mathbf{j}^{\psi_t} \cdot d\mathbf{S} dt$.

If we further introduce the random variables $T_e$, the first exit time from $G$, $T_e := \inf\{t \geq 0 | \mathbf{X}(t) \notin G\}$, and $\mathbf{X}_e$, the position of first exit, $\mathbf{X}_e = \mathbf{X}(T_e)$, we obtain a very natural and principled explanation of what we arrived at in a heuristic and suggestive manner in our treatment of scattering theory and the statistics of the first exit time and position. For Bohmian mechanics the CPC implies that every trajectory crosses the boundary of $G$ at most once, and in this case we have

$$\mathbb{E}(N(dS, dt)) = \mathbb{E}((N_s(dS, dt)) =$$
$$0 \cdot \mathbb{P}(T_e \notin dt \text{ or } \mathbf{X}_e \notin dS) + 1 \cdot \mathbb{P}(\mathbf{X}_e \in dS \text{ and } T_e \in dt)$$

and we find for the joint *probability* of exit through $dS$ in time $dt$

$$\mathbb{P}(\mathbf{X}_e \in dS \text{ and } T_e \in dt) = \mathbf{j}^{\psi_t} \cdot d\mathbf{S} dt. \quad (6.9)$$

In principle one could compute the first exit statistics also when the CPC fails to be satisfied. These are in fact given by the same formula (6.9) as before, provided one replaces $\mathbf{j}^{\psi_t}$ by the truncated probability current $\tilde{\mathbf{j}}$ arising from killing the particle when it reaches the boundary of $G$. This is simply given, on the boundary of $G$, by

$$\tilde{\mathbf{j}}^{\psi_t}(t,\mathbf{x}) = \begin{cases} \mathbf{j}^{\psi_t}(\mathbf{x}) & \text{if } (t,\mathbf{x}) \text{ is a first exit from } G \\ 0 & \text{otherwise} \end{cases} \quad (6.10)$$

where $(t, \mathbf{x})$ is a first exit from $G$ if the Bohmian trajectory passing through $\mathbf{x}$ at time $t$ leaves $G$ at this time, for the first time since $t = 0$. Thus, we have generally that

$$\mathbb{P}(\mathbf{X}_e \in dS \text{ and } T_e \in dt) = \tilde{\mathbf{j}}^{\psi_t} \cdot d\mathbf{S} dt. \quad (6.11)$$

However, there is an important difference between the CPC probability formula (6.9), involving the usual current, and the formula (6.11), involving the truncated current. The usual current is well defined in orthodox quantum theory, even if it is true, as we argue, that its full significance can only be appreciated from a Bohmian perspective. The truncated current cannot even be *defined* without reference to Bohmian mechanics, since whether or not $(t, \mathbf{x})$ is a first exit from $G$ depends upon the full and detailed trajectory up to time $t$. (In particular, a different choice of dynamics, as for example given by stochastic mechanics [26, 27], would yield a different truncated current. It is natural to wonder whether the truncated current given by Bohmian mechanics provides in the general case the best fit to the measured escape statistics expressible without reference to the measuring apparatus.)

Finally, we note that in the context of scattering theory our definition (6.3) of $\sigma_{\text{flux}}$ captures exactly what it should once one has real trajectories, namely the asymptotic probability distribution of exit positions,

$$\sigma_{\text{flux}}(\Sigma) = \lim_{R \to \infty} \mathbb{P}(\mathbf{X}_e \in R\Sigma).$$

This follows from the fact that the expected number of backward crossings of the sphere of radius $R$ vanishes as $R \to \infty$ (see [9]).

# References

1. Y. Aharonov and L. Vaidman. Measurement of the Schrödinger Wave of a Single Particle. *Physical Letters A*, 178:38–42, 1993.
2. B. Simon. *Quantum Mechanics for Hamiltonians Defined as Quadratic Forms*. Princeton University Press, Princeton, New Jersey, 1971.
3. M. Reed and B. Simon. *Methods of Modern Mathematical Physics III*. Academic Press, New York, 1979.
4. V. Enss and B. Simon. Finite Total Cross-Sections in Nonrelativistic Quantum Mechanics. *Communications in Mathematical Physics*, 76:177–209, 1980.
5. J. D. Dollard. Scattering into Cones i: Potential Scattering. *Communications in Mathematical Physics*, 12:193–203, 1969.
6. R. G. Newton. *Scattering Theory of Waves and Particles*. Springer, New York, 1982.
7. J.-M. Combes, R. G. Newton, and R. Shtokhamer. Scattering Into Cones and Flux Across Surfaces. *Physical Review D*, 11:366–372, 1975.
8. W. Brenig and R. Haag. Allgemeine Quantentheorie der Stoßprozesse. *Fortschritte der Physik*, 7:183–242, 1959.
9. M. Daumer, D. Dürr, S. Goldstein, and N. Zanghì. On the Flux-Across-Surfaces Theorem. *Letters in Mathematical Physics*, 38(103–116), 1996.

10. M. Daumer, D. Dürr, S. Goldstein, and N. Zanghì. Scattering and the Role of Operators in Bohmian Mechanics. In M. Fannes, C. Maes, and A. Verbeure, editors, *On Three Levels: The Micro-, Meso-, and Macroscopic Approaches in Physics*, volume 324 of NATO ASI Series B: Physics, pages 331–338. Plenum Press, 1994.
11. P. Busch, P. Lahti, and P. Mittelstaedt. *The Quantum Theory of Measurement*. Springer, New York-Heidelberg-Berlin, 1991.
12. R. Werner. Screen Observables in Relativistic and Nonrelativistic Quantum Mechanics. *Journal of Mathematical Physics*, 27:793–803, 1986.
13. G. R. Allcock. The Time of Arrival in Quantum Mechanics I-III. *Annals of Physics*, 53:253–348, 1969.
14. M. Büttiker and R. Landauer. Traversal Time for Tunneling. *Physical Review Letters*, 49:1739–1742, 1982.
15. W. Pauli. In S. Flügge, editor, *Encyclopedia of Physics*, volume 60. Springer, New York-Heidelberg-Berlin, 1958.
16. R. Landauer and Th. Martin. Barrier Interaction Time in Tunneling. *Reviews of Modern Physics*, 66:217–228, 1994.
17. E. H. Hauge and J. A. Støvneng. Tunneling Times: A Critical Review. *Reviews of Modern Physics*, 61:917–936, 1989.
18. B. Misra and E. C. G. Sudarshan. The Zeno's Paradox in Quantum Theory. *Journal of Mathematical Physics*, 18:756–763, 1977.
19. A. M. Steinberg. How Much Time Does a Tunneling Particle Spend in the Barrier Region? *Physical Review Letters*, 74:2405–2409, 1995.
20. H. Ekstein and A. J. F. Siegert. On a Reinterpretation of Decay Experiments. *Annals of Physics*, 68:509–520, 1971.
21. Ph. Martin. Time Delay of Quantum Scattering Processes. *Acta Physica Austriaca, Suppl. XXIII*, pages 157–208, 1981.
22. C. R. Leavens. Transmission, Reflection and Dwell Times Within Bohm's Causal Interpretation of Quantum Mechanics. *Solid State Communications*, 74:923–928, 1990.
23. J. T. Cushing. Quantum Tunneling Times: A Crucial Test for the Causal Program? *Foundations of Physics*, 25:269–280, 1995.
24. D. Dürr, S. Goldstein, and N. Zanghì. Quantum Equilibrium and the Origin of Absolute Uncertainty. *Journal of Statistical Physics*, **67**:843–907, 1992.
25. E. B. Davies. *Quantum Theory of Open Systems*. Academic Press, London-New York-San Francisco, 1976.
26. E. Nelson. *Quantum Fluctuations*. Princeton University Press, Princeton, N.J., 1985.
27. S. Goldstein. Stochastic Mechanics and Quantum Theory. *Journal of Statistical Physics*, 47:645–667, 1987.

# Chapter 7
# On the Weak Measurement of Velocity in Bohmian Mechanics

## 7.1 Introduction

According to the uncertainty principle, it is impossible to simultaneously measure both the position and the velocity of a quantum particle, at least not to arbitrary accuracy. The basic reason for this limitation is that in quantum mechanics a measurement of the position of a particle to a given accuracy produces a corresponding narrowing of its wave function and hence a corresponding increase in the uncertainty about its velocity. Moreover, this is as true of Bohmian mechanics, a version of quantum mechanics in which a particle always has a velocity as well as a position, as it is of orthodox quantum theory, in which it does not. This suggests that in order to measure the velocity of a Bohmian particle at a given position, it might be good to exploit a measurement procedure that somehow does not significantly affect the wave function of the particle.

Such a procedure, a so-called *weak measurement*, has been developed by Aharonov, Albert, and Vaidman [1]. And Howard Wiseman has indeed proposed in a recent article [2] that weak measurements be used to measure the velocity of a particle at a given position.

More precisely, Wiseman invokes the theory of weak measurements to provide an "*operational definition*" for the velocity for a particle at position $x$"[1]:

$$v(x) \equiv \lim_{\tau \to 0} \tau^{-1} \, \mathrm{E}[x_{\text{strong}}(\tau) - x_{\text{weak}} | x_{\text{strong}}(\tau) = x]. \tag{7.1}$$

He then observes that this quantity is precisely the velocity that defines Bohmian mechanics and uses this fact to respond to some objections that have been raised against it.

In this formula $x_{\text{weak}}$ and $x_{\text{strong}}$ denote respectively the results of a weak measurement of the position of the particle at some time, say $t = 0$, and a strong measurement of the position a short time $\tau$ later. The expectation symbol E in the formula refers

---

[1] For ease of reading, in this chapter we do not use the boldface notation for three-dimensional vectors.

to the average over a large ensemble of systems, all prepared in the same initial state $\psi$ at time 0, and for all of which the result of a strong measurement of position at time $\tau$, following the weak measurement at time 0, is $x$.

A strong measurement of an observable $\hat{A}$ is just a standard quantum measurement of the observable—one which collapses the wave function of the system involved to the eigenstate of $\hat{A}$ corresponding to the eigenvalue found in the measurement. The average of such values for a large ensemble of systems in the state $\psi$ is, of course, $\langle\psi|\hat{A}|\psi\rangle$. In contrast, a weak measurement of $\hat{A}$ (which will be described in more detail in the next section) does not collapse the wave function of the system, and in fact is such that the change in the wave function that is produced by the procedure can be made arbitrarily small. The price to be paid for this desirable feature is that very little information about the system is obtained in a single such measurement, the result found reflecting mainly the effect of noise introduced by the procedure rather than any property of the system itself. Nonetheless, the ensemble average for such a procedure is $\langle\psi|\hat{A}|\psi\rangle$, just as for a strong measurement.

Weak measurements are most interesting when combined with post-selection: Consider the subensemble for which, after the weak measurement, the system is found in state $\varphi$ at time $\tau > 0$. When $\tau = 0+$, the average over this subensemble of the result of the weak measurement at time 0 is of course still $\langle\psi|\hat{A}|\psi\rangle$ when $\varphi = \psi$, and in general is given by the so-called *weak value* [1][2]

$$_{\langle\varphi|}\!\left\langle\hat{A}_w\right\rangle_{|\psi\rangle} = \mathrm{Re}\frac{\langle\varphi|\hat{A}|\psi\rangle}{\langle\varphi|\psi\rangle}. \tag{7.2}$$

For $\tau > 0$ the value of the subensemble average of course involves the unitary evolution operator $U(\tau)$ for time $\tau$ and is given by the weak value

$$_{\langle\varphi|U(\tau)}\!\left\langle\hat{A}_w\right\rangle_{|\psi\rangle} = \mathrm{Re}\frac{\langle\varphi|U(\tau)\hat{A}|\psi\rangle}{\langle\varphi|U(\tau)|\psi\rangle}. \tag{7.3}$$

In terms of this, the velocity definition (7.1) becomes

$$v(x) = \lim_{\tau\to 0} \tau^{-1}\left[x - \mathrm{Re}\frac{\langle x|U(\tau)\hat{X}|\psi\rangle}{\langle x|U(\tau)|\psi\rangle}\right], \tag{7.4}$$

where $\hat{X}$ is the position operator of the particle. One easily computes with $U(\tau) = \exp(-i\hat{H}\tau/\hbar)$ and $\hat{H} = \hat{p}^2/2m + V(\hat{x})$ that (7.4) becomes

$$v(x) = v^\psi(x) \equiv \frac{j^\psi(x)}{|\psi(x)|^2} \tag{7.5}$$

---

[2] Our usage here is that of Wiseman [2]. It is a bit different from that of [1], which refers to the ratio following "Re" in equation (7.2), which could be complex, as the weak value.

## 7.1 Introduction

with

$$j^\psi(x) = (\hbar/m)\text{Im } \psi(x)\nabla\psi(x), \tag{7.6}$$

the usual quantum flux. (7.5) is the expression for the Bohmian velocity:

$$\dot{X}(t) = v^\psi(X(t), t) = \frac{j^\psi(X(t); t)}{|\psi(t, X(t))|^2}. \tag{7.7}$$

This equation, together with Schrödinger's equation for the wave function, is the defining dynamical equation of Bohmian mechanics for a single particle, with a similar equation for the Bohmian mechanics of a many-particle system.

Wiseman does not claim, either in the above quotation or anywhere else in his article, that his weak measurement procedure, providing an "operational definition for the velocity," actually *measures* the Bohmian velocity. There is a good reason for this: There are variants of Bohmian mechanics based on velocity formulas different from (7.5, 7.6) that yield theories empirically equivalent to Bohmian mechanics. The existence of a procedure to measure the velocity would seem to contradict this empirical equivalence.

We elaborate. Recall the quantum continuity equation

$$\partial_t |\psi(x, t)|^2 = -\text{div } j^\psi(x, t). \tag{7.8}$$

From this equation $j^\psi$ is not uniquely defined: A divergence free vector can be added without affecting the continuity equation. In [3] some "physically reasonable" additions are discussed, giving rise to (empirically equivalent) variants of Bohmian mechanics which have different velocity fields (7.7) with $j^\psi$ replaced by the new $j'^\psi$. With respect to this Wiseman [2] describes his findings as follows:

> ... a particular **j** is singled out if one requires that **j** be determined *experimentally* as a *weak value*, using a technique that would make sense to a physicist with no knowledge of quantum mechanics. This "naively observable" **j** seems the most natural way to define **j** operationally. Moreover, I show that this operationally defined **j** equals the standard **j**, so, assuming $\dot{\mathbf{x}} = \mathbf{j}/P$ one obtains the dynamics of BM. It follows that the possible Bohmian paths are naively observable from a large enough ensemble.

Notice that Wiseman claims only that the Bohmian paths (or Bohmian velocities) are "naively observable," but not that they are *genuinely* observable. He claims not that the current (and the velocity associated with it) found in his procedure is ipso facto the actual current, but only that it is "the most natural way to define **j** operationally." In short, Wiseman does not claim that his procedure, which we shall call a "weak measurment of velocity," provides a genuine measurement of velocity.

But, as we shall argue, it does—despite the apparent contradiction. In more detail, we shall be concerned in this chapter with the following statements:

1. A "weak measurement of velocity" in Bohmian mechanics is, in a reasonable sense, a *genuine* measurement of velocity.
2. The same thing is true for the variants of Bohmian mechanics based on a velocity formula different from the Bohmian one mentioned above.

3. Bohmian mechanics and the variants referred to in (2) are empirically equivalent to each other—and to standard quantum mechanics. In particular, for all of them the result of a "weak measurement of velocity" is given by the Aharonov-Albert-Vaidman formula given above, and hence by the formula for velocity in Bohmian mechanics.
4. It is impossible to measure the velocity in Bohmian mechanics (See Sect. 3.7.2).

There is of course an obvious contradiction between the first three of these statements. A genuine measurement of velocity must reveal the velocity and hence could be used to empirically distinguish the theories based on different velocity formulas. And if a theory is based on a velocity formula different from the Bohmian one, a genuine velocity measurement for the theory can't yield the Bohmian velocity.

At least one of these three statements must be false. However (3) is well established, and true. (The reason for this is basically that in Bohmian mechanics and its variants the statistics for the results of experiments—including weak measurements—are determined by the same $|\Psi|^2$ probabilities as for orthodox quantum theory.) In Sect. 7.2 we shall show that (1) is also correct. We shall do this by an analysis that seems to apply to the variants of Bohmian mechanics referred to in (2), as well as to Bohmian mechanics itself. Thus in Sect. 7.2 it shall seem as if we establish (2) as well as (1). In Sect. 7.3 we shall explain why the analysis in Sect. 7.2 is in fact incorrect for the alternatives to Bohmian mechanics, so that (2) is not established by the analysis yielding (1)—a good thing since (2) is false.

We shall also examine, in Sect. 7.4, the crucial condition responsible for the success of Bohmian mechanics here, showing directly that this condition indeed uniquely characterizes Bohmian mechanics. Finally, in Sect. 7.5, we address the apparent contradiction between statements (1) and (4).

## 7.2 Bohmian Analysis of Weak Measurement of Velocity

We now consider a Bohmian particle with wave function $\psi$ at time 0. We model the measurement apparatus which measures weakly the position of the Bohmian particle by a pointer and we denote the actual Bohmian pointer position by $Y$. We denote by $X(t)$ the position of the Bohmian particle. We measure $X = X(0)$ weakly at time 0 and very shortly after that, at time $\tau$, we perform a strong measurement of $X(\tau)$.

Let us spell out what this means. Let $\Phi = \Phi(y)$ be the wave function of the apparatus in its ready state, with the pointer-variable $Y$ centered at $Y = 0$: $\Phi$ is a real wave packet of spread $\sigma$—i.e., such that $\Phi(y) = \phi(y/\sigma)$ with $\phi$ fixed as $\sigma \to \infty$—for which the expected value of $Y$ vanishes,[3]

$$\int dy\, y\, |\Phi(y)|^2 = 0, \tag{7.9}$$

---

[3] This condition on the initial apparatus state $\Phi$, assumed for our analysis of the weak measurement of velocity in Bohmian mechanics, is weaker than what is in general needed for the result of a weak measurement, with post-selection and averaging, to be given by (7.2), namely that $\phi$ be real and even, or real and odd.

## 7.2 Bohmian Analysis of Weak Measurement of Velocity

for example,

$$\Phi(y) \sim e^{-\frac{y^2}{4\sigma^2}}. \tag{7.10}$$

The weak measurement begins at time 0 with an interaction between system and apparatus that leads to the following instantaneous transition from initial quantum state to final quantum state:

$$\psi(x)\Phi(y) \to \psi(x)\Phi(y-x) \tag{7.11}$$

(corresponding in ket notation to

$$\int dx\, \psi(x)|x\rangle|\Phi\rangle \to \int dx\, \psi(x)|x\rangle|\Phi\rangle_x \tag{7.12}$$

where $|\ \rangle_x$ indicates translation by $x$). Immediately after this, the pointer position is measured and recorded. When the measured value is $Y$, (up to normalization) the system wave function after the measurement is, by the projection postulate,

$$\psi_{0+}(x) = \psi_Y(x) \equiv \psi(x)\Phi(Y-x). \tag{7.13}$$

Note that since the result $Y$ is random, with probability distribution

$$\rho^Y(y) = \int dx |\psi(x)|^2 |\Phi(y-x)|^2, \tag{7.14}$$

the system wave function $\psi_{0+}(x)$ is random as well.

For a standard von Neumann measurement of position, the spread $\sigma$ of the apparatus wave function $\Phi$ is taken very small so that the wave function (7.13) is an approximate eigenstate of the position operator concentrated near the value $x = Y$. But in a weak measurement the pointer wave function $\Phi(y)$ is very spread out ($\sigma$ is very large), varying on the scale $\sigma$, whereas $\psi(x)$ varies on a scale of order unity (near say 0). We thus should have from (7.13) that

$$\psi_Y(x) \approx \Phi(Y)\psi(x) \tag{7.15}$$

and that up to normalization

$$\psi_{0+}(x) \approx \psi(x), \tag{7.16}$$

with small error, of order $1/\sigma$. Although a single weak measurement does not measure the actual position of the particle, by averaging over a large sample of identical experiments one obtains information about the mean value of position; we have, observing (7.10)

$$\mathbb{E}(Y) \equiv \int y\rho^Y(y)dy = \int x\rho^X(x)dx \equiv \mathbb{E}(X)$$

where $\rho^X(x) = |\psi(x)|^2$.

The conditional probability density of $Y$ given $X = x$ is

$$\rho^Y(y|X=x) = \frac{\rho^{X,Y}(x,y)}{\rho^X(x)} = \frac{|\psi(x)|^2|\Phi(y-x)|^2}{|\psi(x)|^2} = |\Phi(y-x)|^2, \quad (7.17)$$

and hence in a weak measurement

$$\mathbb{E}(Y|X=x) \equiv \int y\rho^Y(y|X=x) = x. \quad (7.18)$$

The "weak measurement of velocity" is completed by performing at time $\tau$, on each member of the ensemble, a (strong) measurement of the position of the particle, and taking the conditional average indicated by (7.1). Conditioning on the event that $X(\tau) = x$, the Bohmian version of this conditional average is

$$\lim_{\tau \to 0} \frac{1}{\tau} \mathbb{E}(x - Y|X(\tau) = x) = \lim_{\tau \to 0} \frac{1}{\tau} (x - \mathbb{E}(Y|X(\tau) = x)). \quad (7.19)$$

To see how the Bohmian velocity comes in we note, writing $X$ for $X(0)$, that

$$X(\tau) \approx X + v^{\psi_{0+}}\tau. \quad (7.20)$$

When $\tau \to 0$ the error of this approximation is of smaller order than $\tau$. By (7.16) we have that

$$v^{\psi_{0+}} \approx v^\psi(x) \quad (7.21)$$

and hence that

$$X(\tau) \approx X + v^\psi(X)\tau \approx X + v^\psi(X(\tau))\tau. \quad (7.22)$$

With this approximation we can identify the event $X(\tau) = x$ with the event $X = x - v^\psi(x)\tau$. Therefore by (7.18)

$$\mathbb{E}(Y|X(\tau) = x) \approx \mathbb{E}(Y|X = x - v^\psi(x)\tau) = x - v^\psi(x)\tau \quad (7.23)$$

and thus for (7.19) we obtain

$$\lim_{\tau \to 0} \frac{1}{\tau} (x - \mathbb{E}(Y|X(\tau) = x)) \approx v^\psi(x), \quad (7.24)$$

with the approximation becoming exact for a weak measurement, i.e., in the limit $\sigma \to \infty$.

The details of this analysis, in particular equations (7.18), (7.22) and (7.24), show that in a "weak measurement of velocity" in Bohmian mechanics the result of the averaging is $v^\psi$ precisely because the Bohmian particle had velocity $v^\psi$. We are thus justified in asserting that for Bohmian mechanics this procedure of weak measurement genuinely measures the Bohmian velocity. Now we ask: What is specifically

"Bohmian" in formulas (7.18), (7.22) and (7.24)? The answer, it would seem, is nothing. These formulas seem to hold for all variants of Bohmian mechanics which have differentiable paths. But that is incompatible with the (correct) weak measurement formula (7.4) that yields the Bohmian velocity. But where is the mistake? The answer is in fact not easy to find and lies in scrutinizing more carefully weak measurements.

## 7.3 A More Careful Analysis

Now here is the catch: Since (7.16) is only approximately satisfied for $\sigma$ large, where $\sigma$ is the spread of $\Phi(y)$, we cannot in general dismiss the possibility that $v$ depends also on $Y$. The $Y$ dependence of the conditional wave function $\psi_{0+}(x) = \psi(x)\Phi(Y - x)$ yields in general a $Y$ dependence of the induced velocity field.[4] Therefore after the weak measurement we have truthfully now $v^{\psi_{0+}} = v(x, Y)$ and thus for (7.20) we have

$$X(\tau) \approx X + v(X(\tau), Y)\tau. \qquad (7.25)$$

as a better approximation than (7.22), exact to order $\tau$, and the left hand sides of (7.24) becomes

$$\lim_{\tau \to 0} \frac{1}{\tau} (x - \mathbb{E}(Y|X = x - v(x, Y)\tau)). \qquad (7.26)$$

This expression is not anymore easy to handle and in general it is not equal to $v^\psi$. Before entering into more details we make the trivial observation that if the weak measurement were such that $v(x, Y) = v^\psi(x)$, i.e., if

$$v^{\psi \Phi_y} = v^\psi, \qquad (7.27)$$

where $\Phi_y(x) = \Phi(y - x)$, the analysis of Sect. 7.2 would be correct and the "weak measurement of velocity" would indeed be a genuine measurement of velocity yielding the result $v^\psi$. When (7.27) is satisfied, as is the case for Bohmian mechanics, we have in fact by (7.17) that, to order $\tau$,

$$\rho^Y(y|X(\tau) = x) = \rho^Y(y|X = x - v^\psi(x)\tau) = |\Phi|^2 \left(y - [x - v^\psi(x)\tau]\right). \qquad (7.28)$$

With a variant of Bohmian mechanics, however, the velocity need not (in fact, will not, see below) obey (7.27) exactly. Rather (7.27) will hold only approximately, presumably with an error of order $1/\sigma$ since $\Phi$ varies on scale $\sigma$. At first sight one

---

[4] We note that the $Y$ dependence of the velocity of the particle can be simply computed from the velocity formula which the Bohmian type theory provides given the wave function of the entire system consisting of particle and apparatus. There is thus no need to introduce the conditional wave function. The conditional wave function focuses however on the source of the $Y$-dependence: The weak measurement does affect the wave function—if only a tiny bit. That tiny bit changes the velocity a tiny bit, having possibly a big effect in weak measurements.

might be inclined to ignore this error. However one must be careful here, since in a weak measurement a large quantity (here $Y$, of order $\sigma$) is averaged to yield (because of near-perfect cancellation) a result of order unity. And a small change in the probability distribution involved could lead to an effect of order unity as well.

Indeed, an order $1/\sigma$ error in (7.27) would be expected to yield an additional contribution to (7.28) of order $\tau/\sigma$, yielding a contribution of order $\tau$ to $\mathbb{E}(Y|X(\tau) = x)$ (since $Y$ is of order $\sigma$), and hence a contribution of order unity in (7.19). These expectations are correct.

Indeed, writing now $v_B^\psi$ for the velocity (7.5) in Bohmian mechanics, we have that (to order $\tau$)

$$\rho^Y(y|X(\tau) = x) = |\Phi|^2 \left(y - [x - v_B^\psi(x)\tau]\right)$$
$$= |\Phi|^2 \left(y - [x - v^\psi(x)\tau] - [v^\psi(x) - v_B^\psi(x)]\tau]\right) \quad (7.29)$$
$$\approx |\Phi|^2 \left(y - [x - v^\psi(x)\tau]\right) + (v^\psi(x) - v_B^\psi(x))\tau \cdot \nabla_x |\Phi|^2 (y - x)$$

and since $\Phi(y)$ varies on scale $\sigma$ the second term on the last line is of order $\tau/\sigma$. Thus whenever the velocity $v^\psi$ is non-Bohmian, and in particular whenever (7.27) is not obeyed, the conditional distribututon of $Y$ given $X(\tau)$ is sufficiently affected so as to vitiate the analysis of Sect. 7.2. When the velocity is non-Bohmian the "weak measurement of velocity" fails because of these errors to be a genuine measurement of velocity.

## 7.4 Bohmian Mechanics and the Crucial Condition

There is one issue that might still be puzzling here. We have seen that a "weak measurement of velocity" is in fact a genuine measurement of velocity whenever the condition (7.27) is satisfied. And in this case the velocity found must be the Bohmian velocity. This implies that (7.27) can be satisfied only for Bohmian mechanics—that it characterizes Bohmian mechanics among all of its variants. We shall now provide a more general and precise formulation of this conclusion, as well as a direct argument for it:

Suppose $v^\psi$ defines a variant of Bohmian mechanics for which the condition

$$v^{\psi\phi} = v^\psi \quad (7.30)$$

holds for all (differentiable) real-valued functions $\phi$, or at least for a collection of such functions that is "gradient-total," i.e., such that at every point $x \in \mathbb{R}^3$, the collection of vectors $\nabla \phi(x)$ spans $\mathbb{R}^3$. Then $v^\psi = v_B^\psi$.

A similar conclusion holds for a more general configuration space than $\mathbb{R}^3$, for example for $\mathbb{R}^{3N}$. We note that for any fixed (differentiable) real-valued function $\Phi$ that vanishes at $\infty$ (but is not identically 0), the collection of functions $\Phi_y(x) = \Phi(y - x)$, $y \in \mathbb{R}^3$, is gradient-total, since otherwise there would be a direction in which $\Phi$ does not vary. [Note also that for particles without spin, i.e., when $\psi$

## 7.5 The Impossibility of Measuring the Velocity in Bohmian Mechanics

is complex-scalar-valued, then the condition (7.30) amounts basically to requiring that $v^\psi$ depend only on the phase $S$ of $\psi$ (arising from the polar decomposition $\psi = Re^{iS/\hbar}$).]

Here is the proof: Recall that any variant of Bohmian mechanics, with velocity $v^\psi$, must obey the continuity equation, see (7.8),

$$\partial_t |\psi(x,t)|^2 = -\mathrm{div}\,(v^\psi(x,t)|\psi(x,t)|^2). \tag{7.31}$$

Consider two such velocity functionals, $v_1^\psi$ and $v_2^\psi$. Since they both are such that (7.31) is obeyed, we have that

$$\mathrm{div}\,(j_1^\psi - j_2^\psi) = 0, \tag{7.32}$$

where $j_i^\psi = |\psi|^2 v_i^\psi$. If they both also obey (7.30), we have that

$$j_i^{\psi\phi} = |\phi|^2 j_i^\psi. \tag{7.33}$$

Then from (7.32), with $\psi$ replaced by $\psi' = \psi\phi$, we have that

$$\mathrm{div}\,[|\phi|^2 (j_1^\psi - j_2^\psi)] = 0, \tag{7.34}$$

and since

$$\mathrm{div}\,[|\phi|^2(j_1^\psi - j_2^\psi)] = |\phi|^2 \mathrm{div}\,(j_1^\psi - j_2^\psi) + \nabla|\phi|^2 \cdot (j_1^\psi - j_2^\psi) = \nabla|\phi|^2 \cdot (j_1^\psi - j_2^\psi) \tag{7.35}$$

it follows that

$$\nabla|\phi|^2 \cdot (j_1^\psi - j_2^\psi) = 0. \tag{7.36}$$

Thus if the relevant collection of functions $\phi$ is gradient-total, we have that

$$j_1^\psi = j_2^\psi \tag{7.37}$$

and hence that

$$v_1^\psi = v_2^\psi. \tag{7.38}$$

Since $v_B^\psi$ is a possible choice for $v^\psi$ the conclusion follows.

## 7.5 The Impossibility of Measuring the Velocity in Bohmian Mechanics

We have argued that by using weak measurements it is possible to measure the velocity of a particle in Bohmian mechanics. We will now reconcile this with the proven impossibility of measuring the velocity in Bohmian mechanics (see Chap. 3), in the sense that no measurement procedure involving an interaction between a particle and any sort of apparatus can yield a result that conveys (with arbitrary precision) the velocity of the particle just prior to the beginning of the procedure.

Let us first note that our weak measurement of velocity can, it seems, be regarded as just such a procedure. It involves an ensemble of systems, each with the same wave function. We can regard one member of the ensemble as the special particle whose velocity is to be measured, with the other members of the ensemble constituting (part of) the apparatus. The selection of subensemble and averaging corresponding to (7.24), with $x$ the position of the special particle at time $\tau$, then conveys the velocity of the special particle to arbitrary accuracy.

This procedure, however, is not of the sort contemplated in Sect. 3.7.2. It is assumed there that neither the initial state $\Psi_{app}$ of the apparatus nor the interaction $H_{int}$ between system and apparatus depends on the initial state $\psi$ of the system. (This is reasonable since the whole point of a measurement is to obtain information about a system that would not otherwise be available.) Such measurements have been called linear measurements, in contrast with the nonlinear measurements in which either the initial state of the apparatus $\Psi_{app} = \Psi^\psi$ or the interaction $H_{int} = H^\psi$ depends upon the state $\psi$ of the system.

The weak measurement of velocity discussed in this chapter is clearly nonlinear,[5] and is thus not precluded by the impossibility claim of Sect. 3.7.2. It should nonetheless be contrasted with another nonlinear measurement of velocity for Bohmian mechanics: perform a standard position measurement and plug the result into the formula (7.5) to obtain the corresponding velocity. While it is difficult to take the latter "measurement of velocity" seriously, and to regard it as anything more than cheating, the weak measurement of velocity in Bohmian mechanics is, as we have argued, a genuine measurement of velocity, even though it is nonlinear.

## 7.6 Conclusion

Measurement is a tricky and complicated business. Even when, as with Bohmian mechanics and its variants, there is something to be measured, one must be careful. With orthodox quantum theory and the "measurement" of operators as observables, the situation is even more dangerous. We conclude by quoting Bell [4][p. 166] on this:

> A final moral concerns terminology. Why did such serious people take so seriously axioms which now seem so arbitrary? I suspect that they were misled by the pernicious misuse of the word 'measurement' in contemporary theory. This word very strongly suggests the ascertaining of some preexisting property of some thing, any instrument involved playing a purely passive role. Quantum experiments are just not like that, as we learned especially from Bohr. The results have to be regarded as the joint product of 'system' and 'apparatus,' the complete experimental set-up. But the misuse of the word 'measurement' makes it easy

---

[5] Were it possible to clone the wave function $\psi$, the ensemble could have been produced as part of an overall linear measurement, but cloning is not possible [5].

to forget this and then to expect that the 'results of measurements' should obey some simple logic in which the apparatus is not mentioned. The resulting difficulties soon show that any such logic is not ordinary logic. It is my impression that the whole vast subject of 'Quantum Logic' has arisen in this way from the misuse of a word. I am convinced that the word 'measurement' has now been so abused that the field would be significantly advanced by banning its use altogether, in favour for example of the word 'experiment.'

# References

1. Y. Aharonov, D. Z. Albert, and L. Vaidman. How the Result of a Measurement of a Component of the Spin of a Spin-1/2 Particle Can Turn Out to Be 100. *Physical Review Letters*, 60:1351–1354, 1988.
2. H. M. Wiseman. Grounding Bohmian Mechanics in Weak Values and Bayesianism. *New Journal of Physics*, 9:165, 2007.
3. E. Deotto and G. C. Ghirardi. Bohmian Mechanics Revisited. *Foundations of Physics*, 28:1–30, 1998.
4. J. S. Bell. *Speakable and Unspeakable in Quantum Mechanics*. Cambridge University Press, Cambridge, 1987.
5. G. C. Ghirardi. Private communication, 1981.

# Chapter 8
# Topological Factors Derived From Bohmian Mechanics

## 8.1 Introduction

We consider here a novel approach based on Bohmian mechanics towards topological effects in quantum mechanics. These effects arise when the configuration space $\mathscr{Q}$ of a quantum system is a multiply-connected Riemannian manifold and involve *topological factors* forming a representation (or holonomy-twisted representation) of the fundamental group $\pi_1(\mathscr{Q})$ of $\mathscr{Q}$. The use of Bohmian paths allows a derivation of the link between homotopy and quantum mechanics that is essentially different from derivations based on path integrals.

The topological factors we derive are equally relevant and applicable in orthodox quantum mechanics, or any other version of quantum mechanics. Bohmian mechanics, however, provides a sharp mathematical justification of the dynamics with these topological factors that is absent in the orthodox framework. Different topological factors give rise to different Bohmian dynamics, and thus to different quantum theories, for the same configuration space $\mathscr{Q}$ (whose metric we regard as incorporating the "masses of the particles"), the same potential, and the same value space of the wave function.

The motion of the configuration in a Bohmian system of $N$ distinguishable particles can be regarded as corresponding to a dynamical system in the configuration space $\mathscr{Q} = \mathbb{R}^{3N}$, defined by a time-dependent vector field $v^{\psi_t}$ on $\mathscr{Q}$ which in turn is defined, by the Bohmian law of motion, in terms of $\psi_t$. We are concerned here with the analogues of the Bohmian law of motion when $\mathscr{Q}$ is, instead of $\mathbb{R}^{3N}$, an arbitrary Riemannian manifold.[1] The main result is that, if $\mathscr{Q}$ is multiply connected, there are several such analogues: several dynamics, which we will describe in detail, corresponding to different choices of the topological factors.

It is easy to overlook the multitude of dynamics by focusing too much on just one, the simplest one, which we will define in Sect. 8.2: the *immediate generalization* of the Bohmian dynamics from $\mathbb{R}^{3N}$ to a Riemannian manifold, or, as we shall briefly call

---

[1] Manifolds will throughout be assumed to be Hausdorff, paracompact, connected, and $C^\infty$. They need not be orientable.

it, the *immediate Bohmian dynamics*. Of the other kinds of Bohmian dynamics, the simplest involve phase factors associated with non-contractible loops in $\mathcal{Q}$, forming a character[2] of the fundamental group $\pi_1(\mathcal{Q})$. In other cases, the topological factors are given by matrices or endomorphisms, forming a unitary representation of $\pi_1(\mathcal{Q})$ or, in the case of a vector bundle, a holonomy-twisted representation (see the end of Sect. 8.4 for the definition). As we shall explain, the dynamics of bosons is an "immediate" one, but not the dynamics of fermions (except when using a certain not entirely natural vector bundle). The Aharonov–Bohm effect can be regarded as an example of a non-immediate dynamics on the accessible region of 3-space.

It is not obvious what "other kinds of Bohmian dynamics" should mean. We will investigate one approach here, while others will be studied in forthcoming works. The present approach is based on considering wave functions $\psi$ that are defined not on the configuration space $\mathcal{Q}$ but on its universal covering space $\widehat{\mathcal{Q}}$. We then investigate which kinds of periodicity conditions, relating the values on different levels of the covering fiber by a topological factor, will ensure that the Bohmian velocity vector field associated with $\psi$ is projectable from $\widehat{\mathcal{Q}}$ to $\mathcal{Q}$. This is carried out in Sect. 8.3 for scalar wave functions and in Sect. 8.4 for wave functions with values in a complex vector space (such as a spin-space) or a complex vector bundle. In the case of vector bundles, we derive a novel kind of topological factor, given by a holonomy-twisted representation of $\pi_1(\mathcal{Q})$.

The notion that multiply-connected spaces give rise to different topological factors is not new. The most common approach is based on path integrals and began largely with the work of Schulman [1, 2] and Laidlaw and DeWitt [3]; see [4] for details. Nelson [5] derives the topological phase factors for scalar wave functions from stochastic mechanics. There is also the current algebra approach of Goldin et al. [6].

## 8.2 Bohmian Mechanics in Riemannian Manifolds

Bohmian mechanics can be formulated by appealing only to the Riemannian structure $g$ of the configuration space $\mathcal{Q}$ of a physical system: the state of the system in Bohmian mechanics is given by the pair $(\mathcal{Q}, \psi)$; $\mathcal{Q} \in \mathcal{Q}$ is the configuration of the system and $\psi$ is a (standard quantum mechanical) wave function on the configuration space $\mathcal{Q}$, taking values in some *Hermitian vector space W*, i.e., a finite-dimensional complex vector space endowed with a positive-definite Hermitian (i.e., conjugate-symmetric and sesqui-linear) inner product $(\cdot, \cdot)$.

The state of the system changes according to the guiding equation and Schrödinger's equation [7]:

---

[2] By a *character* of a group we refer to what is sometimes called a unitary multiplicative character, i.e., a one-dimensional unitary representation of the group.

## 8.2 Bohmian Mechanics in Riemannian Manifolds

$$\frac{dQ_t}{dt} = v^{\psi_t}(Q_t) \tag{8.1a}$$

$$i\hbar \frac{\partial \psi_t}{\partial t} = -\frac{\hbar^2}{2}\Delta\psi_t + V\psi_t, \tag{8.1b}$$

where the Bohmian velocity vector field $v^\psi$ associated with the wave function $\psi$ is

$$v^\psi := \hbar \operatorname{Im} \frac{(\psi, \nabla\psi)}{(\psi, \psi)}. \tag{8.2}$$

In the above equations $\Delta$ and $\nabla$ are, respectively, the Laplace-Beltrami operator and the gradient on the configuration space equipped with this Riemannian structure; $V$ is the potential function with values given by Hermitian matrices (endomorphisms of $W$). Thus, given $\mathcal{Q}$, $W$, and $V$, we have specified a Bohmian dynamics, the *immediate Bohmian dynamics*.[3]

The empirical agreement between Bohmian mechanics and standard quantum mechanics is grounded in equivariance [7, 8]. In Bohmian mechanics, if the configuration is initially random and distributed according to $|\psi_0|^2$, then the evolution is such that the configuration at time $t$ will be distributed according to $|\psi_t|^2$. This property is called the equivariance of the $|\psi|^2$ distribution. It follows from comparing the transport equation arising from (8.1a)

$$\frac{\partial \rho_t}{\partial t} = -\nabla \cdot (\rho_t v^{\psi_t}) \tag{8.3}$$

for the distribution $\rho_t$ of the configuration $\mathcal{Q}_t$, where $v^\psi = (\mathbf{v}_1^\psi, \ldots, \mathbf{v}_N^\psi)$, to the quantum continuity equation

$$\frac{\partial |\psi_t|^2}{\partial t} = -\nabla \cdot (|\psi_t|^2 v^{\psi_t}), \tag{8.4}$$

which is a consequence of Schrödinger's equation (8.1b). A rigorous proof of equivariance requires showing that almost all (with respect to the $|\psi|^2$ distribution) solutions of (8.1a) exist for all times. This was done in [9, 10]. A more comprehensive introduction to Bohmian mechanics may be found in [11, 12, 13].

An important example (with, say, $W = \mathbb{C}$) is that of several particles moving in a Riemannian manifold $M$, a possibly curved physical space. Then the configuration space for $N$ distinguishable particles is $\mathcal{Q} := M^N$. Let the masses of the particles be $m_i$ and the metric of $M$ be $g$. Then the relevant metric on $M^N$ is

$$g^N(v_1 \oplus \cdots \oplus v_N, w_1 \oplus \cdots \oplus w_N) := \sum_{i=1}^{N} m_i g(v_i, w_i).$$

---

[3] Since the law of motion involves a derivative of $\psi$, the merely measurable functions in $L^2(\mathcal{Q})$ will of course not be adequate for defining trajectories. However, we will leave aside the question, from which dense subspace of $L^2(\mathcal{Q})$ should one choose $\psi$. For a discussion of the global existence question of Bohmian trajectories in $\mathbb{R}^{3N}$, see [9, 10].

Using $g^N$ allows us to write (8.2) and (8.1a) instead of the equivalent equations

$$\frac{d\mathbf{Q}_k}{dt} = \frac{\hbar}{m_k}\mathrm{Im}\frac{(\psi, \nabla_k\psi)}{(\psi, \psi)}(\mathbf{Q}_1, \ldots, \mathbf{Q}_N), \quad k = 1, \ldots, N \tag{8.5}$$

$$i\hbar\frac{\partial \psi}{\partial t} = -\sum_{k=1}^{N}\frac{\hbar^2}{2m_k}\Delta_k\psi + V\psi, \tag{8.6}$$

where $\mathbf{Q}_k$, the $k^{th}$ component of $\mathscr{Q}$, lies in $M$, and $\nabla_k$ and $\Delta_k$ are the gradient and the Laplacian with respect to $g$, acting on the $k^{th}$ factor of $M^N$. Another important example [3] is that of $N$ identical particles in $\mathbb{R}^3$, for which the natural configuration space is the set ${}^N\mathbb{R}^3$ of all $N$-element subsets of $\mathbb{R}^3$,

$${}^N\mathbb{R}^3 := \{S | S \subseteq \mathbb{R}^3, |S| = N\}, \tag{8.7}$$

which inherits a Riemannian metric from $\mathbb{R}^3$. Spin is incorporated by choosing for $W$ a suitable spin space [14]. For one particle moving in $\mathbb{R}^3$, we may take $W$ to be a complex, irreducible representation space of $SU(2)$, the universal covering group[4] of the rotation group $SO(3)$. If it is the spin-$s$ representation then $W = \mathbb{C}^{2s+1}$.

More generally, we can consider a Bohmian dynamics for wave functions taking values in a complex vector bundle $E$ over the Riemannian manifold $\mathscr{Q}$. That is, the value space then depends on the configuration, and wave functions become sections of the vector bundle. Such a case occurs for identical particles with spin $s$, where the bundle $E$ of spin spaces over the configuration space $\mathscr{Q} = {}^N\mathbb{R}^3$ consists of the $(2s+1)^N$-dimensional spaces

$$E_q = \bigotimes_{\mathbf{q}\in q}\mathbb{C}^{2s+1}, \quad q \in \mathscr{Q}. \tag{8.8}$$

(The tensor product here is an unconventional one, in which the usual index set $\{1, \ldots, N\}$ is replaced by the natural $N$-element set $\{\mathbf{q}_1, \ldots, \mathbf{q}_N\}$ defining the configuration $q$ itself.)

We introduce now some notation and terminology.

**Definition 1** *A Hermitian vector bundle, or Hermitian bundle, over $\mathscr{Q}$ is a finite-dimensional complex vector bundle $E$ over $\mathscr{Q}$ with a connection and a positive-definite, Hermitian local inner product $(\cdot, \cdot) = (\cdot, \cdot)_q$ on $E_q$, the fiber of $E$ over $q \in \mathscr{Q}$, which is parallel.*

Our bundle, the one of which $\psi$ is a section, will always be a Hermitian bundle. Note that since a Hermitian bundle consists of a vector bundle and a connection, it can be nontrivial even if the vector bundle is trivial: namely, if the connection is

---

[4] The universal covering space of a Lie group is again a Lie group, the *universal covering group*. It should be distinguished from another group also called the *covering group*: the group $Cov(\widehat{\mathscr{Q}}, \mathscr{Q})$ of the covering (or deck) transformations of the universal covering space $\widehat{\mathscr{Q}}$ of a manifold $\mathscr{Q}$, which will play an important role later.

nontrivial. The *trivial Hermitian bundle* $\mathcal{Q} \times W$, in contrast, consists of the trivial vector bundle with the trivial connection, whose parallel transport $P_\beta$, in general a unitary endomorphism from $E_q$ to $E_{q'}$ for $\beta$ a path from $q$ to $q'$, is always the identity on $W$. The case of a $W$-valued function $\psi : \mathcal{Q} \to W$ corresponds to the trivial Hermitian bundle $\mathcal{Q} \times W$.

The global inner product on the Hilbert space of wave functions is the local inner product integrated against the Riemannian volume measure associated with the metric $g$ of $\mathcal{Q}$,

$$\langle \phi, \psi \rangle = \int_{\mathcal{Q}} dq \, (\phi(q), \psi(q)).$$

The Hilbert space equipped with this inner product, denoted $L^2(\mathcal{Q}, E)$, contains the square-integrable, measurable (not necessarily smooth) sections of $E$ modulo equality almost everywhere. The covariant derivative $D\psi$ of a section $\psi$ is an "$E$-valued 1-form," i.e., a section of $\mathbb{C}T\mathcal{Q}^* \otimes E$ (with $T\mathcal{Q}^*$ the cotangent bundle), while we write $\nabla\psi$ for the section of $\mathbb{C}T\mathcal{Q} \otimes E$ metrically equivalent to $D\psi$. The potential $V$ is now a self-adjoint section of the endomorphism bundle $E \otimes E^*$ acting on the vector bundle's fibers. The equations defining the Bohmian dynamics are, *mutatis mutandis*, the same equations (8.1) and (8.2) as before.

We wish to introduce now further Bohmian dynamics beyond the immediate one. To this end, we will consider wave functions on $\widehat{\mathcal{Q}}$, the universal covering space of $\mathcal{Q}$. This idea is rather standard in the literature on quantum mechanics in multiply-connected spaces [3, 15, 16, 17, 18]. However, the complete specification of the possibilities that we give in Sect. 8.4 includes some, corresponding to what we call *holonomy-twisted representations* of $\pi_1(\mathcal{Q})$, that have not yet been considered. Each possibility has locally the same Hamiltonian $-\frac{\hbar^2}{2}\Delta + V$, with the same potential $V$, and each possibility is equally well defined and equally reasonable. While in orthodox quantum mechanics it may seem more or less axiomatic that the configuration space $\mathcal{Q}$ is the space on which $\psi_t$ is defined, $\mathcal{Q}$ appears in Bohmian mechanics also in another role: as the space in which $\mathcal{Q}_t$ moves. It is therefore less surprising from the Bohmian viewpoint, and easier to accept, that $\psi_t$ is defined not on $\mathcal{Q}$ but on $\widehat{\mathcal{Q}}$. In the next section all wave functions will be complex-valued; in Sect. 8.4 we shall consider wave functions with higher-dimensional value spaces.

## 8.3 Scalar Wave Functions on the Covering Space

The motion of the configuration $\mathcal{Q}_t$ in $\mathcal{Q}$ is determined by a velocity vector field $v_t$ on $\mathcal{Q}$, which may arise from a wave function $\psi$ not on $\mathcal{Q}$ but instead on $\widehat{\mathcal{Q}}$, the universal covering space of $\mathcal{Q}$, in the following way: Suppose we are given a complex-valued map $\gamma$ on the covering group $Cov(\widehat{\mathcal{Q}}, \mathcal{Q})$, $\gamma : Cov(\widehat{\mathcal{Q}}, \mathcal{Q}) \to \mathbb{C}$, and suppose that a wave function $\psi : \widehat{\mathcal{Q}} \to \mathbb{C}$ satisfies the *periodicity condition associated with the topological factors* $\gamma$, i.e.,

$$\psi(\sigma\hat{q}) = \gamma_\sigma \psi(\hat{q}) \qquad (8.9)$$

for every $\hat{q} \in \widehat{\mathcal{Q}}$ and $\sigma \in Cov(\widehat{\mathcal{Q}}, \mathcal{Q})$. For (8.9) to be possible for a $\psi$ that does not identically vanish, $\gamma$ must be a representation of the covering group, as was first emphasized in [15]. To see this, let $\sigma_1, \sigma_2 \in Cov(\widehat{\mathcal{Q}}, \mathcal{Q})$. Then we have the following equalities

$$\gamma_{\sigma_1\sigma_2}\psi(\hat{q}) = \psi(\sigma_1\sigma_2\hat{q}) = \gamma_{\sigma_1}\psi(\sigma_2\hat{q}) = \gamma_{\sigma_1}\gamma_{\sigma_2}\psi(\hat{q}). \qquad (8.10)$$

We thus obtain the fundamental relation

$$\gamma_{\sigma_1\sigma_2} = \gamma_{\sigma_1}\gamma_{\sigma_2}, \qquad (8.11)$$

establishing (since $\gamma_{\mathrm{Id}} = 1$) that $\gamma$ is a representation.

Let $\pi_1(\mathcal{Q}, q)$ denote the *fundamental group of $\mathcal{Q}$ at a point $q$* and let $\pi$ be the covering map (a local diffeomorphism) $\pi : \widehat{\mathcal{Q}} \to \mathcal{Q}$, also called the projection (the *covering fiber* for $q \in \mathcal{Q}$ is the set $\pi^{-1}(q)$ of points in $\widehat{\mathcal{Q}}$ that project to $q$ under $\pi$). The 1-dimensional representations of the covering group are, via the canonical isomorphisms $\varphi_{\hat{q}} : Cov(\widehat{\mathcal{Q}}, \mathcal{Q}) \to \pi_1(\mathcal{Q}, q)$, $\hat{q} \in \pi^{-1}(q)$, in canonical correspondence with the 1-dimensional representations of any fundamental group $\pi_1(\mathcal{Q}, q)$: The different isomorphisms $\varphi_{\hat{q}}$, $\hat{q} \in \pi^{-1}(q)$, will transform a representation of $\pi_1(\mathcal{Q}, q)$ into representations of $Cov(\widehat{\mathcal{Q}}, \mathcal{Q})$ that are conjugate. But the 1-dimensional representations are homomorphisms to the *abelian* multiplicative group of $\mathbb{C}$ and are thus invariant under conjugation.

From (8.9) it follows that $\nabla\psi(\sigma\hat{q}) = \gamma_\sigma \sigma^* \nabla\psi(\hat{q})$, where $\sigma^*$ is the (pushforward) action of $\sigma$ on tangent vectors, using that $\sigma$ is an isometry. Thus, the velocity field $\hat{v}^\psi$ on $\widehat{\mathcal{Q}}$ associated with $\psi$ according to

$$\hat{v}^\psi(\hat{q}) := \hbar \, \mathrm{Im}\, \frac{\nabla\psi}{\psi}(\hat{q}) \qquad (8.12)$$

is projectable, i.e.,

$$\hat{v}^\psi(\sigma\hat{q}) = \sigma^* \hat{v}^\psi(\hat{q}), \qquad (8.13)$$

and therefore gives rise to a velocity field $v^\psi$ on $\mathcal{Q}$,

$$v^\psi(q) = \pi^* \hat{v}^\psi(\hat{q}) \qquad (8.14)$$

where $\hat{q}$ is an arbitrary element of $\pi^{-1}(q)$.

If we let $\psi$ evolve according to the Schrödinger equation on $\widehat{\mathcal{Q}}$,

$$i\hbar \frac{\partial \psi}{\partial t}(\hat{q}) = -\frac{\hbar^2}{2}\Delta\psi(\hat{q}) + \widehat{V}(\hat{q})\psi(\hat{q}) \qquad (8.15)$$

with $\widehat{V}$ the lift of the potential $V$ on $\mathcal{Q}$, then the periodicity condition (8.9) is preserved by the evolution, since, according to

## 8.3 Scalar Wave Functions on the Covering Space

$$i\hbar \frac{\partial \psi}{\partial t}(\sigma\hat{q}) \stackrel{(8.15)}{=} -\frac{\hbar^2}{2}\Delta\psi(\sigma\hat{q}) + \widehat{V}(\sigma\hat{q})\psi(\sigma\hat{q}) = -\frac{\hbar^2}{2}\Delta\psi(\sigma\hat{q}) + \widehat{V}(\hat{q})\psi(\sigma\hat{q})$$
(8.16)

(note the different arguments in the potential), the functions $\psi \circ \sigma$ and $\gamma_\sigma \psi$ satisfy the same evolution equation (8.15) with, by (8.9), the same initial condition, and thus coincide at all times.

Therefore we can let the Bohmian configuration $\mathcal{Q}_t$ move according to $v^{\psi_t}$,

$$\frac{dQ_t}{dt} = v^{\psi_t}(Q_t) = \hbar \pi^*\left(\operatorname{Im}\frac{\nabla\psi}{\psi}\right)(Q_t) = \hbar \pi^*\left(\operatorname{Im}\frac{\nabla\psi}{\psi}\Big|_{\hat{q}\in\pi^{-1}(Q_t)}\right). \quad (8.17)$$

One can also view the motion in this way: Given $Q_0$, choose $\widehat{Q}_0 \in \pi^{-1}(Q_0)$, let $\widehat{Q}_t$ move in $\widehat{\mathcal{Q}}$ according to $\hat{v}^{\psi_t}$, and set $Q_t = \pi(\widehat{Q}_t)$. Then the motion of $Q_t$ is independent of the choice of $\widehat{Q}_0$ in the fiber over $Q_0$, and obeys (8.17).

If, as we shall assume from now on, $|\gamma_\sigma| = 1$ for all $\sigma \in Cov(\widehat{\mathcal{Q}}, \mathcal{Q})$, i.e., if $\gamma$ is a *unitary* representation (in $\mathbb{C}$) or a *character*, then the motion (8.17) also has an equivariant probability distribution, namely

$$\rho(q) = |\psi(\hat{q})|^2. \quad (8.18)$$

To see this, note that we have

$$|\psi(\sigma\hat{q})|^2 \stackrel{(8.9)}{=} |\gamma_\sigma|^2 |\psi(\hat{q})|^2 = |\psi(\hat{q})|^2, \quad (8.19)$$

so that the function $|\psi(\hat{q})|^2$ is projectable to a function on $\mathcal{Q}$ which we call $|\psi|^2(q)$ in this paragraph. From (8.15) we have that

$$\frac{\partial |\psi_t(\hat{q})|^2}{\partial t} = -\nabla \cdot \left(|\psi_t(\hat{q})|^2 \, \hat{v}^{\psi_t}(\hat{q})\right)$$

and, by projection, that

$$\frac{\partial |\psi_t|^2(q)}{\partial t} = -\nabla \cdot \left(|\psi_t|^2(q) \, v^{\psi_t}(q)\right),$$

which coincides with the transport equation for a probability density $\rho$ on $\mathcal{Q}$,

$$\frac{\partial \rho_t(q)}{\partial t} = -\nabla \cdot \left(\rho_t(q) \, v^{\psi_t}(q)\right).$$

Hence,

$$\rho_t(q) = |\psi_t|^2(q) \quad (8.20)$$

for all times if it is so initially; this is equivariance.

The relevant wave functions are those with

$$\int_{\mathcal{Q}} dq \, |\psi(\hat{q})|^2 = 1 \tag{8.21}$$

where the choice of $\hat{q} \in \pi^{-1}(q)$ is arbitrary by (8.19). The relevant Hilbert space, which we denote $L^2(\widehat{\mathcal{Q}}, \gamma)$, thus consists of the measurable functions $\psi$ on $\widehat{\mathcal{Q}}$ (modulo changes on null sets) satisfying (8.9) with

$$\int_{\mathcal{Q}} dq \, |\psi(\hat{q})|^2 < \infty. \tag{8.22}$$

It is a Hilbert space with the scalar product

$$\langle \phi, \psi \rangle = \int_{\mathcal{Q}} dq \, \overline{\phi(\hat{q})} \, \psi(\hat{q}). \tag{8.23}$$

Note that the value of the integrand at $q$ is independent of the choice of $\hat{q} \in \pi^{-1}(q)$ since, by (8.9) and the fact that $|\gamma_\sigma| = 1$,

$$\overline{\phi(\sigma\hat{q})} \, \psi(\sigma\hat{q}) = \overline{\gamma_\sigma \phi(\hat{q})} \, \gamma_\sigma \, \psi(\hat{q}) = \overline{\phi(\hat{q})} \, \psi(\hat{q}).$$

We summarize the results of our reasoning.

**Assertion 1** *Given a Riemannian manifold $\mathcal{Q}$ and a smooth function $V : \mathcal{Q} \to \mathbb{R}$, there is a Bohmian dynamics in $\mathcal{Q}$ with potential $V$ for each character $\gamma$ of the fundamental group $\pi_1(\mathcal{Q})$; it is defined by (8.9), (8.15), and (8.17), where the wave function $\psi_t$ lies in $L^2(\widehat{\mathcal{Q}}, \gamma)$ and has norm one.*

Assertion 1 provides as many dynamics as there are characters of $\pi_1(\mathcal{Q})$ because different characters $\gamma' \neq \gamma$ always define different dynamics. In particular, for the trivial character $\gamma_\sigma = 1$, we obtain the immediate dynamics, as defined by (8.2) and (8.1).

An important application of Assertion 1 is provided by identical particles without spin. The natural configuration space $^N\mathbb{R}^3$ for identical particles has fundamental group $S_N$, the group of permutations of $N$ objects, which possesses two characters, the trivial character, $\gamma_\sigma = 1$, and the alternating character, $\gamma_\sigma = \text{sgn}(\sigma) = 1$ or $-1$ depending on whether $\sigma \in S_N$ is an even or an odd permutation. The Bohmian dynamics associated with the trivial character is that of bosons, while the one associated with the alternating character is that of fermions. However, in a two-dimensional world there would be more possibilities since $\pi_1(^N\mathbb{R}^2)$ is the braid group, whose generators $\sigma_i$, $i = 1, \ldots, N-1$, are a certain subset of braids that exchange two particles and satisfy the defining relations

$$\sigma_i \sigma_j = \sigma_j \sigma_i \quad \text{for} \quad i \leq N-3, j \geq i+2,$$
$$\sigma_i \sigma_{i+1} \sigma_i = \sigma_{i+1} \sigma_i \sigma_{i+1} \quad \text{for} \quad i \leq N-2.$$

Thus, a character of the braid group assigns the same complex number $e^{i\beta}$ to each generator, and therefore, according to Assertion 1, each choice of $\beta$ corresponds to a Bohmian dynamics; two-dimensional bosons correspond to $\beta = 0$ and two-dimensional fermions to $\beta = \pi$. The particles corresponding to the other possibilities are usually called *anyons*. They were first suggested in [16], and their investigation began in earnest with [6, 9]. See [17] for some more details and references.

## 8.4 Vector-Valued Wave Functions on the Covering Space

The analysis of Sect. 8.3 can be carried over with little change to the case of vector-valued wave functions, $\psi(q) \in W$. In this case, however, the topological factors may be given by any endomorphisms $\Gamma_\sigma$ of $W$ that form a representation of $Cov(\widehat{\mathcal{Q}}, \mathcal{Q})$ and need not be restricted to characters, a possibility first mentioned in [4], Notes to Sect. 23.3. Rather than directly considering this case, we focus instead on one that is a bit more general and that will require a new sort of topological factor, that of wave functions that are sections of a vector bundle. The topological factors for this case will be expressed as *periodicity sections*, i.e., parallel unitary sections of the endomorphism bundle indexed by the covering group and satisfying a certain composition law, or, equivalently, as *holonomy-twisted representations* of $\pi_1(\mathcal{Q})$.

If $E$ is a vector bundle over $\mathcal{Q}$, then the lift of $E$, denoted by $\widehat{E}$, is a vector bundle over $\widehat{\mathcal{Q}}$; the fiber space at $\hat{q}$ is defined to be the fiber space of $E$ at $q$, $\widehat{E}_{\hat{q}} := E_q$, where $q = \pi(\hat{q})$. It is important to realize that with this construction, it makes sense to ask whether $v \in \widehat{E}_{\hat{q}}$ is equal to $w \in \widehat{E}_{\hat{r}}$ whenever $\hat{q}$ and $\hat{r}$ are elements of the same covering fiber. Equivalently, $\widehat{E}$ is the pull-back of $E$ through $\pi : \widehat{\mathcal{Q}} \to \mathcal{Q}$. As a particular example, the lift of the tangent bundle of $\mathcal{Q}$ to $\widehat{\mathcal{Q}}$ is canonically isomorphic to the tangent bundle of $\widehat{\mathcal{Q}}$. Sections of $E$ or $E \otimes E^*$ can be lifted to sections of $\widehat{E}$ respectively $\widehat{E} \otimes \widehat{E}^*$.

If $E$ is a Hermitian vector bundle, then so is $\widehat{E}$. The wave function $\psi$ that we consider here is a section of $\widehat{E}$, so that $\psi(\hat{q})$ is a vector in the $\hat{q}$-dependent Hermitian vector space $\widehat{E}_{\hat{q}}$. $V$ is a section of the bundle $E \otimes E^*$, i.e., $V(q)$ is an element of $E_q \otimes E_q^*$. To indicate that every $V(q)$ is a Hermitian endomorphism of $E_q$, we say that $V$ is a Hermitian section of $E \otimes E^*$.

Since $\psi(\sigma\hat{q})$ and $\psi(\hat{q})$ lie in the same space $E_q = \widehat{E}_{\hat{q}} = \widehat{E}_{\sigma\hat{q}}$, a periodicity condition can be of the form

$$\psi(\sigma\hat{q}) = \Gamma_\sigma(\hat{q})\psi(\hat{q}) \tag{8.24}$$

for $\sigma \in Cov(\widehat{\mathcal{Q}}, \mathcal{Q})$, where $\Gamma_\sigma(\hat{q})$ is an endomorphism $E_q \to E_q$. By the same argument as in (8.10), the condition for (8.24) to be possible, if $\psi(\hat{q})$ can be any element of $\widehat{E}_{\hat{q}}$, is the composition law

$$\Gamma_{\sigma_1\sigma_2}(\hat{q}) = \Gamma_{\sigma_1}(\sigma_2\hat{q})\,\Gamma_{\sigma_2}(\hat{q}). \tag{8.25}$$

Note that this law differs from the one $\Gamma(\hat{q})$ would satisfy if it were a representation, which reads $\Gamma_{\sigma_1\sigma_2}(\hat{q}) = \Gamma_{\sigma_1}(\hat{q})\Gamma_{\sigma_2}(\hat{q})$, since in general $\Gamma(\sigma\hat{q})$ need not be the same as $\Gamma(\hat{q})$.

For periodicity (8.24) to be preserved under the Schrödinger evolution,

$$i\hbar \frac{\partial \psi}{\partial t}(\hat{q}) = -\frac{\hbar^2}{2}\Delta\psi(\hat{q}) + \widehat{V}(\hat{q})\psi(\hat{q}), \qquad (8.26)$$

we need that multiplication by $\Gamma_\sigma(\hat{q})$ commute with the Hamiltonian. Observe that

$$[H, \Gamma_\sigma]\psi(\hat{q}) = -\frac{\hbar^2}{2}(\Delta\Gamma_\sigma(\hat{q}))\psi(\hat{q}) - \hbar^2(\nabla\Gamma_\sigma(\hat{q}))\cdot(\nabla\psi(\hat{q})) + [\widehat{V}(\hat{q}), \Gamma_\sigma(\hat{q})]\psi(\hat{q}). \qquad (8.27)$$

Since we can choose $\psi$ such that, for any one particular $\hat{q}$, $\psi(\hat{q}) = 0$ and $\nabla\psi(\hat{q})$ is any element of $\mathbb{C}T_{\hat{q}}\widehat{\mathcal{Q}} \otimes E_q$ we like, we must have that

$$\nabla\Gamma_\sigma(\hat{q}) = 0 \qquad (8.28)$$

for all $\sigma \in Cov(\widehat{\mathcal{Q}}, \mathcal{Q})$ and all $\hat{q} \in \widehat{\mathcal{Q}}$, i.e., that $\Gamma_\sigma$ is parallel. Inserting this in (8.27), the first two terms on the right hand side vanish. Since we can choose for $\psi(\hat{q})$ any element of $E_q$ we like, we must have that

$$[\widehat{V}(\hat{q}), \Gamma_\sigma(\hat{q})] = 0 \qquad (8.29)$$

for all $\sigma \in Cov(\widehat{\mathcal{Q}}, \mathcal{Q})$ and all $\hat{q} \in \widehat{\mathcal{Q}}$. Conversely, assuming (8.28) and (8.29), we obtain that $\Gamma_\sigma$ commutes with $H$ for every $\sigma \in Cov(\widehat{\mathcal{Q}}, \mathcal{Q})$, so that the periodicity (8.24) is preserved.

From (8.24) and (8.28) it follows that $\nabla\psi(\sigma\hat{q}) = (\sigma^* \otimes \Gamma_\sigma(\hat{q}))\nabla\psi(\hat{q})$. If every $\Gamma_\sigma(\hat{q})$ is *unitary*, as we assume from now on, the velocity field $\hat{v}^\psi$ on $\widehat{\mathcal{Q}}$ associated with $\psi$ according to

$$\hat{v}^\psi(\hat{q}) := \hbar\, \mathrm{Im}\, \frac{(\psi, \nabla\psi)}{(\psi, \psi)}(\hat{q}) \qquad (8.30)$$

is projectable, $\hat{v}^\psi(\sigma\hat{q}) = \sigma^*\hat{v}^\psi(\hat{q})$, and gives rise to a velocity field $v^\psi$ on $\mathcal{Q}$. We let the configuration move according to $v^{\psi_t}$,

$$\frac{dQ_t}{dt} = v^{\psi_t}(Q_t) = \hbar\,\pi^*\Big(\mathrm{Im}\,\frac{(\psi, \nabla\psi)}{(\psi, \psi)}\Big)(Q_t). \qquad (8.31)$$

**Definition 2** *Let E be a Hermitian bundle over the manifold $\mathcal{Q}$. A periodicity section $\Gamma$ over E is a family indexed by $Cov(\widehat{\mathcal{Q}}, \mathcal{Q})$ of unitary parallel sections $\Gamma_\sigma$ of $\widehat{E} \otimes \widehat{E}^*$ satisfying the composition law (8.25).*

Since $\Gamma_\sigma(\hat{q})$ is unitary, one sees as before that the probability distribution

$$\rho(q) = (\psi(\hat{q}), \psi(\hat{q})) \qquad (8.32)$$

does not depend on the choice of $\hat{q} \in \pi^{-1}(q)$ and is equivariant.

## 8.4 Vector-Valued Wave Functions on the Covering Space

As usual, we define for any periodicity section $\Gamma$ the Hilbert space $L^2(\widehat{\mathcal{Q}}, \widehat{E}, \Gamma)$ to be the set of measurable sections $\psi$ of $\widehat{E}$ (modulo changes on null sets) satisfying (8.24) with

$$\int_{\mathcal{Q}} dq \, (\psi(\hat{q}), \psi(\hat{q})) < \infty, \tag{8.33}$$

endowed with the scalar product

$$\langle \phi, \psi \rangle = \int_{\mathcal{Q}} dq \, (\phi(\hat{q}), \psi(\hat{q})). \tag{8.34}$$

As before, the value of the integrand at $q$ is independent of the choice of $\hat{q} \in \pi^{-1}(q)$. We summarize the results of our reasoning.

**Assertion 2** *Given a Hermitian bundle $E$ over the Riemannian manifold $\mathcal{Q}$ and a Hermitian section $V$ of $E \otimes E^*$, there is a Bohmian dynamics for each periodicity section $\Gamma$ commuting (pointwise) with $\widehat{V}$ (cf. (8.29)); it is defined by (8.24), (8.26), and (8.31), where the wave function $\psi_t$ lies in $L^2(\widehat{\mathcal{Q}}, \widehat{E}, \Gamma)$ and has norm 1.*

Every character $\gamma$ of $Cov(\widehat{\mathcal{Q}}, \mathcal{Q})$ (or of $\pi_1(\mathcal{Q})$) defines a periodicity section by setting

$$\Gamma_\sigma(\hat{q}) := \gamma_\sigma \mathrm{Id}_{\widehat{E}_{\hat{q}}}. \tag{8.35}$$

It commutes with every potential $V$. Conversely, a periodicity section $\Gamma$ that commutes with every potential must be such that every $\Gamma_\sigma(\hat{q})$ is a multiple of the identity, $\Gamma_\sigma(\hat{q}) = \gamma_\sigma(\hat{q}) \mathrm{Id}_{\widehat{E}_{\hat{q}}}$. By unitarity, $|\gamma_\sigma| = 1$; by parallelity (8.28), $\gamma_\sigma(\hat{q}) = \gamma_\sigma$ must be constant; by the composition law (8.25), $\gamma$ must be a homomorphism, and thus a character.

We briefly indicate how a periodicity section $\Gamma$ corresponds to something like a representation of $\pi_1(\mathcal{Q})$. Fix a $\hat{q} \in \widehat{\mathcal{Q}}$. Then $Cov(\widehat{\mathcal{Q}}, \mathcal{Q})$ can be identified with $\pi_1(\mathcal{Q}) = \pi_1(\mathcal{Q}, \pi(\hat{q}))$ via $\varphi_{\hat{q}}$. Since the sections $\Gamma_\sigma$ of $\widehat{E} \otimes \widehat{E}^*$ are parallel, $\Gamma_\sigma(\hat{r})$ is determined for every $\hat{r}$ by $\Gamma_\sigma(\hat{q})$. (Note in particular that the parallel transport $\Gamma_\sigma(\tau\hat{q})$ of $\Gamma_\sigma(\hat{q})$ from $\hat{q}$ to $\tau\hat{q}, \tau \in Cov(\widehat{\mathcal{Q}}, \mathcal{Q})$, may differ from $\Gamma_\sigma(\hat{q})$.) Thus, the periodicity section $\Gamma$ is completely determined by the endomorphisms $\Gamma_\sigma := \Gamma_\sigma(\hat{q})$ of $E_q$, $\sigma \in Cov(\widehat{\mathcal{Q}}, \mathcal{Q})$, which satisfy the composition law

$$\Gamma_{\sigma_1 \sigma_2} = h_{\alpha_2} \Gamma_{\sigma_1} h_{\alpha_2}^{-1} \Gamma_{\sigma_2}, \tag{8.36}$$

where $\alpha_2$ is any loop in $\mathcal{Q}$ based at $\pi(\hat{q})$ whose lift starting at $\hat{q}$ leads to $\sigma_2 \hat{q}$, and $h_{\alpha_2}$ is the associated holonomy endomorphism of $E_q$. Since (8.36) is not the composition law $\Gamma_{\sigma_1 \sigma_2} = \Gamma_{\sigma_1} \Gamma_{\sigma_2}$ of a representation, the $\Gamma_\sigma$ form, not a representation of $\pi_1(\mathcal{Q})$, but what we call a *holonomy-twisted representation*.

The situation where the wave function assumes values in a fixed Hermitian space $W$, instead of a bundle, corresponds to the trivial Hermitian bundle $E = \mathcal{Q} \times W$ (i.e., with the trivial connection, for which parallel transport is the identity on $W$). Then, parallelity (8.28) implies that $\Gamma_\sigma(\hat{r}) = \Gamma_\sigma(\hat{q})$ for any $\hat{r}, \hat{q} \in \widehat{\mathcal{Q}}$, or $\Gamma_\sigma(\hat{q}) = \Gamma_\sigma$, so that (8.25) becomes the usual composition law $\Gamma_{\sigma_1\sigma_2} = \Gamma_{\sigma_1}\Gamma_{\sigma_2}$ and $\Gamma$ is a unitary representation of $Cov(\widehat{\mathcal{Q}}, \mathcal{Q})$.

The most important case of topological factors that are characters is provided by identical particles *with spin*. In fact, for this case, Assertion 2 entails the same conclusions we arrived at the end of Sect. 8.3, even for particles with spin. To understand how this comes about, consider the potential occurring in the Pauli equation for $N$ identical particles with spin,

$$V(q) = -\mu \sum_{\mathbf{q} \in q} \mathbf{B}(\mathbf{q}) \cdot \boldsymbol{\sigma}_\mathbf{q} \qquad (8.37)$$

on the spin bundle (8.8) over $^N\mathbb{R}^3$, with $\boldsymbol{\sigma}_\mathbf{q}$ the vector of spin matrices acting on the spin space of the particle at $\mathbf{q}$. Clearly, the algebra generated by $\{V(q)\}$ arising from all possible choices of the magnetic field $\mathbf{B}$ is $\text{End}(E_q)$. Thus the only holonomy-twisted representations that define a dynamics for all magnetic fields are those given by a character.[5]

An example of a topological factor that is not a character is provided by the Aharonov–Casher variant [20] of the Aharonov–Bohm effect, according to which a neutral spin-1/2 particle that carries a magnetic moment $\mu$ acquires a nontrivial phase while encircling a charged wire $\mathscr{C}$. A way of understanding how this effect comes about is in terms of the non-relativistic Hamiltonian $-\frac{\hbar^2}{2}\Delta + V$ based on a nontrivial connection $\nabla = \nabla_{\text{trivial}} - \frac{i\mu}{\hbar}\mathbf{E} \times \boldsymbol{\sigma}$ on the vector bundle $\mathbb{R}^3 \times \mathbb{C}^2$. Suppose the charge density $\rho(\mathbf{q})$ is invariant under translations in the direction $\mathbf{e} \in \mathbb{R}^3$, $\mathbf{e}^2 = 1$ in which the wire is oriented. Then the charge per unit length $\lambda$ is given by the integral

$$\lambda = \int_D \rho(\mathbf{q}) \, dA \qquad (8.38)$$

over the cross-section disk $D$ in any plane perpendicular to $\mathbf{e}$. The restriction of this connection, outside of $\mathscr{C}$, to any plane $\Sigma$ orthogonal to the wire turns out to be flat[6] so that its restriction to the intersection $\mathscr{Q}$ of $\mathbb{R}^3 \setminus \mathscr{C}$ with the orthogonal plane can be replaced, as in the Aharonov–Bohm case, by the trivial connection if we introduce a periodicity condition on the wave function with the topological factor

$$\Gamma_1 = \exp\left(-\frac{4\pi i \mu \lambda}{\hbar} \mathbf{e} \cdot \boldsymbol{\sigma}\right). \qquad (8.39)$$

In this way we obtain a representation $\Gamma: \pi_1(\mathscr{Q}) \to SU(2)$ that is not given by a character.

---

[5] In fact, it can be shown [21] that the only holonomy-twisted representations for a magnetic field $\mathbf{B}$ that is not parallel must be a character.

[6] The curvature is $\Omega = d_{\text{trivial}}\omega + \omega \wedge \omega$ with $\omega = -i\frac{\mu}{\hbar}\mathbf{E} \times \boldsymbol{\sigma}$. The 2-form $\Omega$ is dual to the vector $\nabla_{\text{trivial}} \times \omega + \omega \times \omega = i\frac{\mu}{\hbar}(\nabla \cdot \mathbf{E})\boldsymbol{\sigma} - i\frac{\mu}{\hbar}(\boldsymbol{\sigma} \cdot \nabla)\mathbf{E} - 2i(\frac{\mu}{\hbar})^2(\boldsymbol{\sigma} \cdot \mathbf{E})\mathbf{E}$. Outside the wire, the first term vanishes and, noting that $\mathbf{E} \cdot \mathbf{e} = 0$, the other two terms have vanishing component in the direction of $\mathbf{e}$ and thus vanish when integrated over any region within an orthogonal plane.

## 8.4 Vector-Valued Wave Functions on the Covering Space

Another example of a topological factor that is not a character and which can be generalized to a nonabelian representation is provided by a higher-dimensional version of the Aharonov–Bohm effect: one may replace the vector potential in the Aharonov–Bohm setting by a non-abelian gauge field (à la Yang–Mills) whose field strength (curvature) vanishes outside a cylinder $\mathscr{C}$ but not inside; the value space $W$ (now corresponding not to spin but to, say, quark color) has dimension greater than one, and the difference between two wave packets that have passed $\mathscr{C}$ on different sides is given in general, not by a phase, but by a unitary endomorphism $\Gamma$ of $W$. In this example, involving one cylinder, the representation $\Gamma$, though given by matrices that are not multiples of the identity, is nonetheless abelian, since $\pi_1(\mathscr{Q}) \cong \mathbb{Z}$ is an abelian group. However, when two or more cylinders are considered, we obtain a non-abelian representation $\Gamma$, since when $\mathscr{Q}$ is $\mathbb{R}^3$ minus two disjoint solid cylinders its fundamental group is isomorphic to the non-abelian group $\mathbb{Z} * \mathbb{Z}$, where $*$ denotes the free product of groups, generated by loops $\sigma_1$ and $\sigma_2$ surrounding one or the other of the cylinders. One can easily arrange that the matrices $\Gamma_{\sigma_i}$ corresponding to loops $\sigma_i$, $i = 1, 2$, fail to commute, so that $\Gamma$ is nonabelian.

Our last example involves a holonomy-twisted representation $\Gamma$ that is not a representation in the ordinary sense. Consider $N$ fermions, each as in the previous examples, moving in $M = \mathbb{R}^3 \setminus \cup_i \mathscr{C}_i$, where $\mathscr{C}_i$ are one or more disjoint solid cylinders. More generally, consider $N$ fermions, each having 3-dimensional configuration space $M$ and value space $W$ (which may incorporate spin or "color" or both). Then the configuration space $\mathscr{Q}$ for the $N$ fermions is the set $^N M$ of all $N$-element subsets of $M$, with universal covering space $\widehat{\mathscr{Q}} = \widehat{^N M} = \widehat{M}^N \setminus \Delta$ with $\Delta$ the extended diagonal, the set of points in $\widehat{M}^N$ whose projection to $M^N$ lies in its coincidence set. Every diffeomorphism $\sigma \in Cov(^N M, {^N M})$ can be expressed as a product

$$\sigma = p\tilde{\sigma} \tag{8.40}$$

where $p \in S_N$ and

$$\tilde{\sigma} = (\sigma^{(1)}, \ldots, \sigma^{(N)}) \in Cov(\widehat{M}, M)^N$$

and these act on $\hat{q} = (\hat{\mathbf{q}}_1, \ldots, \hat{\mathbf{q}}_N) \in \widehat{M}^N$ as follows:

$$\tilde{\sigma}\hat{q} = (\sigma^{(1)}\hat{\mathbf{q}}_1, \ldots, \sigma^{(N)}\hat{\mathbf{q}}_N) \tag{8.41}$$

and

$$p\hat{q} = (\hat{\mathbf{q}}_{p^{-1}(1)}, \ldots, \hat{\mathbf{q}}_{p^{-1}(N)}). \tag{8.42}$$

Thus

$$\sigma\hat{q} = (\sigma^{(p^{-1}(1))}\hat{\mathbf{q}}_{p^{-1}(1)}, \ldots, \sigma^{(p^{-1}(N))}\hat{\mathbf{q}}_{p^{-1}(N)}). \tag{8.43}$$

Moreover, the representation (8.40) of $\sigma$ is unique. Thus, since

$$\sigma_1\sigma_2 = p_1\tilde{\sigma}_1 p_2\tilde{\sigma}_2 = (p_1 p_2)(p_2^{-1}\tilde{\sigma}_1 p_2\tilde{\sigma}_2) \tag{8.44}$$

with $p_2^{-1}\tilde{\sigma}_1 p_2 = (\sigma_1^{(p_2(1))}, \ldots, \sigma_1^{(p_2(N))}) \in Cov(\widehat{M}, M)^N$, we find that $Cov(\widehat{^NM}, {^NM})$ is a semidirect product of $S_N$ and $Cov(\widehat{M}, M)^N$, with product given by

$$\sigma_1\sigma_2 = (p_1, \tilde{\sigma}_1)(p_2, \tilde{\sigma}_2) = (p_1 p_2, p_2^{-1}\tilde{\sigma}_1 p_2\tilde{\sigma}_2). \tag{8.45}$$

Wave functions for the $N$ fermions are sections of the lift $\widehat{E}$ to $\widehat{\mathcal{Q}}$ of the bundle $E$ over $\mathcal{Q}$ with fiber

$$E_q = \bigotimes_{\mathbf{q} \in q} W \tag{8.46}$$

and (nontrivial) connection inherited from the trivial connection on $M \times W$. If the dynamics for $N = 1$ involves wave functions on $\widehat{M}$ obeying (8.24) with topological factor $\Gamma_\sigma(\hat{\mathbf{q}}) = \Gamma_\sigma$ given by a unitary representation of $\pi_1(M)$ (i.e., independent of $\hat{\mathbf{q}}$), then the $N$ fermion wave function obeys (8.24) with topological factor

$$\Gamma_\sigma(\hat{q}) = \text{sgn}(p) \bigotimes_{\mathbf{q} \in \pi(\hat{q})} \Gamma_{\sigma^{(i_{\hat{q}}(\mathbf{q}))}} \equiv \text{sgn}(p)\Gamma_{\tilde{\sigma}}(\hat{q}) \tag{8.47}$$

where for $\hat{q} = (\hat{\mathbf{q}}_1, \ldots, \hat{\mathbf{q}}_N)$, $\pi(\hat{q}) = \{\pi_M(\hat{\mathbf{q}}_1), \ldots, \pi_M(\hat{\mathbf{q}}_N)\}$ and $i_{\hat{q}}(\pi_M(\hat{\mathbf{q}}_j)) = j$. Since

$$\Gamma_{\tilde{\sigma}_1\tilde{\sigma}_2}(\hat{q}) = \Gamma_{\tilde{\sigma}_1}(\hat{q})\,\Gamma_{\tilde{\sigma}_2}(\hat{q}) \tag{8.48}$$

we find, using (8.45) and (8.48), that

$$\Gamma_{\sigma_1\sigma_2}(\hat{q}) = \text{sgn}(p_1 p_2)\Gamma_{p_2^{-1}\tilde{\sigma}_1 p_2\tilde{\sigma}_2}(\hat{q}) \tag{8.49a}$$
$$= \text{sgn}(p_1)\Gamma_{p_2^{-1}\tilde{\sigma}_1 p_2}(\hat{q})\text{sgn}(p_2)\Gamma_{\tilde{\sigma}_2}(\hat{q}) \tag{8.49b}$$
$$= P_2\Gamma_{\sigma_1}(\hat{q})P_2^{-1}\Gamma_{\sigma_2}(\hat{q}), \tag{8.49c}$$

which agrees with (8.36) since the holonomy on the bundle $E$ is given by permutations $P$ acting on the tensor product (8.46).

## 8.5 Conclusions

We have investigated the possible quantum theories on a topologically nontrivial configuration space $\mathcal{Q}$ from the point of view of Bohmian mechanics, which is fundamentally concerned with the motion of matter in physical space, represented by the evolution of a point in configuration space.

Our goal was to find all Bohmian dynamics in $\mathscr{Q}$, where the wave functions may be sections of a Hermitian vector bundle $E$. What "all" Bohmian dynamics means is not obvious; we have followed one approach to what it can mean; other approaches will be described in future works. The present approach uses wave functions $\psi$ that are defined on the universal covering space $\widehat{\mathscr{Q}}$ of $\mathscr{Q}$ and satisfy a periodicity condition ensuring that the Bohmian velocity vector field on $\widehat{\mathscr{Q}}$ defined in terms of $\psi$ can be projected to $\mathscr{Q}$. We have arrived in this way at a natural class of Bohmian dynamics beyond the immediate Bohmian dynamics. Such a dynamics is defined by a potential and some information encoded in "topological factors," which form either a character (one-dimensional unitary representation) of the fundamental group of the configuration space, $\pi_1(\mathscr{Q})$, or a more general algebraic-geometrical object, a holonomy-twisted representation $\Gamma$. Only those dynamics associated with characters are compatible with *every* potential, as one would desire for what could be considered a version of quantum mechanics in $\mathscr{Q}$. We have thus arrived at the known fact that for every character of $\pi_1(\mathscr{Q})$ there is a version of quantum mechanics in $\mathscr{Q}$. A consequence of this is the symmetrization postulate for identical particles. These different quantum theories emerge naturally when one contemplates the possibilities for defining a Bohmian dynamics in $\mathscr{Q}$.

# References

1. L. S. Schulman. A Path Itegral for Spin. *Physical Review*, 176:1558–1569, 1968.
2. L. S. Schulman. Approximate Topologies. *Journal of Mathematical Physics*, 12:304–308, 1971.
3. M. G. Laidlaw and C. M. DeWitt. Feynman Functional Integrals for Systems of Indistinguishable Particles. *Physical Review D*, 3:1375–1378, 1971.
4. L. S. Schulman. *Techniques and Applications of Path Integration*. John Wiley & Sons, New York, 1981.
5. E. Nelson. *Quantum Fluctuations*. Princeton University Press, Princeton, N.J., 1985.
6. G. A. Goldin, R. Menikoff, and D. H. Sharp. Representations of a Local Current Algebra in Nonsimply Connected Space and the Aharonov-Bohm Effect. *Journal of Mathematical Physics*, 22:1664–1668, 1981.
7. D. Dürr, S. Goldstein, and N. Zanghì. Quantum Equilibrium and the Origin of Absolute Uncertainty. *Journal of Statistical Physics*, 67:843–907, 1992.
8. D. Dürr, S. Goldstein, and N. Zanghì. Quantum Equilibrium and the Role of Operators as Observables in Quantum Theory. *Journal of Statistical Physics*, 116:959–1055, 2004.
9. K. Berndl, D. Dürr, S. Goldstein, G. Peruzzi, and N. Zanghì. On the Global Existence of Bohmian Mechanics. *Communications in Mathematical Physics*, 173:647–673, 1995.
10. S. Teufel and R. Tumulka. Simple Proof for Global Existence of Bohmian Trajectories. *Communications in Mathematical Physics*, 258:349–365, 2005.
11. S. Goldstein. Bohmian Mechanics. In E. N. Zalta, editor, Stanford Encyclopedia of Philosophy. Published online by Stanford University, 2001.
12. K. Berndl, M. Daumer, D. Dürr, S. Goldstein, and N. Zanghì. A Survey of Bohmian Mechanics. *Il Nuovo Cimento*, 110B:737–750, 1995.
13. D. Dürr, S. Goldstein, and N. Zanghì. Bohmian Mechanics as the Foundation of Quantum Mechanics. In J. T. Cushing, A. Fine, and S. Goldstein, editors, Bohmian Mechanics and Quantum Theory: an Appraisal, volume 184 of Boston Studies in the Philosophy of Science, pages 21–44. Kluwer Acad. Publ., Dordrecht, 1996.

14. J. S. Bell. On the Problem of Hidden Variables in Quantum Mechanics. *Reviews of Modern Physics*, 38:447–452, 1966. Reprinted in [211] and in [26].
15. J. S. Dowker. Quantum Mechanics and Field Theory on Multiply Connected and on Homogeneous Spaces. *Journal of Physics A*, 5:936–943, 1972.
16. J. Leinaas and J. Myrheim. On the Theory of Identical Particles. *Il Nuovo Cimento B*, 37:1–23, 1977.
17. G. Morandi. *The Role of Topology in Classical and Quantum Physics, volume 7 of Lecture Notes in Physics. New Series m: Monographs*. Springer-Verlag, New York-Heidelberg-Berlin, 1992.
18. V. B. Ho and M. J. Morgan. Quantum Mechanics in Multiply-Connected Spaces. *Journal of Physics A*, 29:1497–1510, 1996.
19. F. Wilczek. Quantum Mechanics of Fractional-Spin Particles. *Physical Review Letters*, 49:957–959, 1982.
20. Y. Aharonov and A. Casher. Topological Quantum Effects for Neutral Particles. *Physical Review Letters*, 53:319–321, 1984.
21. D. Dürr, S. Goldstein, J. Taylor, R. Tumulka, and N. Zanghì. Quantum Mechanics in Multiply-Connected Spaces. *Journal of Physics A: Mathematical and Theoretical*, 40:2997–3031, 2007.

# Part III
# Quantum Relativity

# Chapter 9
# Hypersurface Bohm-Dirac Models

## 9.1 Introduction

Among the different approaches to resolving the conceptual problems of quantum theory, Bohm's approach is perhaps the simplest. In a nutshell, it consists in adding the most basic dynamical variables, obeying additional evolution equations, to the description of a quantum system provided by its wave function $\psi$. For nonrelativistic quantum theory the additional variables are the positions of the particles, which evolve according to a "guiding equation" naturally suggested by the Schrödinger evolution. This theory—usually called Bohmian mechanics or the pilot-wave theory—is well understood. It has been analyzed, and its connection with the predictions of orthodox quantum theory explained, in the original papers of Bohm [1, 2] as well as in later works (see, e.g., [3, 4, 5]). One of the main problems remaining for the Bohmian (or any other) approach is to find a *satisfactory* relativistic quantum theory, a theory that is fully Lorentz invariant while avoiding the profound conceptual difficulties of orthodox quantum theory.

In his original papers, Bohm had an outline for a "Bohmian" field theory, with fields on space-time as the additional variables. A year later he proposed a "Bohmian" model for one Dirac particle [6], which was subsequently extended by Bohm and coworkers to $N$ Dirac particles [7]. For this $N$-particle model the additional variables are, as in Bohmian mechanics, the positions $\mathbf{Q}_k$, $k = 1, \ldots, N$, of the particles. However, in contrast with Bohmian mechanics, the guiding equation for this theory

$$\frac{d\mathbf{Q}_k}{dt} = \frac{\psi^* \boldsymbol{\alpha}_k \psi}{\psi^* \psi} \tag{9.1}$$

is ultralocal on configuration space: The right hand side of (9.1) depends only upon the value of $\psi$ at the positions of the particles and not upon spatial derivatives of $\psi$ there. Here $\psi = \psi(\mathbf{q}_1, \ldots, \mathbf{q}_N, t)$, taking values in the $N$-particle spin space $(\mathbb{C}^4)^{\otimes N}$, solves the $N$-particle Dirac equation ($\hbar = c = 1$)

$$i\frac{\partial \psi}{\partial t} = \sum_{k=1}^{N}(-i\boldsymbol{\alpha}_k \cdot \boldsymbol{\nabla}_k - e\boldsymbol{\alpha}_k \cdot \mathbf{A}(\mathbf{q}_k,t)$$
$$+e\Phi(\mathbf{q}_k,t) + \beta_k m)\psi, \tag{9.2}$$

where $\boldsymbol{\alpha}_k = (\alpha_k^1, \alpha_k^2, \alpha_k^3)$, $\alpha_k^i = I \otimes \cdots \otimes I \otimes \alpha^i \otimes I \otimes \cdots \otimes I$, with the $i$-th Dirac $\alpha$ matrix $\alpha^i$ at the $k$-th of the $N$ places, and $\beta_k$ is defined analogously. $\Phi$ and $\mathbf{A}$ are external electromagnetic potentials. (We may of course consider particle-dependent masses $m_k$, charges $e_k$, and external potentials $\Phi_k$ and $\mathbf{A}_k$, but for simplicity we shall not do so.) We shall call this model the Bohm-Dirac model (BD model). Just as with Bohm's proposal for a field theory, the BD model requires for its formulation the specification of a distinguished frame of reference—in terms of which the actual configuration $(\mathbf{Q}_1, \ldots, \mathbf{Q}_N)$ and the generic configuration $(\mathbf{q}_1, \ldots, \mathbf{q}_N)$ at time $t$ is defined—and in fact the model is not Lorentz invariant if $N > 1$ [7].

However, for $N = 1$ this model is Lorentz invariant, and may be formulated in a covariant way: Writing $X = X(\tau)$ for the space-time point along a trajectory, with (scalar) parametrization $\tau$, the guiding equation may be written as

$$\frac{dX}{d\tau} = j \equiv \overline{\psi}\gamma\psi \tag{9.3}$$

with $\psi$ satisfying the Dirac equation

$$(i\gamma \cdot \partial - e\gamma \cdot A - m)\psi = 0, \tag{9.4}$$

where $\gamma \cdot \partial \equiv \gamma^\mu \partial_\mu$ and $\gamma \cdot A \equiv \gamma^\mu A_\mu(x)$. Note that the right hand side of (9.3), the Dirac current $j = j^\mu \equiv \overline{\psi}\gamma^\mu\psi$, is the simplest 4-vector that can be constructed from the Dirac spinor $\psi$.

Note also that the parameter $\tau$ has no intrinsic physical significance, so that Eq. (9.3) is equivalent to

$$\frac{dX}{d\tau} = aj$$

with arbitrary positive scalar field $a = a(x)$. It is not the field of 4-vectors $j$ (having direction and length) that determines the particle motion, but rather the field of directions defined by $j$. In other words, the law for the particle motion could be formulated in a purely geometrical manner as the condition that the Dirac current $j$ at every point along the trajectory be tangent to the trajectory at that point.

Because the Dirac current is time-like and divergence free,[1]

$$\partial \cdot j = 0,$$

there is a dynamically distinguished probability distribution on the set of particle paths $X(\tau)$ arising from (9.3). Any distribution on this space of paths can be defined

---

[1] The claims in this and the next paragraph follow directly from the application of the divergence theorem (or Stokes' theorem) to an infinitesimally thin tube of paths between $\Sigma_0$ (see below) and the relevant hypersurface $\Sigma$.

## 9.1 Introduction

by specifying for the path the crossing probability for some given equal-time surface $\Sigma_0$ in some Lorentz frame. (By this crossing probability we mean the distribution of the point through which the path crosses $\Sigma_0$, which is the same thing as the probability distribution for the position of the particle in this frame at the given time.) The distinguished distribution is then defined by the crossing probability for $\Sigma_0$ given by $\rho = j^0 = \psi^*\psi$ on $\Sigma_0$ (with $\psi$ suitably normalized), which can be written in a covariant manner as $j \cdot n$ where $n$ in the future-oriented unit normal to the surface. For this distribution the crossing probability for any other equal-time surface will also be given by $j \cdot n$, both for the original frame and any other Lorentz frame. We may roughly summarize the situation by saying that for the distinguished probability distribution, quantum equilibrium holds in all Lorentz frames at all times, with the quantum equilibrium distribution given by $\rho = \psi^*\psi$.

More generally, the crossing probability for any space-like hypersurface $\Sigma$ will also be given by $j \cdot n$, with $n = n(x)$ the future-oriented unit normal field to $\Sigma$. Moreover, for any oriented hypersurface $\Sigma$, the crossing measure (a signed measure that need not be normalized), which describes the expected number of signed crossings through area elements of $\Sigma$, with negatively oriented crossings counted negatively, is, for the distinguished distribution, also given by $j \cdot n$, with $n = n(x)$ now the positively oriented unit normal field to $\Sigma$.[2]

The $N$-particle BD model (9.1) also has a dynamically distinguished probability distribution on paths. As a consequence of (9.2) $\rho = \psi^*\psi$ satisfies, in the Lorentz frame in which the dynamics is defined, the continuity equation

$$\frac{\partial \rho}{\partial t} + \sum_{k=1}^{N} \nabla_k \cdot \mathbf{J}_k = 0, \tag{9.5}$$

where

$$\mathbf{J}_k = \rho \mathbf{v}_k = \psi^* \boldsymbol{\alpha}_k \psi. \tag{9.6}$$

Thus, if the joint probability distribution for the positions of the $N$ particles is given by $\rho = \psi^*\psi$ at some time $t = t_0$, then, for the corresponding distribution on paths, it will be given by $\rho = \psi^*\psi$ at all times $t$. However, even for this distinguished distribution, quantum equilibrium will not in general hold in other Lorentz frames: The joint distribution of crossings of equal-time surfaces for other frames will in general not be given by $\psi'^*\psi'$ (where $\psi'$ is the wave function in the relevant Lorentz frame) [8, 9]. Nonetheless, Bohm and coworkers have argued that the observational content of this model is as Lorentz invariant as the covariant formalism of relativistic quantum theory: Since the predictions for results of measurements for this model can be regarded as reflected in the configuration of various devices and registers—and hence can be derived from probabilities for positions given by $\rho = \psi^*\psi$—at a

---

[2] In this regard it is perhaps worth noting the following: In Minkowski space there is a natural duality between divergence-free vector fields and closed 3-forms. Such a vector field defines a "deterministic" random path, whose "law" is given directly by the vector field, as in (9.3), and whose statistics are governed by the dual 3-form, in the manner just described.

common time in the distinguished frame, these predictions must agree with those of the usual interpretation. Thus no violation of Lorentz invariance can be detected in experiments [7]. (In particular, the identity of the distinguished Lorentz frame cannot be ascertained by means of any possible observation.)

Lorentz invariance is, however, a delicate issue. Indeed, any theory can be made trivially Lorentz invariant (or invariant under any other space-time symmetry), even on the microscopic level, by the incorporation of suitable additional structure [9]. For this reason Bell has stressed that one should consider what he has called "serious Lorentz invariance," a notion, however, that is extremely difficult to make precise in an adequate way [3]. Lacking a general criterion, we may nonetheless begin to get a handle on "serious Lorentz invariance" by analyzing some specific models. If the models involve additional structure, then whether or not we have serious Lorentz invariance will depend, of course, upon the detailed nature of this structure.

In [9] we have considered a model for which the additional structure for a system of $N$ (noninteracting) Dirac particles is provided by a global synchronization among the particles: The trajectories of the particles are such that each one of them at some given space-time point is tangent to a vector field determined, given the wave function, by that point and those points along the trajectories of the other particles with which that point has been "synchronized." This additional synchronization structure is defined implicitly by the equation of motion and the model is not amenable to a statistical analysis in any obvious way. In other words, this model is not statistically transparent (see Section IV of [9]). Nonetheless, even this model provides a counterexample to the widely held belief that a Lorentz invariant Bohmian theory for many particles is impossible (unless only product states are allowed). In this regard, see also the local model of Squires [10].

In this chapter we shall analyze a statistically transparent counterexample, the "hypersurface Bohm-Dirac model" (HBD model). The basic idea was proposed in [11] in the context of bosonic quantum field theory: In addition to the wave function and field variables, a distinguished foliation of space-time—a new element of geometrical structure defining simultaneity surfaces—is suggested as an additional dynamical variable of the theory. These surfaces need not be hyperplanes. The defining (Lorentz invariant) equations of the theory should describe the evolution of the wave function, the field variables, and the simultaneity surfaces. For a careful philosophical discussion of how this may be compatible with some appropriate notion of relativity, even if the simultaneity surfaces should turn out to be unobservable, see Maudlin [12].

Here we shall consider such a theory, not for fields but for $N$ (noninteracting) Dirac particles. We shall discuss an as yet incomplete hypersurface Bohm-Dirac model: The law for the evolution of the foliation is not specified, beyond the requirement that it not involve the positions of the particles. We present no hypothesis concerning the origin of the foliation, but have in mind that the foliation should ultimately be governed by a Lorentz invariant law, one that may, for example, involve the $N$-particle wave function. (For definiteness we shall give some very tentative and less than compelling examples of laws for the foliation in Sect. 9.4.) However, we show in Sect. 9.3.1 that, regardless of how the foliation is determined, the dynamics of the HBD model preserves the quantum equilibrium distribution on the leaves of the foliation. Thus the model is amenable to the same sort of statistical analysis as for nonrelativistic Bohmian mechanics. This is discussed briefly in Sect. 9.3.2.

## 9.2 The Hypersurface Bohm-Dirac Model

A general foliation $\mathscr{F}$ of codimension one on Minkowski space $M$ can approximately be thought of as a partition of $M$ into 3-dimensional hypersurfaces. These hypersurfaces are the leaves of the foliation. The simplest way to obtain a foliation is by a smooth function $f : M \to \mathbb{R}$ without critical points, i.e., $df \neq 0$ everywhere. The level sets $f^{-1}(s)$ are smooth hypersurfaces and form a foliation of $M$. With the one-form $df_x$, which vanishes on the tangent space of the hypersurface through $x \in M$, we may associate by the Lorentz metric the normal vector field $\partial f(x)$. If this is time-like everywhere, and thus the foliation hypersurfaces space-like, we may normalize $\partial f(x)$ to obtain a unit normal vector field $n(x)$ associated with the foliation $\mathscr{F}$.

We shall consider in this chapter only space-like foliations, i.e., foliations by space-like hypersurfaces. While obviously different $f$'s may generate the same foliation $\mathscr{F}$, the future-oriented unit normal vector field $n$ is uniquely determined by $\mathscr{F}$. When does a vector field $v(x)$ determine a foliation $\mathscr{F}$ such that for all $x \in M$, $v(x)$ is normal to the tangent space of the foliation hypersurface through $x$? If we denote by $V$ the one-form associated with $v$ by the Lorentz metric, then, by Frobenius' theorem, the necessary and sufficient condition is that $V$ be completely integrable, $V \wedge dV = 0$.

Apart from the foliation, the other dynamical variables of the hypersurface Bohm-Dirac model are the usual ones: the wave function $\psi$, here for $N$ Dirac particles, and the $N$-path, the $N$-tuple of (everywhere either time-like or light-like) space-time paths, which describes the trajectories of the $N$ Dirac particles. Covariant laws for these dynamical variables suggest themselves when we write those of the Bohm-Dirac model, defined by (9.1) and (9.2), in a coordinate-free, i.e., covariant manner.

To achieve this we consider first of all the $\psi$-function in the multi-time formalism: For $N$ Dirac particles the wave function $\psi = \psi(x_1, x_2, \ldots, x_N)$, $x_k \in M$, takes values in the $N$-particle spin space $(\mathbb{C}^4)^{\otimes N}$ and satisfies $N$ Dirac equations

$$(i\gamma_k \cdot \partial_k - e\gamma_k \cdot A(x_k) - m)\psi = 0, \tag{9.7}$$

$k = 1, \ldots, N$. Here $\gamma_k = I \otimes \cdots \otimes I \otimes \gamma \otimes I \otimes \cdots \otimes I$, with $\gamma$ at the $k$-th of the $N$ places, and $A$ is an external electromagnetic potential. (Just as with (9.2), we may of course consider particle-dependent masses $m_k$, charges $e_k$, and external potentials $A_k$.) The system of Eq. (9.7) is a covariant version of (9.2); in this multi-time form the Lorentz invariance of the law for $\psi$ is manifest [3].[3] The $N$ Dirac particles are coupled by the common wave function $\psi$. If this is entangled, we have

---

[3] Note that in the single-time form (9.2) we can easily add an explicit interaction potential $V(\mathbf{q}_1, \ldots, \mathbf{q}_N, t)$ for the $N$ Dirac particles, while in the multi-time form this is impossible.

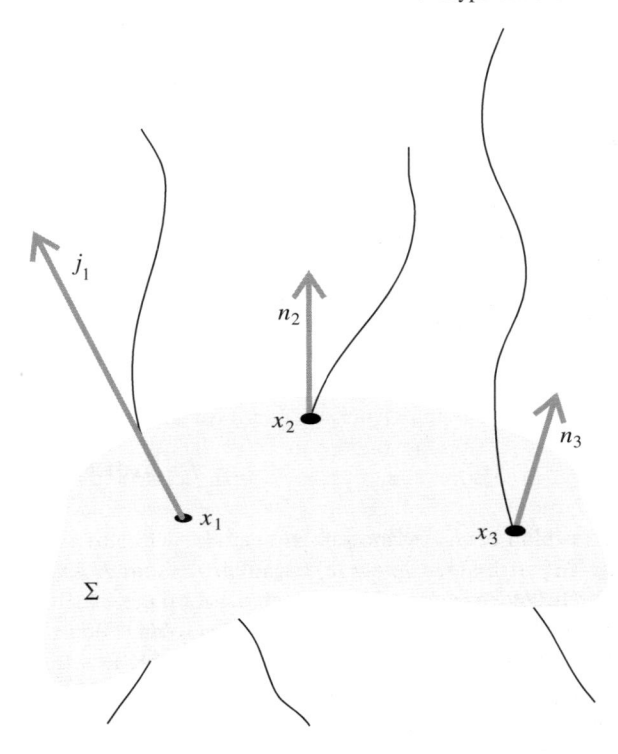

**Fig. 9.1** Geometrical formulation of the dynamics for a system of three particles: For each particle the path of that particle, say particle 1 at $x_1$, must be tangent to the 4-vector $j_1$ which is determined by: 1) the intersections $x_2$ and $x_3$ of the trajectories of the other two particles with the hypersurface $\Sigma$ containing $x_1$, 2) the future-oriented unit normals $n_2$ and $n_3$ at these points, and 3) the wave function of the system evaluated at $x_1$, $x_2$ and $x_3$: $j_1 = \overline{\psi}(x_1, x_2, x_3)\gamma_1(\gamma_2 \cdot n_2)(\gamma_3 \cdot n_3)\psi(x_1, x_2, x_3)$

nonlocal correlations between the $N$ particles, despite the fact that the particles are noninteracting.

We shall now develop the guiding law for the $N$-path. Note that the numerator of the right hand side of (9.1) is given by a current $j_k$,

$$j_k = \overline{\psi}\gamma_1^0 \ldots \gamma_k \ldots \gamma_N^0 \psi,$$

that involves matrix elements of an operator having as factors the 0-component $\gamma^0$ of a 4-vector for all but the $k$-th particle. Therefore $j_k$ can be expressed in a covariant manner by replacing $\gamma_i^0$ in the above expression with $\gamma_i \cdot n$, where $n$ is the future-oriented unit normal to the $t = $ const hyperplanes,

$$j_k = \overline{\psi}(\gamma_1 \cdot n) \ldots \gamma_k \ldots (\gamma_N \cdot n)\psi. \tag{9.8}$$

## 9.2 The Hypersurface Bohm-Dirac Model

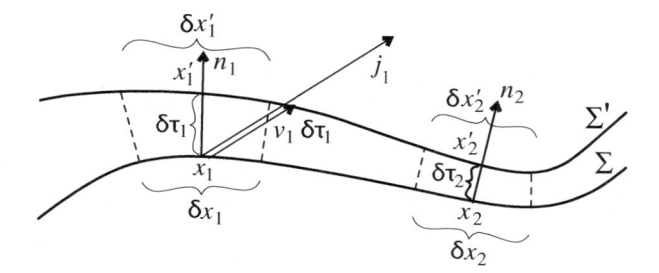

**Fig. 9.2** Motion of two particles in one space dimension from hypersurface $\Sigma$ to $\Sigma'$: space-time view. We have indicated the positions of the primed points $x'_k$ obtained from $x_k$ via displacement from $\Sigma$ to $\Sigma'$ in the normal direction, and the images $\delta x'_k$ of the regions $\delta x_k$ under this correspondence. The point on $\Sigma'$ to which particle $k$ moves when starting at $x_k \in \Sigma$ is given (to leading order) by $x_k + v_k \delta\tau_k$ with $v_k = j_k/(j_k \cdot n_k)$, where $\delta\tau_k$ is the Minkowski distance between $x_k$ and $x'_k$.

Moreover, the denominator of the right hand side of (9.1) can be expressed covariantly as $j_k \cdot n$. Then the covariant velocity of the $k$-th particle—with respect to the time of a Lorentz frame with $n$ as time axis—is

$$\frac{dX_k}{dt} = \frac{j_k}{j_k \cdot n}. \tag{9.9}$$

Since $j_k \cdot n = \overline{\psi}(\gamma_1 \cdot n)\ldots(\gamma_N \cdot n)\psi$ is independent of $k$, we may reparametrize the paths with a parameter $s$ so related to $t$ that $t'(s) = j_k \cdot n$ to obtain

$$\frac{dX_k}{ds} = j_k. \tag{9.10}$$

More generally, by further reparametrization, we may obtain $dX_k/d\tau = aj_k$, where $a$ is any positive scalar field. The physical particle dynamics—i.e., the $N$ space-time paths defined by the equations of motion (and initial conditions)—is invariant under reparametrization.

A manifestly "parametrization invariant" formulation of the dynamics—that is, such that a time parameter plays no role—is easily obtained: The space-time paths for the $N$ particles are constrained by the currents $j_k$ by requiring that the path for the $k$-th particle at the point $x_k$ be tangent to the current $j_k$ evaluated at $x_k$ and at the intersection points of the paths of the $N-1$ other particles with the $t = $ const-hyperplane $\Sigma_t$ containing $x_k$. If we denote by $X_k(\Sigma_t)$ the intersection point of the path $X_k$ with the hyperplane $\Sigma_t$, and by $\dot{X}_k(\Sigma_t)$ a tangent of (or the tangent line to) the path $X_k$ at $X_k(\Sigma_t)$, we may write the law for the $N$-path as

$$\dot{X}_k(\Sigma_t) \parallel j_k(X_1(\Sigma_t), \ldots, X_N(\Sigma_t)), \tag{9.11}$$

using the symbol $\parallel$ for "is parallel to." In this geometric formulation the Bohm-Dirac dynamics depends upon the Lorentz frame only via its associated foliation into simultaneity hypersurfaces $\Sigma_t$, and thus naturally extends to an arbitrary foliation $\mathscr{F}$ of Minkowski space-time $M$ by curved space-like hypersurfaces:[4]

Given such a foliation $\mathscr{F}$ and $\Sigma \in \mathscr{F}$, let $X_k(\Sigma)$ be the intersection of the path $X_k$ with $\Sigma$,[5] and let $\dot{X}_k(\Sigma)$ be a tangent of (or the tangent line to) the path $X_k$ at $X_k(\Sigma)$. The law of the $N$-path $X = (X_1, \ldots, X_N)$ for the hypersurface Bohm-Dirac model is defined by the currents $j_k$ naturally extending (9.8)

$$j_k = \overline{\psi}(\gamma_1 \cdot n_1) \ldots \gamma_k \ldots (\gamma_N \cdot n_N)\psi, \tag{9.12}$$

where $n_1 \equiv n(x_1), \ldots, n_N \equiv n(x_N)$, with $n$ the future-oriented unit normal vector field associated with $\mathscr{F}$, via the HBD tangency condition (see also Fig. 9.1)

$$\dot{X}_k(\Sigma) \parallel j_k(X_1(\Sigma), \ldots, X_N(\Sigma)). \tag{9.13}$$

(By considering the action of a suitable Lorentz transformation on $\gamma^0 \gamma \cdot n$ for arbitrary time-like unit vector $n$ (transforming $n$ to $(1, 0, 0, 0)$), one sees that $\gamma^0 \gamma \cdot n$ is a positive operator in spin space $\mathbb{C}^4$. Hence $(\gamma_1^0 \gamma_1 \cdot n_1) \ldots (\gamma_k^0 \gamma_k \cdot n) \ldots (\gamma_N^0 \gamma_N \cdot n_N)$ is also positive, i.e., $j_k \cdot n \geq 0$ with "=" only if $\psi = 0$. This means that, where it is nonzero, $j_k$ is future-oriented and, like the path $X_k$, nowhere space-like.)

We may also write down the equations of motion in the parametrized form analogous to (9.9) or (9.10). To do so it is convenient to label the hypersurfaces of the foliation using a function $f : M \to \mathbb{R}$ that generates the foliation as described above, and use this hypersurface labeling as the parameter for the particle trajectories—so that $X_k(s)$ is on the hypersurface $f^{-1}(s)$. From the geometrical characterization of the dynamics (9.13) we know that $dX_k/ds$ is parallel to $j_k(X_1(s), \ldots, X_N(s))$, and the scale factor required to ensure $f(X_k(s)) = s$ for all $k$ and $s$ is easily seen to be $1/(\partial f \cdot j_k)$. Therefore

$$\frac{dX_k}{ds} = \frac{j_k(X_1(s), \ldots, X_N(s))}{\partial f(X_k(s)) \cdot j_k(X_1(s), \ldots, X_N(s))}. \tag{9.14}$$

For a flat foliation we may choose a Lorentz frame such that the foliation hyperplanes are the $x^0 = $ const-planes, i.e., $f(x) = x^0$ for all $x$. Then $n = \partial f = (1, 0, 0, 0)$ and (9.14) reduces to the Bohm-Dirac law (9.1).

---

[4] This is in marked contrast with the parametrized dynamics such as given by Eqs. (9.9 or 9.10), which need not extend in anything like the same form to a general foliation since the parametrized paths generated by the dynamics need not, in general, respect the foliation.

[5] Note that the paths $X_k$ comprising an $N$-path, since they are nowhere space-like, can intersect $\Sigma$ at most once. This is the main reason why it is important that the foliation $\mathscr{F}$ be space-like. Of course, also from the physical point of view a synchronization along space-like hypersurfaces yields a picture which perhaps makes most sense. We shall assume, without further ado, global existence: that a fragment of an $N$-path locally satisfying the HBD tangency condition, see (9.13), can be continued in such a manner that each of its paths $X_k$ intersects every $\Sigma \in \mathscr{F}$.

## 9.3 Statistical Analysis of the HBD Model

### 9.3.1 Quantum Equilibrium

We shall show now that for the hypersurface Bohm-Dirac model, with foliation $\mathscr{F}$, there is a distinguished probability measure on $N$-paths $X$ satisfying the HBD tangency condition (9.13), one for which the distribution of hypersurface crossings $X_1(\Sigma), \ldots, X_N(\Sigma)$ for $\Sigma \in \mathscr{F}$ depends only upon $\psi$ restricted to $\Sigma$ (or, more precisely, to $\Sigma^N$) for $\psi$ satisfying (9.7). We shall say that such a distinguished measure, as well as the corresponding hypersurface crossing distribution, is *equivariant*, defining *quantum equilibrium*. The physical significance of the hypersurfaces $\Sigma \in \mathscr{F}$ is thus twofold: They serve (via (9.13)) to define the motion of the particles, and, for a quantum equilibrium $N$-path, it is "on these hypersurfaces" that, manifestly, the "particles are in quantum equilibrium."

The natural candidate for the equivariant crossing probability density $\rho$ of the HBD model is given by the obvious covariant extension of the equivariant density $\psi^*\psi \, (= \overline{\psi}\gamma_1^0 \ldots \gamma_N^0 \psi)$ of the BD model:

$$\rho = \overline{\psi}(\gamma_1 \cdot n_1)\ldots(\gamma_N \cdot n_N)\psi. \tag{9.15}$$

To see that this is in fact equivariant, note the following: In view of (9.12), (i) $\rho = j_k \cdot n_k$ and

$$j_k \cdot n_k \text{ is independent of } k. \tag{9.16}$$

Furthermore, (ii) the currents $j_k$ are divergence free:

$$\partial_k \cdot j_k = 0, \tag{9.17}$$

which follows immediately from (9.12) using the Dirac equation (9.7) and its adjoint. These two properties of the currents, (9.16) and (9.17), are the key ingredients for the proof of the equivariance of $\rho$. For any current satisfying (9.16) and (9.17), for the particle dynamics defined by (9.13), $\rho = j_k \cdot n_k$ is an equivariant probability density for crossings of the leaves of the foliation.[6]

The proof of this assertion consists of two steps: First we determine how an arbitrary probability density $R$ on crossings of a foliation hypersurface $\Sigma$ evolves under the dynamics (9.13), i.e., we formulate the continuity equation of the hypersurface dynamics. In the second step, we show that $R = \rho$ solves the continuity equation. It then follows that if the probability distribution of the "positions of the $N$ particles" on $\Sigma \in \mathscr{F}$ is given by $\rho$ restricted to $\Sigma$, then for any other hypersurface $\Sigma' \in \mathscr{F}$, the probability distribution of the "positions of the $N$ particles" on $\Sigma'$ which emerges by transport according to the dynamics (9.13) is given by $\rho$ restricted to $\Sigma'$. Thus $\rho$ is equivariant.

---

[6] In contrast, the current $j_k = \overline{\psi}\gamma_k\psi$ we considered in [9] satisfies (9.17) but not (9.16).

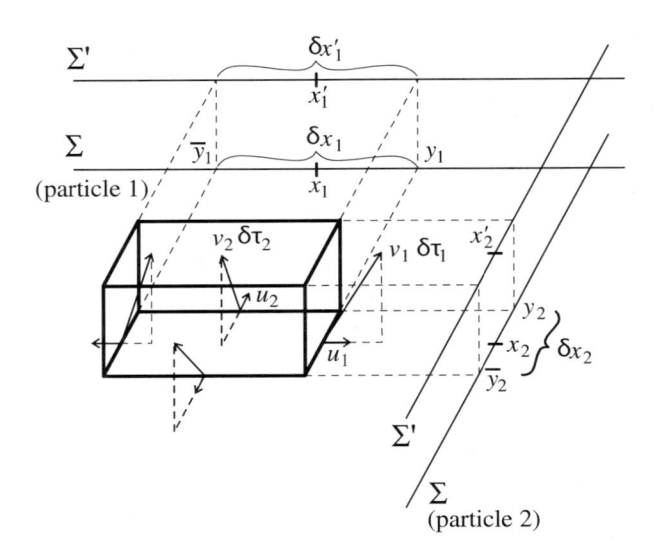

**Fig. 9.3** Conservation of probability for a system of two particles in one space dimension: configuration-space-time view with, for simplicity, the hypersurfaces drawn straightened out. (Note that the figure fails to convey the fact—displayed in Fig. 9.2—that the areas $\delta x_k$ and $\delta x'_k$ may differ, and that also $\delta \tau(y_k)$ may differ from $\delta \tau(\bar{y}_k)$, where $y_k$ and $\bar{y}_k$ are the boundary points of $\delta x_k$.) The change of the probability of particle 1 being in $\delta x_1$ and particle 2 being in $\delta x_2$ from hypersurface $\Sigma$ to $\Sigma'$ is accounted for by the single particle fluxes through the lateral sides of the configuration-space-time box between $\delta x_1 \cdot \delta x_2 \subset \Sigma^2$ and the corresponding set of primed points on $(\Sigma')^2$, i.e., $R_{\Sigma'}(x'_1, x'_2)\delta x'_1 \delta x'_2 - R_\Sigma(x_1, x_2)\delta x_1 \delta x_2 = -((R_\Sigma v_1)(\bar{y}_1, x_2) \cdot (u_1 \delta \tau)(\bar{y}_1) + (R_\Sigma v_1)(y_1, x_2) \cdot (u_1 \delta \tau)(y_1))\delta x_2 - ((R_\Sigma v_2)(x_1, \bar{y}_2) \cdot (u_2 \delta \tau)(\bar{y}_2) + (R_\Sigma v_2)(x_1, y_2) \cdot (u_2 \delta \tau)(y_2))\delta x_1$. Equation (9.18) is the natural extension of this formula to $N$ particles in Minkowski space

Consider thus two infinitesimally close hypersurfaces $\Sigma$ and $\Sigma'$ belonging to the foliation $\mathscr{F}$. The probability distribution of the positions of the $N$ particles on $\Sigma$ is given by a density $R_\Sigma : \Sigma^N \to \mathbb{R}$ such that

$$\text{Prob(particle } i \text{ crosses } \Sigma \text{ in } \delta x_i, \ i = 1, \ldots, N)$$
$$= R_\Sigma(x_1, \ldots, x_N)\delta x_1 \cdots \delta x_N.$$

By $\delta x$ we denote simultaneously an infinitesimal region on $\Sigma$ around $x$ and its area (i.e., 3-volume). Now we compare $R_\Sigma$ evaluated at $(x_1, \ldots, x_N) \in \Sigma^N$ with $R_{\Sigma'}$ evaluated at $(x'_1, \ldots, x'_N) \in (\Sigma')^N$, where $x' \in \Sigma'$ is obtained from $x \in \Sigma$ via displacement from $\Sigma$ to $\Sigma'$ in the normal direction, see Fig. 9.2. Let $\delta x'$ be the area of the image of the region $\delta x$ under this correspondence. (Since the projection of the Lorentz metric on $\Sigma'$ need not agree with the image, under $x \mapsto x'$, of its projection on $\Sigma$, $\delta x$ and $\delta x'$ need not agree.)

Recall from elementary physics that a continuity equation such as (9.5) is an expression of a local conservation law that, on the infinitesimal level, can be stated

## 9.3 Statistical Analysis of the HBD Model

as follows: The difference between the probability densities $R_\Sigma$ on $\Sigma^N$ and $R_{\Sigma'}$ on $(\Sigma')^N$ (with $\Sigma'$ infinitesimally later than $\Sigma$) is accounted for by the flux through the lateral sides—to which the hypersurface normals are tangent—of the configuration-space-time box between $\delta x_1 \times \cdots \times \delta x_N \subset \Sigma^N$ and the corresponding set of (primed) points in $(\Sigma')^N$, see Fig. 9.3;

$$R_{\Sigma'}(x'_1,\ldots,x'_N)\delta x'_1 \cdots \delta x'_N - R_\Sigma(x_1,\ldots,x_N)\delta x_1 \cdots \delta x_N$$
$$= -\sum_{k=1}^{N} \delta x_1 \ldots \widehat{\delta x_k} \cdots \delta x_N \int_{\partial(\delta x_k)} (R_\Sigma v_k)(x_1,\ldots,x_{k-1},y,x_{k+1},\ldots,x_N) \cdot$$
$$(u_k \delta \tau)(y)\, dS_k, \tag{9.18}$$

where the $\widehat{\ }$ on $\widehat{\delta x_k}$ indicates that this term should be omitted from the product. Here $y$ is the integration variable on $\partial(\delta x_k)$, the (2-dimensional) boundary of $\delta x_k$ regarded as a region in $\Sigma$, $dS_k$ is the area element of $\partial(\delta x_k)$, $u_k$ is the outward unit normal vector field in $\Sigma$ to $\partial(\delta x_k)$, $\delta \tau(y)$ is the Minkowski distance between $y \in \Sigma$ and the corresponding $y' \in \Sigma'$ (so that $y' = y + \delta \tau(y) n(y)$) and

$$v_k = \frac{j_k}{j_k \cdot n_k} \tag{9.19}$$

is the covariant velocity of the $k$-th particle relative to $\Sigma$, see Fig. 9.2.

Equation (9.18) is the continuity equation for the HBD model in the "infinitesimally integrated form." It is valid for any hypersurface dynamics defined by (9.13), regardless of whether the currents $j_k$ satisfy (9.16) and (9.17). However, as we shall now show, if the currents do satisfy (9.16) and (9.17), then $R_\Sigma = \rho|_\Sigma = (j_k \cdot n_k)|_\Sigma$ satisfies (9.18).

Since[7]

$$\rho(x'_1,\ldots,x'_N)\delta x'_1 \cdots \delta x'_N - \rho(x_1,\ldots,x_N)\delta x_1 \cdots \delta x_N$$
$$= \rho(x'_1,\ldots,x'_N)\delta x'_1 \cdots \delta x'_N$$
$$-\rho(x_1,x'_2,\ldots,x'_N)\delta x_1 \delta x'_2 \cdots \delta x'_N$$
$$+\rho(x_1,x'_2,\ldots,x'_N)\delta x_1 \delta x'_2 \cdots \delta x'_N$$
$$-\rho(x_1,x_2,x'_3,\ldots,x'_N)\delta x_1 \delta x_2 \delta x'_3 \cdots \delta x'_N$$
$$+\cdots + \rho(x_1,\ldots,x_{N-1},x'_N)\delta x_1 \cdots \delta x_{N-1} \delta x'_N$$
$$-\rho(x_1,\ldots,x_N)\delta x_1 \cdots \delta x_N, \tag{9.20}$$

we obtain in this case for the left hand side of (9.18) (to leading order)

---

[7] Note that this decomposition is possible because $\rho$ is defined on $M^N$ (with $M$ Minkowski space), in contrast with an arbitrary $R = (R_\Sigma)_{\Sigma \in \mathscr{F}}$, defined only for $N$-tuples belonging to $\Sigma^N$ for some $\Sigma \in \mathscr{F}$, for which therefore such a decomposition is impossible.

$$\sum_{k=1}^{N} \delta x_1 \ldots \widehat{\delta x_k} \cdots \delta x_N \big(j_k(x_1,\ldots,x'_k,\ldots,x_N) \cdot n(x'_k)\delta x'_k$$
$$- j_k(x_1,\ldots,x_k,\ldots,x_N) \cdot n(x_k)\delta x_k\big), \tag{9.21}$$

while the integrand on the right hand side of (9.18) becomes $(j_k \cdot u_k)\delta\tau \, dS_k$. Thus, subtracting the right hand side of (9.18) from (9.21), we obtain (to leading order) the sum over $k$ of the integral of $j_k$ over the (outward oriented) boundary of the space-time region above $\delta x_k$ between $\Sigma$ and $\Sigma'$. But since $j_k$ is divergence-free (9.17), each such term, and hence the sum, vanishes. Thus (9.18) is satisfied, establishing the equivariance of $\rho$.

We may also write the continuity equation (9.18) in a purely local form: Writing
$$\delta R_\Sigma(x_1,\ldots,x_N) = R'_{\Sigma'}(x'_1,\ldots,x'_N) - R_\Sigma(x_1,\ldots,x_N),$$
where
$$R'_{\Sigma'}(x'_1,\ldots,x'_N)\delta x_1 \cdots \delta x_N$$
$$= R_{\Sigma'}(x'_1,\ldots,x'_N)\delta x'_1 \cdots \delta x'_N, \tag{9.22}$$

and applying Gauss' theorem to the right hand side of (9.18)
$$\int_{\partial(\delta x_k)} R_\Sigma v_k \cdot u_k \delta\tau \, dS_k = \operatorname{div}^\Sigma_k (R_\Sigma v^\Sigma_k \delta\tau(x_k)) \, \delta x_k,$$

where $\operatorname{div}^\Sigma_k$ is the divergence with respect to the $k$-th coordinate $x_k$ on the Riemannian manifold $\Sigma$ and $v^\Sigma_k$ is the projection of $v_k$ on $\Sigma$, yields

$$\delta R_\Sigma + \sum_{k=1}^{N} \operatorname{div}^\Sigma_k (R_\Sigma v^\Sigma_k \delta\tau_k) = 0, \tag{9.23}$$

where $\delta\tau_k \equiv \delta\tau(x_k)$.

Using this form we may also check the equivariance of $\rho$. To do so, we first "smoothly" label the hypersurfaces of the foliation $\mathscr{F}$ by a parameter $s \in \mathbb{R}$, increasing in the future direction, which may be called a "time parameter," in terms of which (9.23) becomes a standard differential equation. The function $f : M \to \mathbb{R}$ that maps any point $x \in M$ to the label $s$ of the hypersurface $\Sigma_s$ to which $x$ belongs generates the foliation in the manner described in Sect 9.2. In particular, $\partial f = \|\partial f\| n$, where $n$ is the future-oriented unit normal vector field of $\mathscr{F}$. With $\delta s = \|\partial f_k\| \delta\tau_k$, where $\partial f_k \equiv \partial f(x_k)$, we get from (9.19) that

$$v_k \delta\tau_k = \frac{j_k}{j_k \cdot \partial f_k} \delta s \equiv \hat{v}_k \delta s, \tag{9.24}$$

with $\hat{v}_k = dX_k/ds$ the velocity of the $k$-th particle in the parametrized formulation of the dynamics (9.14).

## 9.3 Statistical Analysis of the HBD Model

Consider now a coordinate system adapted to our parametrized foliation $\Sigma_s$: one coordinate is clearly given by $s$, and on one foliation hypersurface we introduce an (arbitrary) coordinate system $p$, which is transported to the other foliation hypersurfaces by the flow along the normal field, yielding the system of coordinates $(s, p)$, allowing us to write $x = (s, p)$ for $x \in M$. Then $x_k = (s_k, p_k) \in \Sigma_s \Leftrightarrow s_k = s$, and the relation between $x = (s, p)$ and $x' = (s', p')$ from Fig. 9.2 becomes $p = p'$. Let $\delta p$ be the volume element defined by the $p$-coordinates and let $\delta x = g(p, s)\delta p$. In these adapted coordinates the continuity Eq. (9.23) assumes, using (9.22) and (9.24), the more standard form

$$\frac{1}{g_1 \cdots g_N} \frac{\partial (g_1 \cdots g_N R_s)}{\partial s} + \sum_{k=1}^{N} \operatorname{div}_k^{\Sigma_s} (R_s \hat{v}_k^{\Sigma_s}) = 0, \qquad (9.25)$$

with $R_s(p_1, \ldots, p_N) = R_{\Sigma_s}((s, p_1), \ldots, (s, p_N))$ and $g_k = g(s, p_k)$, $k = 1, \ldots, N$ (and where $\hat{v}_k^{\Sigma_s}$ is the projection of $\hat{v}_k$ on $\Sigma_s$).[8]

For $R_s = \rho_s$, (9.20) is what lies behind the usual (implication of the) chain rule

$$\frac{1}{g_1 \cdots g_N} \frac{\partial (g_1 \cdots g_N \rho_s)}{\partial s}$$
$$= \sum_{k=1}^{N} \frac{1}{g_k} \frac{\partial (g(s_k, p_k)\rho(s_1, p_1, \ldots, s_k, p_k))}{\partial s_k}\bigg|_{s_k=s}. \qquad (9.26)$$

Splitting the 4-divergence into pieces corresponding to variations orthogonal to and variations within $\Sigma_s$, we obtain

$$\operatorname{div} j = \|\partial f\| \left( \frac{1}{g} \frac{\partial}{\partial s} (g j^0) + \operatorname{div}^{\Sigma_s} \left( \|\partial f\|^{-1} j^{\Sigma_s} \right) \right),$$

where $j^0$ is the normal component of $j$, $j^0 = j \cdot n$. Setting $j = j_k$ and using $\operatorname{div} j_k = 0$ (9.17) we then find with (9.24) that

$$\frac{1}{g_k} \frac{\partial (g_k \rho)}{\partial s_k} + \operatorname{div}_k^{\Sigma_s} \left( \rho \hat{v}_k^{\Sigma_s} \right) = 0$$

for all $k$. Therefore, in view of (9.26), summation over $k$ establishes that $R_s = \rho_s$ satisfies the HBD continuity equation (9.25).

### 9.3.2 Comparison with Quantum Mechanics

The statistical analysis of the hypersurface Bohm-Dirac model can be based on the assumption that the probability distribution on $N$-paths is given by the equivariant

---

[8] This evolution equation depends upon $g$ only through the area-expansion factor arising from the normal flow between hypersurfaces, and thus does not really depend upon the choice of coordinates on the hypersurfaces.

density $\rho$ (9.15) on some simultaneity surface $\Sigma$ belonging to the foliation $\mathscr{F}$. Then, by equivariance, the statistical predictions of the HBD model (i.e., the crossing probabilities) agree with the quantum predictions for positions for any hypersurface in $\mathscr{F}$. But what can be said about the statistical predictions concerning a hypersurface which is not part of a member of $\mathscr{F}$?

For one particle the situation is very simple: From the geometrical formulation of the HBD model (Sect. 9.2) it follows immediately that the HBD model for one particle is foliation-independent, and in fact is the usual one-particle Bohm-Dirac theory given by Eqs. (9.3 and 9.4), with current $j = \overline{\psi}\gamma\psi$. Thus in this case the statistical predictions of the model agree with the quantum predictions for position along *any* hypersurface.

The situation is analogous for $N$ independent particles: If the wave function $\psi$ is a product wave function, $\psi = \psi_1(x_1)\cdots\psi_N(x_N)$, then it follows from the multi-time Dirac equation (9.7) that $\psi_k$ satisfies the usual one-particle Dirac equation. Furthermore, the path of the $k$-th particle is tangent to the one-particle current $\overline{\psi}_k\gamma\psi_k$ and thus independent of the paths of the other particles. Moreover $\rho$ is the product of the corresponding 1-particle distributions. Therefore, a product wave function indeed generates a foliation-independent motion, the motion of $N$ independent Bohm-Dirac particles, and we thus have agreement with all the quantum position distributions in this case.

In the general case the situation is more subtle: If the $N$-particle wave function is entangled, it will not in general be the case that the distribution of crossings of hypersurfaces not belonging to the foliation agree with the corresponding quantum position distributions [8, 9] (which, in fact, may be incompatible with the crossing statistics for any trajectory model whatsoever). However, this disagreement does not entail violations of the quantum predictions, as has been discussed for the case of the multi-time translation invariant Bohmian theory in [9]. In fact, insofar as results of measurement are concerned, the predictions of our model are the same as those of orthodox quantum theory, for positions or any other quantum observables, regardless of whether or not these observables refer to a common hypersurface belonging to $\mathscr{F}$.[9]

This is because the outcomes of all quantum measurements can ultimately be reduced to the orientations of instrument pointers, counter readings, or the ink distribution of computer printouts, if necessary brought forward in time to a common hypersurface in $\mathscr{F}$, or even to a single common location, for which agreement is assured. Nonetheless, this situation may seem paradoxical if we forget the non-passive character of measurement in quantum mechanics. The point is that for Bohmian quantum theory, measurement can effect even distant systems, so that the resulting positions—and hence their subsequently measured values—are different from what they would have been had no measurement occurred.

---

[9] This conclusion requires the rather dubious assumption that the relevant measurements can be understood in terms of noninteracting Dirac particles. However, in order to talk coherently about the quantum predictions for a model, it must be possible to understand measurement processes in terms of that model. The remarks we are making here would also be appropriate for the more realistic models for which this would be true.

## 9.4 Perspective

We have presented a hypersurface Bohm-Dirac model for $N$ entangled but noninteracting Dirac particles. This model is a covariant extension of the Bohm-Dirac model, which involves a foliation by equal-time (flat) hypersurfaces, to arbitrarily shaped (smooth) hypersurfaces. How natural is this model?

When looking for a relativistic extension of nonrelativistic Bohmian mechanics one inevitably encounters two central, very different problems: that such an extension must involve a mechanism for nonlocal interactions between the particles, and that quantum equilibrium cannot hold in all Lorentz frames. For both of these problems the additional space-time structure provided by a foliation yields the most obvious solution: The motion of each particle at a point $x \in M$ depends upon the paths of the other particles via the points at which they intersect the leaf of the foliation containing $x$, and we have an equivariant density on the leaves of the foliation.

And the simplest way to achieve this, in a covariant manner, for a Dirac wave function $\psi$, is via the current (9.12): Form the natural tensor $\overline{\psi}\gamma_1 \ldots \gamma_N \psi$, evaluated at $x$ and the other intersection points, and contract in the slots corresponding to the other particles with the $N-1$ unit normals to the hypersurface at the corresponding points, to obtain the divergence-free 4-vector $j_k$, the tangent to the trajectory at $x$. Thus, *the dynamics of the HBD model is the simplest Lorentz invariant dynamics compatible with the structure at hand, namely, the Dirac wave function and the foliation.* Furthermore, the simultaneous normal component $\rho = j_k \cdot n_k$ is an equivariant density on the leaves of the foliation.

It should be stressed, however, that the Lorentz invariance of the HBD model is—in Bell's sense—"serious" only if the foliation can be regarded as an objective *dynamical*—in contrast to absolute—structure in the theory (and in the world, if the theory is to describe the world). It is this structure that is the innovation of what has been proposed here and in [11], not the model per se, which is indeed a rather straightforward covariant extension of the BD model.

However, in this chapter we have not tried to find a "serious" law for the foliation $\mathscr{F}$ or, what amounts to the same thing, its normal vector field $n$. As a toy example, however, the foliation law could be given by an autonomous equation for $n$, such as $\partial_\nu n_\mu = 0$. (A better proposal, that the unit normal vector field $n$ obey $\partial_\nu n_\mu - \partial_\mu n_\nu = 0$, has been made by Roderich Tumulka; see Eq. (26) of [13].) Another class of toy examples involves a vector field $n$ constructed from the wave function $\psi(x_1, \ldots, x_N)$: Consider the space-time vector fields $v_{kl}^\mu(x) = (\overline{\psi}\gamma_k^\mu \psi)(\widehat{x_1}, \ldots, \widehat{x_{l-1}}, x, \widehat{x_{l+1}}, \ldots, \widehat{x_N})$, where $(\widehat{x_1}, \ldots, \widehat{x_N})$ is a point fixed in a Lorentz invariant way, for example as a maximum of $\overline{\psi}\psi$. (Simply considering $v_k^\mu(x) = (\overline{\psi}\gamma_k^\mu \psi)(x, \ldots, x)$ is not a good idea, since this will be zero for antisymmetric (fermion) wave functions.) Now one may set $n$ equal to the integrable[10] part

---

[10] For an arbitrary vector field $v^\mu(x)$, the Fourier transformed $\hat{v}^\mu(k)$ may be split into $\hat{v}_\parallel^\mu(k) = \hat{v}^\nu(k)k_\nu k^\mu/(k_\lambda k^\lambda)$ and $\hat{v}_\perp^\mu(k) = \hat{v}^\mu(k) - \hat{v}_\parallel^\mu(k)$. The inverse Fourier transformed $v_\parallel^\mu(x)$ satisfies the integrability condition $\partial_\mu v_{\parallel\nu} - \partial_\nu v_{\parallel\mu} = 0$.

of some $v_{kl}$. (However, the result may not be entirely satisfactory; it may fail to be time-like.)

A further possibility, which may be more serious, is to have, in addition to the particle degrees of freedom, an independent quantum field $\phi_\mu$ that determines the foliation. Assume that for any quantum state $\Phi$ of the field, $(\Phi, \phi_\mu \Phi)$ is time-like and completely integrable. Then for any state $\Psi$ of the particle-field system, set $n_\mu = (\Psi, \phi_\mu \Psi)$. Suppose that the particle and the field degrees of freedom are both dynamically and statistically independent, i.e., that there is neither quantum interaction nor entanglement between these degrees of freedom, so that in particular the full wave function $\Psi = \psi \otimes \Phi$. Then we may define the foliation by the normal field $n_\mu$. The $\phi$-field can be regarded as very roughly analogous to a Higgs field, producing a kind of spontaneous symmetry breaking, where by choice of $\Phi$ a particular foliation is determined, and relativistic invariance thereby broken.

For more recent work along similar lines, see Sect. 12.3.7.

# References

1. D. Bohm. A Suggested Interpretation of the Quantum Theory in Terms of "Hidden" Variables: Part I. *Physical Review*, 85:166–179, 1952. Reprinted in [?].
2. D. Bohm. A Suggested Interpretation of the Quantum Theory in Terms of "Hidden" Variables: Part II. *Physical Review*, 85:180–193, 1952. Reprinted in [?].
3. J. S. Bell. *Speakable and Unspeakable in Quantum Mechanics*. Cambridge University Press, Cambridge, 1987.
4. D. Dürr, S. Goldstein, and N. Zanghì. Quantum Equilibrium and the Origin of Absolute Uncertainty. *Journal of Statistical Physics*, 67:843–907, 1992.
5. J. T. Cushing, A. Fine, and S. Goldstein, editors. *Bohmian Mechanics and Quantum Theory: an Appraisal, volume 184 of Boston Studies in the Philosophy of Science*. Kluwer Acad. Publ., Dordrecht, 1996.
6. D. Bohm. Comments on an Article of Takabayasi Concerning the Formulation of Quantum Mechanics with Classical Pictures. Progress of Theoretical Physics, 9:273–287, 1953.
7. D. Bohm and B. J. Hiley. *The Undivided Universe: An Ontological Intepretation of Quantum Theory*. Routledge & Kegan Paul, London, 1993.
8. T.M. Samols. A Stochastic Model of a Quantum Field Theory. *Journal of Statistical Physics*, 80:793–809, 1995.
9. K. Berndl, D. Dürr, S. Goldstein, and N. Zanghì. Nonlocality, Lorentz Invariance, and Bohmian Quantum Theory. *Physical Review A*, 53:2062–2073, 1996.
10. E. Squires. A Local Hidden-Variable Theory that, FAPP, Agrees with Quantum Theory. *Physics Letters A*, 178:22–26, 1993.
11. D. Dürr, S. Goldstein, and N. Zanghì. On a Realistic Theory for Quantum Physics. In S. Albeverio, G. Casati, U. Cattaneo, D. Merlini, and R. Mortesi, editors, *Stochastic Processes, Geometry and Physics*, pages 374–391. World Scientific, Singapore, 1990.
12. T. Maudlin. Space-time in the Quantum World. In J. T. Cushing, A. Fine, and S. Goldstein, editors, *Bohmian Mechanics and Quantum Theory: An Appraisal, volume 184 of Boston Studies in the Philosophy of Science*, pages 285–307. Kluwer Acad. Publ., Dordrecht, 1996.
13. R. Tumulka. The "Unromantic Pictures" of Quantum Theory. *Journal of Physics A: Mathematical and Theoretical*, 40:3245–3273, 2007.

# Chapter 10
# Bohmian Mechanics and Quantum Field Theory

## 10.1 Introduction

Despite the uncertainty principle, the predictions of nonrelativistic quantum mechanics permit particles to have precise positions at all times. The simplest theory demonstrating that this is so is Bohmian mechanics, in which the position of a particle cannot be known to macroscopic observers more accurately than the $|\psi|^2$ distribution would allow. A frequent complaint about Bohmian mechanics is that, in the words of Steven Weinberg (private communication), "it does not seem possible to extend Bohm's version of quantum mechanics to theories in which particles can be created and destroyed, which includes all known relativistic quantum theories."

To remove the grounds of the concern that such an extension may be impossible, we show how, with (more or less) any regularized quantum field theory (QFT), one can associate a particle theory—describing moving particles—that is empirically equivalent to that QFT. In particular, there is a particle theory that recovers all predictions of regularized QED[1].

However, we will not attempt to achieve full Lorentz invariance; that would lead to quite a different set of questions, orthogonal to those with which we shall be concerned here. But we note that though the theories we present here require a preferred reference frame, there can be no experiment that would allow an observer to determine which frame is the preferred one, provided the corresponding QFTs are such that their empirical predictions are Lorentz invariant.

The theories we present are based on the work of Bell [1] and our own recent results [2, 3, 4]; in [2] we study a simple model QFT, and in [3, 4] we give a detailed account of the mathematics needed for treating other QFTs. While Bell replaced physical 3-space by a lattice, we describe directly what presumably is the continuum

---
[1] One may worry that such a particle theory, in which particles always have actual positions but have no additional actual discrete degrees of freedom corresponding for example to spin, cannot do justice to QFT, for which spin and other discrete degrees of freedom are often regarded as playing a crucial role. However, additional actual discrete degrees of freedom are entirely unnecessary: as Bell has shown in [5], the straightforward extension of Bohmian mechanics to spinor-valued wavefunctions, which we use here, accounts for all phenomena involving spin.

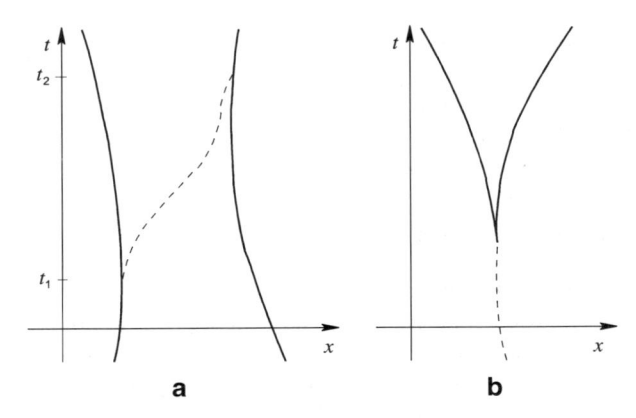

**Fig. 10.1** Two patterns of world lines as they may arise from some Bell-type QFT. **a** The world line of a photon (*dashed curve*) starts at an emission event (at time $t_1$) on the world line of an electron (*bold curve*), and ends at an absorption event (at time $t_2$) on the world line of another electron. **b** An electron–positron pair (*bold curves*) is created at the end point of a photon world line

limit of Bell's model [3, 4, 6, 7]. Since Bell's proposal was the first in this direction, we call these models "Bell-type QFTs". The trajectories we use as the world lines consist of pieces of Bohmian trajectories, or similar ones. A novel element is that the world lines can begin and end. This is essential for describing processes involving particle creation or annihilation, such as, e.g., positron–electron pair creation. Our description of such events is the most naive and natural one: the world line of the particle begins at some space-time point, its creation event, and ends at another (see Fig. 10.1). The models thus involve "particle creation" in the literal sense.

The patterns of world lines are reminiscent of Feynman diagrams, and the possible Feynman diagrams correspond to the possible types of world-line patterns. Note, however, that the role of Feynman diagrams is to aid with computing the evolution of the state vector $\Psi$, while the world lines here are supposed to exist in addition to $\Psi$. Unlike Feynman diagrams, which are computational tools not to be confused with actual particle paths, the world-line patterns of our models are to be regarded as describing the possibilities for what might actually happen (in a universe governed by that model).

## 10.2 Configuration Space

Whatever the pattern of world lines may look like, it can be described by a time-dependent configuration $Q_t = Q(t)$ moving in the configuration space $\mathcal{Q}$ of possible positions for a variable number of particles. In the case of a single particle species, this is the disjoint union of the $n$-particle configuration spaces,

## 10.2 Configuration Space

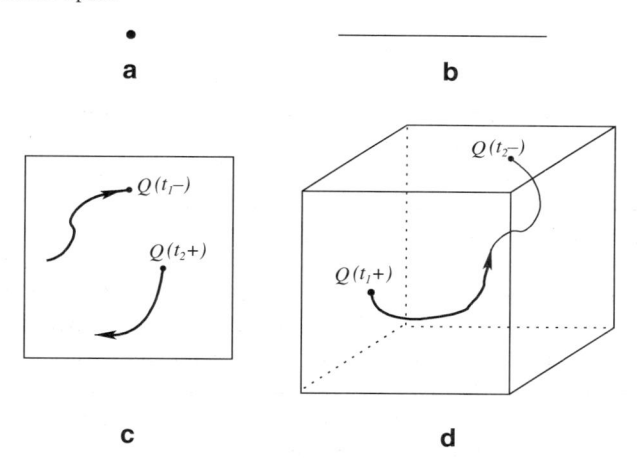

**Fig. 10.2** Schematic representation of the configuration space of a variable number of particles. **a–d** show the sectors $\mathcal{Q}^{[0]}$ through $\mathcal{Q}^{[3]}$. A configurational history $\mathcal{Q}(t)$ jumps to the next higher sector at each creation event, and to the next lower sector at each annihilation event. The history shown corresponds to a world-line pattern like that of Fig. 10.1a

$$\mathcal{Q} = \bigcup_{n=0}^{\infty} \mathcal{Q}^{[n]}. \tag{10.1}$$

Since the particles are identical, the sector $\mathcal{Q}^{[n]}$ is best defined as $\mathbb{R}^{3n}$ modulo permutations, $\mathbb{R}^{3n}/S_n$. For simplicity, we will henceforth pretend that $\mathcal{Q}^{[n]}$ is simply $\mathbb{R}^{3n}$; we discuss $\mathbb{R}^{3n}/S_n$ in [4]. For several particle species, one forms the Cartesian product of several copies of the space (10.1), one for each species. One obtains in this way a configuration space which is, like (10.1), a union of sectors $\mathcal{Q}^{[n]}$ where, however, now $n = (n_1, \ldots, n_\ell)$ is an $\ell$-tuple of particle numbers for the $\ell$ species of particles. For QED, for example, $\mathcal{Q}$ is the product of three copies of the space (10.1), corresponding to electrons, positrons, and photons; thus, a configuration specifies the number and positions of all electrons, positrons, and photons[2].

Let us explore what $Q(t)$ looks like for a typical world line pattern (see Fig. 10.2). $Q(t)$ will typically have discontinuities, even if there is nothing discontinuous in the world line pattern (Fig. 10.1), because it jumps to a different sector at every creation or annihilation event. Between such events, $Q(t)$ moves smoothly within one sector.

It is helpful to note that the bosonic Fock space can be understood as a space of $L^2$ (i.e., square-integrable) functions on $\bigcup_n \mathbb{R}^{3n}/S_n$. The fermionic Fock space consists of $L^2$ functions on $\bigcup_n \mathbb{R}^{3n}$ which are anti-symmetric under permutations.

---

[2] The choice of $\mathcal{Q}$ for a given QFT, such as QED, is however not unique. For instance, it is possible to devise a version of the theory in which the configuration variable $\mathcal{Q}$ represents only the configuration of the *fermions* [1]. We have argued in [2] that it is more natural to include photons among the particles, represented in $\mathcal{Q}$.

## 10.3 The Laws of Motion

A Bell-type QFT specifies such world-line patterns, or histories in configuration space, by specifying three sorts of "laws of motion": when to jump, where to jump, and how to move between the jumps. Before we say more on what precisely the laws are, we elucidate one consequence of the laws: if at $t = 0$, the configuration $Q(0)$ is chosen at random with probability distribution $|\Psi_0|^2$, then at any later time $t$, $Q(t)$ has distribution $|\Psi_t|^2$. This property we call *equivariance*. The main consequence is that these theories are empirically equivalent to their corresponding QFTs. This conclusion has been explained in detail in Chap. 2 for Bohmian mechanics and the predictions of nonrelativistic quantum mechanics, and the same reasoning applies here. It involves a law of large numbers governing the empirical frequencies in a typical universe, and involves the recognition that the variables that record the outcome of an experiment are ultimately particle positions (orientations of meter pointers, ink marks on paper, etc.).

In a Bell-type QFT, the state of a system is described by the pair $(\Psi_t, Q_t)$, where $\Psi_t$ is an (arbitrary) vector in the appropriate Fock space and may well involve a superposition of states of different particle numbers. As remarked before, $\Psi_t$ can thus be viewed as a function $\Psi_t(q)$ on the configuration space $\mathcal{Q}$ of a variable number of particles. (For photons, whose position observable is represented by a positive-operator-valued measure (POVM), $\Psi_t$ can be represented by a wavefunction $\Psi_t(q)$ satisfying a constraint.) $\Psi_t$ evolves according to the appropriate Schrödinger equation

$$i\hbar \frac{d\Psi_t}{dt} = H\Psi_t. \qquad (10.2)$$

Typically $H = H_0 + H_I$ is the sum of a free Hamiltonian $H_0$ and an interaction Hamiltonian $H_I$. It is important to appreciate that although there is an actual particle number, defined by $N(t) = \#Q(t) := $ [number of entries in $Q(t)$] or $Q(t) \in \mathcal{Q}^{[N(t)]}$, $\Psi$ need not be a number eigenstate (i.e., concentrated on one sector). This is similar to the situation in the usual double-slit experiment, in which the particle passes through only one slit although the wavefunction passes through both. And as with the double-slit experiment, the part of the wavefunction that passes through another sector of $\mathcal{Q}$ (or another slit) may well influence the behavior of $Q(t)$ at a later time.

The laws of motion for $Q_t$ depend on $\Psi_t$ (and on $H$). The continuous part of the motion is governed by a first-order ordinary differential equation

$$\frac{dQ_t}{dt} = v^{\Psi_t}(Q_t) = \mathrm{Re}\, \frac{\Psi_t^*(Q_t)\,(\dot{\hat{q}}\Psi_t)(Q_t)}{\Psi_t^*(Q_t)\,\Psi_t(Q_t)} \qquad (10.3)$$

$$\text{where } \dot{\hat{q}} = \frac{d}{d\tau} e^{iH_0\tau/\hbar}\, \hat{q}\, e^{-iH_0\tau/\hbar}\Big|_{\tau=0} = \frac{i}{\hbar}[H_0, \hat{q}] \qquad (10.4)$$

is the time derivative of the $\mathcal{Q}$-valued Heisenberg position operator $\hat{q}$, evolved with $H_0$ alone. Since in the absence of global coordinates on $\mathcal{Q}$, the notion of a "$\mathcal{Q}$-valued

## 10.3 The Laws of Motion

operator" may be somewhat obscure, one should understand (10.3) as saying this: for any smooth function $f : \mathscr{Q} \to \mathbb{R}$,

$$\frac{df(Q_t)}{dt} = \mathrm{Re}\, \frac{\Psi_t^*(Q_t)(\frac{i}{\hbar}[H_0, \hat{f}]\Psi_t)(Q_t)}{\Psi_t^*(Q_t)\Psi_t(Q_t)} \tag{10.5}$$

where $\hat{f}$ is the multiplication operator corresponding to $f$. This expression is of the form $v^\Psi \cdot \nabla f(Q_t)$, as it must be for defining a dynamics for $Q_t$, if the free Hamiltonian is a differential operator of up to second order [4]. The Klein–Gordon operator is not covered by (10.3) or (10.5); its treatment will be discussed in future work. The numerator and denominator of (10.3) resp. (10.5) involve, when appropriate, scalar products in spin space. One may view $v$ as a vector field on $\mathscr{Q}$, and thus as consisting of one vector field $v^{[n]}$ on every manifold $\mathscr{Q}^{[n]}$; it is then $v^{[N(t)]}$ that governs the motion of $Q(t)$ in (10.3).

If $H_0$ were the Schrödinger Hamiltonian $-\sum_{i=1}^n \frac{\hbar^2}{2m_i}\Delta_i + V$ of quantum mechanics, formula (10.3) would yield the velocity proposed by Bohm in [8],

$$v_i^\Psi = \frac{\hbar}{m_i}\, \mathrm{Im}\, \frac{\Psi^* \nabla_i \Psi}{\Psi^*\Psi},\ i = 1,\ldots,n. \tag{10.6}$$

When $H_0$ is the "second quantization" of a one-particle Schrödinger Hamiltonian, (10.3) amounts to (10.6), with equal masses, in every sector $\mathscr{Q}^{[n]}$. Similarly, in case $H_0$ is the second quantization of the Dirac Hamiltonian $-i c\hbar \boldsymbol{\alpha} \cdot \nabla + \beta mc^2$, (10.3) says a configuration $Q(t)$ (with $N$ particles) moves according to (the $N$-particle version of) the known variant of Bohm's velocity formula for Dirac wavefunctions [9],

$$v^\Psi = \frac{\Psi^* \boldsymbol{\alpha} \Psi}{\Psi^*\Psi}\, c. \tag{10.7}$$

The jumps are stochastic in nature, i.e., they occur at random times and lead to random destinations. In Bell-type QFTs, God does play dice. There are no hidden variables which would fully pre-determine the time and destination of a jump. (Note also that a deterministic jump law that prescribes the time and destination of the jump as a (smooth) function of the initial configuration would lack sufficient randomness to be compatible with equivariance, since after a jump from a sector with dimension $d'$ to a sector with dimension $d > d'$ the configuration would have to belong, at any specific time, to a $d'$-dimensional submanifold.)

The probability of jumping, within the next $dt$ seconds, to the volume $dq$ in $\mathscr{Q}$, is $\sigma^\Psi(dq|Q_t)\, dt$ with

$$\sigma^\Psi(dq|q') = \frac{2}{\hbar}\, \frac{[\mathrm{Im}\, \Psi^*(q)\langle q|H_I|q'\rangle \Psi(q')]^+}{\Psi^*(q')\Psi(q')}\, dq, \tag{10.8}$$

where $x^+ = \max(x, 0)$ means the positive part of $x \in \mathbb{R}$. Thus the jump rate $\sigma^\Psi$ depends on the present configuration $\mathscr{Q}_t$, on the state vector $\Psi_t$, which has a "guiding" role similar to that in Bohm's velocity law (10.6), and of course on the overall setup of the QFT as encoded in the interaction Hamiltonian $H_I$. In [2], we spelled out in detail a simple example of a Bell-type QFT.

Together, (10.3) and (10.8) define a Markov process on $\mathcal{Q}$. The "free" part of this process, defined by (10.3), can also be regarded as arising as follows: if $H_0$ is as usual the "second quantization" of a 1-particle Hamiltonian $h$, one can construct the dynamics corresponding to $H_0$ from a given 1-particle dynamics corresponding to $h$ (be it deterministic or stochastic) by an algorithm that one may call the "second quantization" of a Markov process [4]. Moreover, this algorithm can still be used when formula (10.3) fails to define a dynamics (in particular when $H_0$ is the second quantized Klein–Gordon operator).

## 10.4 Field Operators

We now discuss the role of field operators (operator-valued fields on space-time) in a theory of particles. Almost by definition, it would seem that QFT concerns fields, and not particles. But there is less to this than might be expected. The field operators do not function as observables in QFT. It is far from clear how to actually "observe" them, and even if this could somehow, in some sense, be done, it is important to bear in mind that the standard predictions of QFT are grounded in the particle representation, not the field representation: Experiments in high energy physics are scattering experiments, in which what is observed is the asymptotic motion of the outgoing particles. Moreover, for Fermi fields—the matter fields—the field as a whole (at a given time) could not possibly be observable, since Fermi fields anti-commute, rather than commute, at space-like separation. We note, though, that a theory in which $\Psi_t$ guides an actual field can be devised, at least formally [8].

The role of the field operators is to provide a connection, the only connection in fact, between space-time and the abstract Hilbert space containing the quantum states $|\Psi\rangle$, which are usually regarded not as functions but as abstract vectors. For our purpose, what is crucial are the following facts that we shall explain presently: (i) the field operators naturally correspond to the spatial structure provided by a projection-valued (PV) measure on configuration space $\mathcal{Q}$, and (ii) the process we have defined in this chapter can be efficiently expressed in terms of a PV measure.

Consider a PV measure $P$ on $\mathcal{Q}$ acting on $\mathcal{H}$: For $B \subseteq \mathcal{Q}$, $P(B)$ means the projection to the space of states localized in $B$. All our formulas above can be formulated in terms of $P$ and $|\Psi\rangle$: (10.5) becomes

$$\frac{df(Q_t)}{dt} = \text{Re} \frac{\langle \Psi | P(dq) \frac{i}{\hbar}[H_0, \hat{f}] | \Psi \rangle}{\langle \Psi | P(dq) | \Psi \rangle} \Big|_{q=Q_t} \qquad (10.9)$$

$$\text{with } \hat{f} = \int_{q \in \mathcal{Q}} f(q) \, P(dq), \qquad (10.10)$$

for any smooth function $f : \mathcal{Q} \to \mathbb{R}$, and (10.8) becomes

$$\sigma^{\Psi}(dq|q') = \frac{2}{\hbar} \frac{[\operatorname{Im} \langle \Psi|P(dq)H_I P(dq')|\Psi\rangle]^+}{\langle \Psi|P(dq')|\Psi\rangle}. \tag{10.11}$$

Note that $\langle \Psi|P(dq)|\Psi\rangle$ is the probability distribution analogous to $|\Psi(q)|^2 dq$.

We now turn to (i): how we obtain the PV measure $P$ from the field operators. For the configuration space $\mathcal{Q} = \bigcup_n \mathbb{R}^{3n}/S_n$ of a variable number of identical particles, a configuration can be specified by giving the number of particles $n(R)$ in every region $R \subseteq \mathbb{R}^3$. A PV measure $P$ on $\mathcal{Q}$ is mathematically equivalent to a family of number operators: an additive operator-valued set function $N(R)$, $R \subseteq \mathbb{R}^3$, such that the $N(R)$ commute pairwise and have spectra in the nonnegative integers. Indeed, $P$ is the joint spectral decomposition of the $N(R)$ [4]. And the easiest way to obtain such a family of number operators is by setting

$$N(R) = \int_R \phi^*(\mathbf{x}) \phi(\mathbf{x}) d^3\mathbf{x},$$

exploiting the canonical commutation or anti-commutation relations for the field operators $\phi(\mathbf{x})$. These observations suggest that field operators are just what the doctor ordered for the efficient construction of a theory describing the creation, motion, and annihilation of particles.

(It is only the positive-energy one-particle states that are used for constructing the Fock space $\mathcal{H}$, so that $\mathcal{H}$ is really a subspace of a larger Hilbert space $\mathcal{H}_0$ which contains also unphysical states (with contributions from one-particle states of negative energy). Since position operators may fail to map positive energy states into positive energy states, the PV measure $P$ is typically defined on $\mathcal{H}_0$ but not on $\mathcal{H}$, in which case (10.9) and (10.11) have to be read as applying in $\mathcal{H}_0$. While $H_0$ is defined on $\mathcal{H}_0$, $H_I$ is usually not and needs to be "filled up with zeroes," i.e. replaced by $P'H_I P'$ where $P'$ is the projection $\mathcal{H}_0 \to \mathcal{H}$.)

## 10.5 Conclusions

To sum up, we have shown how the realist view which Bohmian mechanics provides for the realm of nonrelativistic quantum mechanics can be extended to QFT, including creation and annihilation of particles. Those who find the all too widespread positivistic attitude in quantum theory unsatisfactory may find these ideas helpful. But even those who think that Copenhagen quantum theory is just fine may find it interesting to see how the particle picture, ubiquitous in the pictorial lingo and heuristic intuition of QFT, can be made consistent, internally and with the observable facts of QFT, by introducing suitable laws of motion.

# References

1. J. S. Bell. Beables for Quantum Field Theory. *Physics Reports*, 137:49–54, 1986. Reprinted in [26].
2. D. Dürr, S. Goldstein, R. Tumulka, and N. Zanghì. Trajectories and Particle Creation and Annihilation in Quantum Field Theory. *Journal of Physics A*, 36:4143–4150, 2003.
3. D. Dürr, S. Goldstein, R. Tumulka, and N. Zanghì. Quantum Hamiltonians and Stochastic Jumps. *Communications in Mathematical Physics*, 254:129–166, 2005.
4. D. Dürr, S. Goldstein, R. Tumulka, and N. Zanghì. Bell-Type Quantum Field Theories. *Journal of Physics A*, 38:R1–R43, 2005.
5. J. S. Bell. On the Problem of Hidden Variables in Quantum Mechanics. *Reviews of Modern Physics*, 38:447–452, 1966. Reprinted in [211] and in [26].
6. A. Sudbery. Objective Interpretations of Quantum Mechanics and the Possibility of a Deterministic Limit. *Journal of Physics A*, 20:1743–1750, 1987.
7. J. C. Vink. Quantum Mechanics in Terms of Discrete Beables. *Physical Review A*, 48:1808–1818, 1993.
8. D. Bohm. A Suggested Interpretation of the Quantum Theory in Terms of "Hidden" Variables: Part I. *Physical Review*, 85:166–179, 1952. Reprinted in [211].
9. D. Bohm and B. J. Hiley. *The Undivided Universe: An Ontological Intepretation of Quantum Theory*. Routledge & Kegan Paul, London, 1993.

# Chapter 11
# Quantum Spacetime without Observers: Ontological Clarity and the Conceptual Foundations of Quantum Gravity

## 11.1 Introduction

> The term "3-geometry" makes sense as well in quantum geometrodynamics as in classical theory. So does superspace. But space-time does not. Give a 3-geometry, and give its time rate of change. That is enough, under typical circumstances to fix the whole time-evolution of the geometry; enough in other words, to determine the entire four-dimensional space-time geometry, provided one is considering the problem in the context of classical physics. In the real world of quantum physics, however, one cannot give both a dynamic variable and its time-rate of change. The principle of complementarity forbids. Given the precise 3-geometry at one instant, one cannot also know at that instant the time-rate of change of the 3-geometry. ... The uncertainty principle thus deprives one of any way whatsoever to predict, or even to give meaning to, "the deterministic classical history of space evolving in time." *No prediction of spacetime, therefore no meaning for spacetime,* is the verdict of the quantum principle. (Misner et al. [1]).

One of the few propositions about quantum gravity that most physicists in the field would agree upon, that our notion of space-time must, at best, be altered considerably in any theory conjoining the basic principles of quantum mechanics with those of general relativity, will be questioned in this article. We will argue, in fact, that most, if not all, of the conceptual problems in quantum gravity arise from the sort of thinking on display in the preceding quotation.

It is also widely agreed, almost 40 years after the first attempts to quantize general relativity, that there is still no single set of ideas on how to proceed, and certainly no physical theory successfully concluding this program. Rather, there are a great variety of approaches to quantum gravity; for a detailed overview, see, e.g., Rovelli [2]. While the different approaches to quantum gravity often have little in common, they all are intended ultimately to provide us with a consistent quantum theory agreeing in its predictions with general relativity in the appropriate physical domain. Although we will focus here on the conceptual problems faced by those approaches which amount to a canonical quantization of classical general relativity, the main lessons will apply to most of the other approaches as well.

This is because, as we shall argue, many of these difficulties arise from the subjectivity and the ontological vagueness inherent in the very framework of orthodox quantum theory, a framework taken for granted by almost all approaches to quantum gravity. We shall sketch how most, and perhaps all, of the conceptual problems of canonical quantum gravity vanish if we insist upon formulating our cosmological theories in such a manner that it is reasonably clear what they are about—if we insist, that is, upon ontological clarity—and, at the same time, avoid any reference to such vague notions as measurement, observers, and observables.

The earliest approach, canonical quantum gravity, amounts to quantizing general relativity according to the usual rules of canonical quantization. However, to apply canonical quantization to general relativity, the latter must first be cast into canonical form. Since the quantization of the standard canonical formulation of general relativity, the Arnowitt, Deser, Misner formulation [3], has led to severe conceptual and technical difficulties, nonstandard choices of canonical variables, such as in the Ashtekar formulation [4] and in loop quantum gravity [5], have been used as starting points for quantization. While some of the technical problems have been resolved by these new ideas, the basic conceptual problems have not been addressed.

After the great empirical success of the standard model in particle physics, the hope arose that the gravitational interaction could also be incorporated in a similar model. The search for such a unified theory led to string theory, which apparently reproduces not only the standard model but also general relativity in a suitable low energy limit. However, since string theory is, after all, a quantum theory, it retains all the conceptual difficulties of quantum theory. Nonetheless, our focus, again, will be on the canonical approaches, restricted for simplicity to pure gravity, ignoring matter.

This article is organized as follows: In Sect. 11.2 we will sketch the fundamental conceptual problems faced by most approaches to quantum gravity. The seemingly unrelated problems in the foundations of orthodox quantum theory will be touched upon in Sect. 11.3. Our central point will be made in Sect. 11.4, where we indicate how the conceptual problems of canonical quantum gravity disappear when the main insights of the Bohmian approach to quantum theory are applied.

Finally, in Sect. 11.5, we will discuss how the status and significance of the wave function, in Bohmian mechanics as well as in orthodox quantum theory, is radically altered when we adopt a universal perspective. This altered status of the wave function, together with the very stringent symmetry demands so central to general relativity, suggests the possibility—though by no means the inevitability—of finding an answer to the question: Why should the universe be governed by laws so apparently peculiar as those of quantum mechanics?

## 11.2 The Conceptual Problems of Quantum Gravity

In the canonical approach to quantum gravity one must first reformulate general relativity as a Hamiltonian dynamical system. This was done by ADM [3], using the 3-metric $g_{ij}(x^a)$ on a space-like hypersurface $\Sigma$ as the configurational variable and

## 11.2 The Conceptual Problems of Quantum Gravity

the extrinsic curvature of the hypersurface as its conjugate momentum $\pi^{ij}(x^a)$[1]. The real time parameter of usual Hamiltonian systems is replaced by a "multi-fingered time" corresponding to arbitrary deformations $d\Sigma$ of the hypersurface. These deformations are split into two groups: those changing only the three dimensional coordinate system $x^a$ on the hypersurface (with which, as part of what is meant by the hypersurface, it is assumed to be equipped) and deformations of the hypersurface along its normal vector field. While the changes of the canonical variables under both kind of deformations are generated by Hamiltonian functions on phase space, $H_i(g, \pi)$ for spatial diffeomorphisms and $H(g, \pi)$ for normal deformations, their changes under pure coordinate transformations on the hypersurfaces are dictated by their geometrical meaning. The dynamics of the theory is therefore determined by the Hamiltonian functions $H(g, \pi)$ generating changes under normal deformations of the hypersurface.

Denote by $N(x^a)$ the freely specifiable lapse function that determines how far, in terms of proper length, one moves the space-like hypersurface at the point $x = (x^a)$ along its normal vector: This distance is $N(x^a)d\tau$, where $\tau$ is a parameter labeling the successive hypersurfaces arrived at under the deformation (and defined by this equation). The infinitesimal changes of the canonical variables are then generated by the Hamiltonian $H_N$ associated with $N$ (an integral over $\Sigma$ of the product of $N$ with a Hamiltonian density $H(g, \pi; x^a)$):

$$dg_{ij}(x^a) = \frac{\delta H_N(g, \pi)}{\delta \pi^{ij}(x^a)} d\tau$$
$$d\pi^{ij}(x^a) = -\frac{\delta H_N(g, \pi)}{\delta g_{ij}(x^a)} d\tau. \qquad (11.1)$$

In what follows we shall denote by $H(g, \pi)$ the collection $\{H_N(g, \pi)\}$ of all such Hamiltonians (or, what comes pretty much to the same thing, the collection $\{H(g, \pi; x)\}$ for all points $x \in \Sigma$) and similarly for $H_i$.

It is important to stress that the theory can be formulated completely in terms of geometrical objects on a three dimensional manifold, with no a priori need to refer to its embedding into a space-time. A solution of (11.1) is a family of 3-metrics $g(\tau)$ that can be glued together to build up a 4-metric using the lapse function $N$ (to determine the transverse geometry). In this way the space-time metric emerges dynamically when one evolves the canonical variables with respect to multi-fingered time.

However, the initial canonical data cannot be chosen arbitrarily, but must obey certain constraints: Only for initial conditions that lie in the submanifold of phase space on which $H_i(g, \pi)$ and $H(g, \pi)$ vanish do the solutions (space-time metrics $g_{\mu\nu}(x^\mu)$) also satisfy Einstein's equations. In fact, away from this so called constraint manifold the theory is not even well defined, at least not as a theory involving a multi-fingered time, since the solutions would depend on the special way we choose

---

[1] Actually the extrinsic curvature is given by $K_{ij} = G_{ijab}\pi^{ab}$ where $G_{ijab}$ is the so called supermetric, which is itself a function of $g_{ij}$. This distinction is, however, not relevant to our discussion.

to evolve the space-like hypersurface, i.e., on the choice of $N(x^a)$, to build up space-time. Of course, a theory based on a single choice, for example $N(x^a) = 1$, would be well defined, at least formally.

By the same token, the invariance of the theory under space-time diffeomorphisms is no longer so obvious as in the formulation in terms of Einstein's equations: In the ADM formulation 4-diffeomorphism invariance amounts to the requirement that one ends up with the same space-time, up to coordinate transformations, regardless of which path in multi-fingered time is followed, i.e., which lapse function $N$, or $\tau$-dependent sequence of lapse functions $N(\tau)$, is used. This says that for the space-time built up from any particular choice of multi-fingered time, the dynamical Eq. (11.1) will be satisfied for *any* foliation of the resulting space-time into space-like hypersurfaces—using in (11.1) the lapse function $N(\tau)$ associated with that foliation—and not just for the foliation associated with that particular choice.

Formally, it is now straightforward to quantize this constrained Hamiltonian theory using Dirac's rules for the quantization of constrained systems [6]. First one must replace the canonical variables $g_{ij}$ and $\pi^{ij}$ by operators $\hat{g}_{ij}$ and $\hat{\pi}^{ij} = -i\frac{\delta}{\delta g_{ij}}$ satisfying the canonical commutation relations.[2] One then formally inserts these into the Hamiltonian functions $H(g, \pi)$ and $H_i(g, \pi)$ of the classical theory to obtain operators $\widehat{H}(\hat{g}, \hat{\pi})$ and $\widehat{H}_i(\hat{g}, \hat{\pi})$ acting on functionals $\Psi(g)$ on the configuration space of 3-metrics. Since the Hamiltonians were constrained in the classical theory one demands that the corresponding operators annihilate the physical states in the corresponding quantum theory:

$$\widehat{H}\Psi = 0 \tag{11.2}$$
$$\widehat{H}_i\Psi = 0. \tag{11.3}$$

Equation (11.3) has a simple meaning, namely that $\Psi(g)$ be invariant under 3-diffeomorphisms (coordinate changes on the 3-manifold), so that it depends on the 3-metric $g$ only through the 3-geometry. However, the interpretation of the *Wheeler-DeWitt equation* (11.2) is not at all clear.

Before discussing the several problems which arise in attempts to give a physical meaning to the approach just described, a few remarks are in order: While we have omitted many technical details and problems from our schematic description of the "Dirac constraint quantization" of gravity, these problems either do not concern, or are consequences of, the main conceptual problems of canonical quantum gravity. Other approaches, such as the canonical quantization of the Ashtekar formulation of classical general relativity and its further development into loop quantum gravity, resolve some of the technical problems faced by canonical quantization in the metric representation, but leave the main conceptual problems untouched.

Suppose now that we have found a solution $\Psi(g)$ to Eqs. (11.2 and 11.3). What physical predictions would be implied? In orthodox quantum theory a solution $\Psi_t$ of the time-dependent Schrödinger equation provides us with a time-dependent probability distribution $|\Psi_t|^2$, as well as with the absolute square of other time-dependent

---

[2] We choose units in which $\hbar$ and $c$ are 1.

## 11.2 The Conceptual Problems of Quantum Gravity

probability amplitudes. The measurement problem and the like aside, the physical meaning of these is reasonably clear: they are probabilities for the results of the measurement of the configuration or of other observables. But any attempt to interpret canonical quantum gravity along orthodox lines immediately faces the following problems:

- **The problem of time**: In canonical quantum gravity there is no time-dependent Schrödinger equation; it was replaced by the time-independent Wheeler-DeWitt equation. The Hamiltonians—the generators of multi-fingered-time evolution in the classical case—annihilate the state vector and therefore cease to generate any evolution at all. The theory provides us with only a timeless wave function on the configuration space of 3-metrics, i.e., on the possible configurations of space, not of space-time. But how can a theory that provides us (at best) with a single fixed probability distribution for configurations of space ever be able to describe the always changing world in which we live? This, in a nutshell, is the problem of time in canonical quantum gravity.
- **The problem of 4-diffeomorphism invariance:** The fundamental symmetry at the heart of general relativity is its invariance under general coordinate transformations of space-time. It is important to stress that almost any theory can be formulated in such a 4-diffeomorphism invariant manner by adding further structure to the theory (e.g., a preferred foliation of space-time as a dynamical object). General relativity has what is sometimes called serious diffeomorphism-invariance, meaning that it involves no space-time structure beyond the 4-metric and, in particular, singles out no special foliation of space-time. In canonical quantum gravity, while the invariance under coordinate transformations of space is retained, it is not at all clear what 4-diffeomorphism invariance could possibly mean. Therefore the basic symmetry, and arguably the essence, of general relativity seems to be lost in quantization.
- **The problem of "no outside observer":** One of the most fascinating applications of quantum gravity is to quantum cosmology. Orthodox quantum theory attains physical meaning only via its predictions about the statistics of outcomes of measurements of observables, performed by observers that are not part of the system under consideration, and seems to make no clear physical statements about the behavior of a closed system, not under observation. The quantum formalism concerns the interplay between—and requires for its very meaning—two kinds of objects: a quantum system and a more or less classical apparatus. It is hardly imaginable how one could make any sense out of this formalism for quantum cosmology, for which the system of interest is the whole universe, a closed system if there ever was one.
- **The problem of diffeomorphism invariant observables:** Even if we pretend for the moment that we are able to give meaning to the quantum formalism without referring to an observer located outside of the universe, we encounter a more subtle difficulty. Classical general relativity is fundamentally diffeomorphism invariant. It is only the space-time geometry, not the 4-metric nor the identity of the individual points in the space-time manifold, that has physical significance. Therefore the physical observables in general relativity should be independent of special

coordinate systems; they must be invariant under 4-diffeomorphisms, which are in effect generated by the Hamiltonians $H$ and $H_i$. Since the quantum observables are constructed, via quantization, from the classical ones, it would seem that they must commute with the Hamiltonians $\widehat{H}$ and $\widehat{H_i}$. But such diffeomorphism invariant quantum observables are extremely hard to come by, and there are certainly far too few of them to even begin to account for the bewildering variety of our experience which it is the purpose of quantum theory to explain. (For a discussion of the question of existence of diffeomorphism invariant observables, see Kuchar [7].)

These conceptual problems, and the attempts to solve them, have lead to a variety of technical problems that are discussed in much detail in, e.g., Kuchar [7], [8] and Isham [9]. However, since we are not aware of any orthodox proposal successfully resolving the conceptual problems, we shall not discuss such details here. Rather, we shall proceed in the opposite direction, toward their common cause, and argue that they originate in a deficiency shared with, and inherited from, orthodox quantum mechanics: the lack of a coherent ontology.

Regarding the first two problems of canonical quantum gravity, it is not hard to discern their origin: the theory is concerned only with configurations of and on space, the notion of a space-time having entirely disappeared. It is true that even with classical general relativity, Newton's external absolute time is abandoned. But a notion of time, for an observer located somewhere in space-time and employing a coordinate system of his convenience, is retained, emerging from space-time. The problem of time in canonical quantum gravity is a direct consequence of the fact that in an orthodox quantum theory for space-time itself we must insist on its nonexistence (compare the quote at the beginning of this article). Similarly, the problem of diffeomorphism invariance, or, better, the problem of not even being able to address this question properly, is an immediate consequence of having no notion of space-time in orthodox quantum gravity.

## 11.3 The Basic Problem of Orthodox Quantum Theory: the Lack of a Coherent Ontology

Despite its extraordinary predictive successes, quantum theory has, since its inception some seventy-five years ago, been plagued by severe conceptual difficulties. The most widely cited of these is the measurement problem, best known as the paradox of Schrödinger's cat. For many physicists the measurement problem is, in fact, not *a* but *the* conceptual difficulty of quantum theory.

In orthodox quantum theory the wave function of a physical system is regarded as providing its complete description. But when we analyze the process of measurement itself in quantum mechanical terms, we find that the after-measurement wave function for system and apparatus arising from Schrödinger's equation for the composite system typically involves a superposition over terms corresponding to what we would like to regard as the various possible results of the measurement—e.g., different pointer orientations. Since it seems rather important that the actual result of the measurement be a part of the description of the after-measurement situation, it is difficult to believe that the wave function alone provides the complete description of that situation.

The usual collapse postulate for quantum measurement solves this problem for all practical purposes, but only at the very steep price of the introduction of an *observer* or *classical measurement apparatus* as an irreducible, unanalyzable element of the theory. This leads to a variety of further problems. The unobserved physical reality becomes drastically different from the observed, even on the macroscopic level of daily life. Even worse, with the introduction at a fundamental level of such vague notions as classical measurement apparatus, the physical theory itself becomes unprofessionally vague and ill defined. The notions of observation and measurement can hardly be captured in a manner appropriate to the standards of rigor and clarity that should be demanded of a fundamental physical theory. And in quantum cosmology the notion of an external observer is of course entirely obscure.

The collapse postulate is, in effect, an unsuccessful attempt to evade the measurement problem without taking seriously its obvious implication: that the wave function does not provide a complete description of physical reality. If we do accept this conclusion, we must naturally inquire about the nature of the more complete description with which a less problematical formulation of quantum theory should be concerned. We must ask, which theoretical entities, in addition to the wave function, might the theory describe? What mathematical objects and structures represent entities that, according to the theory, simply *are*, regardless of whether or not they are observed? We must ask, in other words, about the *primitive ontology* of the theory, what the theory is fundamentally about (see Goldstein [10, 11]). And when we know what the theory is really about, measurement and observation become secondary phenomenological concepts that, like anything else in a world governed by the theory, can be analyzed in terms of the behavior of its primitive ontology.

By far the simplest possibility for the primitive ontology is that of particles described by their positions. The corresponding theory, for non-relativistic particles, is Bohmian mechanics.

## 11.4 Bohmian Quantum Gravity

We shall now turn to what one might call a Bohmian approach to quantum gravity. There are two important lessons to be learned from a Bohmian perspective on quantum theory. First of all, the existence of Bohmian mechanics demonstrates that the characteristic features of quantum theory, usually viewed as fundamental—intrinsic randomness, operators as observables, non-commutativity, and uncertainty—need play no role whatsoever in the formulation of a quantum theory, naturally emerging instead, as a consequence of the theory, in special measurement-like situations. Therefore we should perhaps not be too surprised when approaches to quantum gravity that regard these features as fundamental encounter fundamental conceptual difficulties. Second, the main point of this chapter is made transparent in the simple example of Bohmian mechanics. If we base our theory on a coherent ontology, the conceptual problems may disappear, and, what may be even more important, a genuine understanding of the features that have seemed most puzzling might be achieved.

The transition from quantum mechanics to Bohmian mechanics is very simple, if not trivial: one simply incorporates the actual configuration into the theory as the basic variable, and stipulates that this evolve in a natural way, suggested by symmetry and by Schrödinger's equation. The velocity field $v^{\Psi_t}$ is, in fact, related to the quantum probability current $j^{\Psi_t}$ by

$$v^{\Psi_t} = \frac{j^{\Psi_t}}{|\Psi_t|^2},$$

suggesting, since $\rho^{\Psi} = |\Psi|^2$ satisfies the continuity equation with $j^{\Psi_t} = \rho^{\Psi_t} v^{\Psi_t}$, that the empirical predictions of Bohmian mechanics, for positions and ultimately, in fact, for other "observables" as well, agree with those of quantum mechanics (as in fact they do; see Chap. 2).

Formally, one can follow exactly the same procedure in canonical quantum gravity, where the configuration space is the space of (positive-definite) 3-metrics (on an appropriate fixed manifold). The basic variable in Bohmian quantum gravity is therefore the 3-metric $g$ (representing the geometry on a space-like hypersurface of the space-time to be generated by the dynamics) and its change under (what will become) normal deformations is given by a vector field on configuration space generated by the wave function $\Psi(g)$. Considerations analogous to those for non-relativistic particles lead to the following form for the Bohmian equation of motion:

$$dg_{ij}(x^a) = G_{ijab}(x^a) \operatorname{Im}\left( \Psi(g)^{-1} \frac{\delta \Psi(g)}{\delta g_{ab}(x^a)} \right) N(x^a) d\tau. \qquad (11.4)$$

The wave function $\Psi(g)$ is a solution of the timeless Wheeler-DeWitt Eq. (11.2) and therefore does not evolve. But the vector field on the right hand side of (11.4) that it generates is typically nonvanishing if $\Psi(g)$ is complex, leading to a nontrivial evolution $g(\tau)$ of the 3-metric. Suitably gluing together the 3-metrics $g(\tau)$, we obtain a space-time (see the paragraph after Eq. (11.1)). Interpretations of canonical quantum gravity along these lines have been proposed by, e.g., by Holland [12] and discussed, e.g., by Shtanov [13]. Minisuperspace Bohmian cosmologies have been considered by Kowalski-Glikman and Vink [14], Squires [15] and Callender and Weingard [16].

However, there is a crucial point which is often overlooked or, at least, not made sufficiently clear in the literature. A space-time generated by a solution of (11.2) via (11.4) will in general depend on the choice of lapse function $N$ (or $N(\tau)$). Thus the theory is not well defined as so far formulated. There are essentially two ways to complete the theory. Either one chooses a special lapse function $N$, e.g., $N = 1$, or one employs only special solutions $\Psi$ of (11.2), those yielding a vector field that generates an $N$-independent space-time. In the first case, with special $N$ but general

## 11.4 Bohmian Quantum Gravity

solution $\Psi$ of (11.2), the general covariance of the theory will typically be broken, the theory incorporating a special foliation (see the paragraph before the one containing Eq. (11.2)). The possible existence of special solutions giving rise to a covariant dynamics will be touched upon towards the end of Sect. 11.5. However, most the following discussion, especially in the first part of Sect. 11.5, does not depend upon whether or not the theory incorporates a special foliation.

Let us now examine the impact of the Bohmian formulation of canonical quantum gravity on the basic conceptual problems of orthodox canonical quantum gravity. Since a solution to the equations of Bohmian quantum gravity defines a space-time, the problem of time is resolved in the most satisfactory way: Time plays exactly the same role as in classical general relativity; there is no need whatsoever for an external absolute time, which has seemed so essential to orthodox quantum theory. The problem of diffeomorphism-invariance is ameliorated, in that in this formulation it is at least clear what diffeomorphism-invariance means. But, as explained above, general covariance can be expected at most for special solutions of (11.2). If it should turn out, however, that we must abandon general covariance on the fundamental level by introducing a special foliation of space-time, it may still be possible to retain it on the observational level (see, e.g., Chap. 9, where it is also argued, however, that a special, dynamical, foliation of space-time need not be regarded as incompatible with serious covariance).

A short answer to the problems connected with the role of observers and observables is this: There can be no such problems in the Bohmian formulation of canonical quantum gravity since observers and observables play no role in this formulation. But this is perhaps too short. What, after all, is wrong with the observation that, since individual space-time points have no physical meaning, physically significant quantities must correspond to diffeomorphism-invariant observables, of which there are far too few to describe very much of what we most care about?

The basic answer, we believe, is this: We ourselves are not—or, at least, need not be[3]—diffeomorphism invariant: Most physical questions of relevance to us are not formulated in a diffeomorphism invariant manner because, naturally enough, they refer to our own special location in space-time. Nonetheless, we know very well what they mean—we know, e.g., what it means to ask where and when something happens with respect to our own point of view. Such questions can be addressed, in fact because of diffeomorphism invariance, by taking into account the details of our environment and asking about the local predictions of the theory conditioned on such an environment, past and present.

The observer who sets the frame of reference for his physical predictions is part of and located inside the system—the universe. In classical general relativity this is not at all problematical, since that theory provides us with a coherent ontology, a potentially complete description of space-time and, if we wish, a description taking into account our special point of view in the universe. But once the step to quantum theory is taken, the coherent space-time ontology is replaced by an incoherent "ontology" of quantum observables. In orthodox quantum theory this problem can be talked away

---

[3] In some models of quantum cosmology, e.g., those permitting the definition of a global time function, it may well be possible to pick ourselves out in a diffeomorphism-invariant manner.

by introducing an outside observer actually serving two purposes: the observer sets the frame of reference with respect to which the predictions are to be understood, a totally legitimate and sensible purpose. But of course the main reason for the focus on observers in quantum theory is that it is only with respect to them that the intrinsically incoherent quantum description of the system under observation can be given any meaning. In quantum cosmology, however, no outside observer is at hand, neither for setting a frame of reference nor for transforming the incoherent quantum picture into a coherent one.

In Bohmian quantum gravity, again, both problems disappear. Since we have a coherent description of the system itself, in this case the universe, there is no need for an outside observer in order to give meaning to the theory. Nor do we have to worry about the diffeomorphism invariance of observables, since we are free to refer to observers who are themselves part of the system.

There is however an important aspect of the problem of time that we have not yet addressed. From a Bohmian perspective, as we have seen, a time-dependent wave function, satisfying Schrödinger's equation, is by no means necessary to understand the possibility of what we call change. Nonetheless, a great deal of physics is, in fact, described by such time-dependent wave functions. We shall see in the next section how these also naturally emerge from the structure of Bohmian quantum gravity, which fundamentally has only a timeless universal wave function.

The issue of the absence of diffeomorphism invariant observables is related to another problem: stationarity and equivariance are notions formulated in terms of a dynamics and a time parameter. On the cosmological level, on which there is no physical external time parameter, these notions become rather unclear. It is thus very much an open question how the quantum equilibrium analysis of Chap. 2, based crucially on such notions, can be performed in this setting. We believe that in addressing this matter, the random system, random time analysis of Sect. 2.9, as well as the conditional wave function of the universe relative to the geometrical degrees of freedom registering physical time, for example the radius of the universe, point us in the right direction.

## 11.5 A Universal Bohmian Theory

When Bohmian mechanics is viewed from a universal perspective, the status of the wave function is dramatically altered. To appreciate what we have in mind here, it might help to consider two very common objections to Bohmian mechanics.

Bohmian mechanics violates the action-reaction principle that is central to all of modern physics, both classical and (non-Bohmian) quantum: In Bohmian mechanics there is no back-action of the configuration upon the wave function, which evolves, autonomously, according to Schrödinger's equation. And the wave function, which is part of the state description of—and hence presumably part of the reality comprising—a Bohmian universe, is not the usual sort of physical field on physical space (like the electromagnetic field) to which we are accustomed, but rather a field

## 11.5 A Universal Bohmian Theory

on the abstract space of all possible configurations, a space of enormous dimension, a space constructed, it would seem, by physicists as a matter of convenience.

It should be clear by now what, from a universal viewpoint, the answer to these objections must be: the wave function $\Psi$ of the universe should be regarded as a representation, not of substantial physical reality, but of physical law. In a universal Bohmian theory, $\Psi$ should be a functional of the configurations of all elements of physical reality: geometry, particle positions, field or string configurations, or whatever primitive ontology turns out to describe nature best. As in the case of pure quantum gravity, $\Psi$ should be a (special) solution of some fundamental eq.(such as the Wheeler-DeWitt equation (11.2) with additional terms for particles, fields, etc.). Such a universal wave function would be static—a wave function whose timelessness constitutes the problem of time in canonical quantum gravity—and, insofar as our universe is concerned, unique. But this doesn't mean, as we have already seen, that the world it describes would be static and timeless. No longer part of the state description, the universal wave function $\Psi$ provides a representation of dynamical law, via the vector field on configuration space that it defines. As such, the wave function plays a role analogous to that of the Hamiltonian function $H = H(Q, P) \equiv H(\xi)$ in classical mechanics—a function on phase space, a space even more abstract than configuration space. In fact, the wave function and the Hamiltonian function generate motions in pretty much the same way

$$\frac{d\xi}{dt} = \text{Der}H \longleftrightarrow \frac{dQ}{dt} = \text{Der}(\log \Psi),$$

with Der a derivation. (For more detail, see Sect. 12.3.1 below). And few would be tempted to regard the Hamiltonian function $H$ as a real physical field, or expect any back-action of particle configurations on this Hamiltonian function.

Once we recognize that the role of the wave function is thus nomological, two important questions naturally arise: Why and how should a formalism involving time-dependent wave functions obeying Schrödinger's equation emerge from a theory involving a fixed timeless universal wave function? And which principle singles out the special unique wave function $\Psi$ that governs the motion in our universe? Our answers to these questions are somewhat speculative. But they do provide further insight into the role of the wave function in quantum mechanics and might even explain why, in fact, our world is quantum mechanical. (For another problem that arises when the universal wave function is timeless, see the last paragraph of Sect. 12.3.3.)

In order to understand the emergence of a time-dependent wave function, we must ask the right question, which is this: Is it ever possible to find a simple effective theory governing the behavior of suitable subsystems of a Bohmian universe? Suppose, then, that the configuration of the universe has a decomposition of the form $q = (x, y)$, where $x$ describes the degrees of freedom with which we are somehow most directly concerned (defining the *subsystem*, the "$x$-system") and $y$ describes the remaining degrees of freedom (the subsystem's *environment*, the "$y$-system"). For example, $x$ might be the configuration of all the degrees of freedom governed by standard quantum field theory, describing the fermionic matter fields as well as the bosonic force fields, while $y$ refers to the gravitational degrees of freedom. Suppose further that we have, corresponding to this decomposition, a solution $Q(\tau) = (X(\tau), Y(\tau))$ of the appropriate (yet to be defined) extension of (11.4), where the real continuous parameter $\tau$ labels the slices in a suitable foliation of space-time.

Focus now on the *conditional wave function* (Eq. (2.39) of Chap. 2)

$$\psi_\tau(x) = \Psi(x, Y(\tau)) \tag{11.5}$$

of the subsystem, governing its motion, and ask whether $\psi_\tau(x)$ could be—and might under suitable conditions be expected to be—governed by a simple law that does not refer directly to its environment. (The conditional wave function of the $x$-system should be regarded as defined only up to a factor that does not depend upon $x$.)

Suppose that $\Psi$ satisfies an equation of the form (11.2), with $\widehat{H} = \{\widehat{H}_N\}$. Suppose further that for $y$ in some "$y$-region" of configuration space and for some choice of lapse function $N$ we have that $\widehat{H}_N \simeq \widehat{H}_N^{(x)} + \widehat{H}_N^{(y)}$ and can write

$$\Psi(x, y) = e^{-i\tau \widehat{H}_N} \Psi(x, y) \simeq e^{-i\tau \widehat{H}_N} \sum_\alpha \psi_0^\alpha(x) \phi_0^\alpha(y)$$

$$\simeq \sum_\alpha \left( e^{-i\tau \widehat{H}_N^{(x)}} \psi_0^\alpha(x) \right) \left( e^{-i\tau \widehat{H}_N^{(y)}} \phi_0^\alpha(y) \right)$$

$$=: \sum_\alpha \psi_\tau^\alpha(x) \phi_\tau^\alpha(y) \tag{11.6}$$

where the $\phi_0^\alpha$ are "narrow disjoint wave packets" and remain approximately so as long as $\tau$ is not too large. Suppose (as would be the case for Bohmian mechanics) that the motion is such that if the configuration $Y(0)$ lies in the support of one $\phi_0^{\alpha'}$, then $Y(\tau)$ will keep up with $\phi_\tau^{\alpha'}$ as long as the above conditions are satisfied. It then follows from (11.6) that for the conditional wave function of the subsystem we have

$$\psi_\tau(x) \approx \psi_\tau^{\alpha'},$$

and it thus approximately satisfies the time-dependent Schrödinger equation

$$i \frac{\partial \psi}{\partial \tau} = \widehat{H}_N^{(x)} \psi. \tag{11.7}$$

(In the case of (an extension of) Bohmian quantum gravity with preferred foliation, this foliation must correspond to the lapse function $N$ in (11.6).)

We may allow here for an interaction $\widehat{W}_N(x, y)$ between the subsystem and its environment in the Hamiltonian in (11.6), provided that the influence of the $x$-system on the $y$-system is negligible. In this case we can replace $\widehat{H}_N^{(x)}$ in (11.6) and (11.7) by $\widehat{H}_N^{(x)}(Y(\tau)) \equiv \widehat{H}_N^{(x)} + \widehat{W}_N(x, Y(\tau))$, since the wave packets $\phi^\alpha(y)$ are assumed to be narrow. Think, for the simplest example, of the case in which the $y$-system is the gravitational field and the $x$-system consists of very light particles.

Now one physical situation (which can be regarded as corresponding to a region of configuration space) in which (11.6), and hence the Schrödinger evolution (11.7),

## 11.5 A Universal Bohmian Theory

should obtain is when the $y$-system behaves semiclassically: In the semiclassical regime, one expects an initial collection of narrow and approximately disjoint wave packets $\phi_0^\alpha(y)$ to remain so under their (approximately classical) evolution.

As a matter of fact, the emergence of Schrödinger's equation in the semiclassical regime for gravity can be justified in a more systematic way, using perturbation theory, by expanding $\Psi$ in powers of the gravitational constant $\kappa$. Then for a "semiclassical wave function" $\Psi$, the phase $S$ of $\Psi$, to leading order, $\kappa - 1$, depends only on the 3-metric and obeys the classical Einstein-Hamilton-Jacobi equation, so that the metric evolves approximately classically, with the conditional wave function for the matter degrees of freedom satisfying, to leading (zeroth) order, Schrödinger's equation for, say, quantum field theory on a given evolving background. The relevant analysis was done by Banks [17] for canonical quantum gravity, but the significance of that analysis is rather obscure from an orthodox perspective:

The semiclassical limit has been proposed as a solution to the problem of time in quantum gravity, and as such has been severely criticized by Kuchar [7], who concludes his critique by observing that "the semiclassical interpretation does not solve the standard problems of time. It merely obscures them by the approximation procedure and, along the way, creates more problems." Perhaps the main difficulty is that, within the orthodox framework, the classical evolution of the metric is not really an approximation at all. Rather, it is put in by hand, and can in no way be justified on the basis of an entirely quantum mechanical treatment, even as an approximation. This is in stark contrast with the status of the semiclassical approximation within a Bohmian framework, for which there is no problem of time. In this approach, the classical evolution of the metric is indeed merely an approximation to its exact evolution, corresponding to the exact phase of the wave function (i.e., to Eq. (11.4)). To the extent that this approximation is valid, the appropriate conclusions can be drawn, but the theory makes sense, and suffers from no conceptual problems, even when the approximation is not valid.

Now to our second question. Suppose that we demand of a universal dynamics that it be first-order for the variables describing the primitive ontology (the simplest possibility for a dynamics) and covariant—involving no preferred foliation, no special choice of lapse function $N$, in its formulation. This places a very strong constraint on the vector field defining the law of motion—and on the universal wave function, should this motion be generated by a wave function. The set of wave functions satisfying this constraint should be very small, far smaller than the set of wave functions we normally consider as possible initial states for a quantum system. However, according to our conception of the wave function as nomological, this very fact might well be a distinct virtue.

We have begun to investigate the possibility of a first-order covariant geometrodynamics in connection with Bohmian quantum gravity, and have found that the constraint for general covariance is captured by the Dirac algebra (see also [18]), which expresses the relation between successive infinitesimal deformations of hypersurfaces, taken in different orders: It has been argued by one of us (S. G.) and Stefan Teufel (unpublished) that defining a representation of the Dirac algebra is more or less the necessary and sufficient condition for a vector field on the space of

3-metrics to yield a generally covariant dynamics, generating a 4-geometry involving no dynamically distinguished hypersurfaces.

This work is very much in its infancy. In addition to the problem of finding a mathematically rigorous proof of the result just mentioned, there remains the difficult question of the possible representations of the Dirac algebra, both for pure gravity and for gravity plus matter. For pure gravity it seems that a first-order generally covariant geometrodynamics is achievable, but only with vector fields that generate classical 4-geometries—solutions of the Einstein equations with a possible cosmological constant. How this situation might be affected by the inclusion of matter is not easy to say.

Even a negative result—to the effect that a generally covariant Bohmian theory must involve additional space-time structure—would be illuminating. A positive result—to the effect that a first-order dynamics, for geometry plus matter, that does not invoke additional space-time structure can be generally covariant more or less only when the vector field defining this dynamics arises from an essentially unique wave function of the universe that happens to satisfy an equation like the Wheeler-DeWitt equation (and from which a time-dependent Schrödinger equation emerges, in the manner we've described, as part of a phenomenological description governing the behavior of appropriate subsystems)—would be profound. For then we would know, not just what quantum mechanics is, but why it is.

# References

1. C. Misner, K. Thorne, and J. Wheeler. Gravitation. W. H. Freeman, San Francisco, 1973.
2. C. Rovelli. Strings, Loops and Others: A Critical Survey of the Present Approaches to Quantum Gravity. In N. Dadhich and J. Narlikar, editors, *Proceedings of the 15th International Conference on General Relativity and Gravitation (1997)*, Pune, India, 1998. IUCAA.
3. R. Arnowitt, S. Deser, and C. W. Misner. The Dynamics of General Relativity. In *Gravitation: an Introduction to Current Research*. Wiley, New York, 1962.
4. A. Ashtekar. New Hamiltonian Formulation of General Relativity. *Physical Review D*, 36:1587–1602, 1987.
5. C. Rovelli and L. Smolin. Loop Representation of Quantum General Relativity. *Nuclear Physics B*, 331:80–152, 1990.
6. P. A. M. Dirac. *Lectures on Quantum Mechanics*. Yeshiva University, New York, 1964.
7. K. Kuchar. Time and Interpretations of Quantum Gravity. In *Proceedings of the 4th Canadian Conference on General Relativity and Relativistic Astrophysics (Winnipeg, MB, 1991)*, pages 211–314, River Edge, NJ, 1992. World Sci. Publishing.
8. K. Kuchar. Canonical Quantum Gravity. In R.J. Gleiser, C.N. Kozameh, and O.M. Moreschi, editors, *General Relativity and Gravitation 1992*, pages 119–150, Bristol, Philadelphia, 1993. Institute of Physics Pub.
9. C. J. Isham. Prima Facie Questions in Quantum Gravity. In *Canonical Gravity: From Classical to Quantum*, Lecture Notes in Physics. Springer-Verlag, New York-Heidelberg-Berlin, 1994.
10. S. Goldstein. Quantum Theory Without Observers. Part One. *Physics Today*, 51(3):42–46, 1998.
11. S. Goldstein. Quantum Theory Without Observers. Part Two. *Physics Today*, 51(4):38–42, 1988.
12. P. R. Holland. *The Quantum Theory of Motion*. Cambridge University Press, Cambridge-New York-New Rochelle-Melbourne-Sydney, 1993.

# References

13. Y. Shtanov. On Pilot Wave Quantum Cosmology. *Physical Review D*, 54:2564–2570, 1996.
14. J. Kowalski-Glikman and J. C. Vink. Gravity-Matter Mini-Superspace: Quantum Regime, Classical Regime and in Between. *Classical and Quantum Gravity*, 7:901–918, 1990.
15. E. Squires. A Quantum Solution to a Cosmological Mystery. *Physical Letters A*, 162:35–36, 1992.
16. C. Callender and R. Weingard. The Bohmian Model of Quantum Cosmology. In *Proceedings of the Philosophy of Science Association*, volume 1, pages 228–237, 1994.
17. T. Banks. TCP, Quantum Gravity, the Cosmological Constant, and All That. *Nuclear Physics B*, 249:332–360, 1985.
18. S. A. Hojman, K. Kuchař, and C. Teitelboim. Geometrodynamics Regained. *Annals of Physics*, 96:88–135, 1976.

# Chapter 12
# Reality and the Role of the Wave Function in Quantum Theory

## 12.1 Questions About the Wave Function

We shall be concerned here with the role and status of the wave function in quantum theory, and especially in Bohmian mechanics.

The wave function is arguably the main innovation of quantum theory. Nonetheless the issue of its status has not received all that much attention over the years. What the hell is this strange thing, the wave function, that we have in quantum mechanics? What's going on with that? Who ordered that?

In more detail, is the wave function subjective or epistemic, or is it objective? Does it merely describe our information or does it describe an observer independent reality? What's the deal with collapse? Why does the wave function collapse? What's going on there? And if the wave function is objective, is it some sort of concrete material reality or something else?

Let us say a word about what it means for the wave function to be merely epistemic. To us that means first that there is something else, let's call it $X$, describing some physical quantity, say the result of an experiment, or maybe the whole history up to the present of some variable or collection of variables—things we're primarily interested in. And then to say that the wave function is merely epistemic is to say that it is basically equivalent to a probability distribution on the space of possible values for $X$.

You should note that orthodox quantum theory is not of this form. That's because the $X$ is in effect a hidden variable and there are no hidden variables in orthodox quantum theory—there's just the wave function. So the wave function is certainly not merely epistemic in orthodox quantum theory.

And neither is Bohmian mechanics of this form. In Bohmian mechanics it is indeed the case that the wave function sort of has a probabilistic role to play, since the absolute square of the wave function gives the probability of the configuration of the Bohmian system. However, that's not the only role for the wave function in Bohmian mechanics; it's not its fundamental role and certainly not its most important role.

We should all agree—and maybe this is the only thing we would all agree upon—that there are three possibilities for the wave function: (i) that it is everything, as would seem to be the case with Everett [1], (ii) what would seem to be the most modest possibility, that it is something (but not everything), as with Bohmian mechanics, for example, where there's the wave function and something else, or (iii) maybe it's nothing—that would solve the problem of what's this weird thing, the wave function: if you can get rid of it you don't have to agonize about it.

## 12.2 The Wave Function of a Subsystem

We recall that a crucial ingredient in the extraction of the implications of Bohmian mechanics is the notion of the wave function of a subsystem of a Bohmian universe, a universe of particles governed by the equations of Bohmian mechanics, defining a motion choreographed by the *wave function of the universe* $\Psi$.

In almost all applications of quantum mechanics it is the wave function of a subsystem with which we are concerned, not the wave function of the universe. The latter, after all, must be rather elusive. Most physicists don't deal with the universe as a whole. They deal with subsystems more or less all the time: a hydrogen atom, particles going through Stern-Gerlach magnets, a Bose-Einstein condensate, or whatever. And yet from a fundamental point of view the only genuine Bohmian system in a Bohmian universe—the only system you can be sure is Bohmian—is the universe itself, in its entirety. It can't be an immediate consequence of that that subsystems of a Bohmian universe are themselves Bohmian, with the motion of their particles governed by wave functions in the Bohmian way.

That is, one can't simply demand of subsystems of a Bohmian universe that they be Bohmian systems in their own right. The behavior of the parts of a big system are already determined by the behavior of the whole. And what you have for the whole is the wave function $\Psi$ of the universe, together with its configuration $Q$. That's your data. That's what's objective in a Bohmian universe. The wave function of a subsystem, if it exists at all, must be definable in terms of that data.

Now corresponding to a subsystem of the universe is a splitting $Q = (Q_{sys}, Q_{env}) = (X, Y)$ of its configuration $Q$ into the configuration $Q_{sys} = X$ of the subsystem, the "$x$-system," formed from the positions of the particles of the subsystem, and the configuration $Q_{env} = Y$ of the environment of the subsystem—the configuration of everything else. So the data in terms of which the wave function of a subsystem must be defined are the universal wave function $\Psi(q) = \Psi(x, y)$ and the actual configurations $X$ of the subsystem and $Y$ of its environment.

The first guess people make about what the wave function $\psi(x)$ of the $x$-subsystem should be usually turns out to be wrong. The right guess, and the natural thing to do, is to define the wave function of a subsystem in this way: Remembering that the wave function of a subsystem should be a function on its configuration space, a function, that is, of $x$ alone, you take the universal wave function $\Psi(x, y)$ and plug the actual configuration $Y$ of the environment into the second slot to obtain a function of $x$,

## 12.2 The Wave Function of a Subsystem

$$\psi(x) = \Psi(x, Y). \tag{12.1}$$

If you think about it you see that this is exactly the right definition. The situation is simplest for spin-0 particles, which we will henceforth assume. First of all, it is easy to see that the velocity that the configuration $X$ of the subsystem inherits from the motion of the configuration $Q$ can be expressed in terms of this $\psi$ in the usual Bohmian way. In other words, if $dQ/dt = v^\Psi(X, Y)$ then $dX/dt = v^\psi(X)$ for $\psi(x) = \Psi(x, Y)$.

However, the evolution law for the wave function of the subsystem need not be Bohmian. Explicitly putting in the time dependence, we have for the wave function of the $x$-system at time $t$:

$$\psi_t(x) = \Psi_t(x, Y_t).$$

Thus the wave function of a subsystem has an interesting time dependence. Time appears here in two places: the wave function of the universe depends on $t$ since it evolves according to Schrödinger's equation. And the configuration of the environment $Y$ also evolves and depends on $t$ as part of the evolving configuration of the universe $Q_t = (X_t, Y_t)$.

This suggest a rich variety of ways that the wave function of a subsystem might behave in time. Everyone readily believes, and it is in fact the case, that the wave function of a subsystem evolves just as it should for a Bohmian system, according to Schrödinger's equation for the subsystem, when the subsystem is suitably decoupled from its environment. And it's actually rather easy to see that the wave function of a subsystem collapses according to the usual textbook rules with the usual textbook probabilities in the usual measurement situations. The wave function of the $x$-system thus collapses in just the way wave functions in quantum mechanics are supposed to collapse. This follows more or less directly from standard quantum measurement theory together with the definition of the wave function of the $x$-system and the quantum equilibrium hypothesis, see Chap. 2.

It is a sociological fact, for whatever reason, that even very talented mathematical physicists have a lot of trouble accepting that the wave function of a subsystem collapses as claimed. We guess that's because people know that collapse in quantum mechanics is supposed to be some really problematical, difficult issue, so they think it can't be easy for Bohmian mechanics either. But it is easy for Bohmian mechanics. And the thing everyone is happy to take for granted, that the wave function of a subsystem will evolve according to Schrödinger's equation in the appropriate situations—they do so presumably because nobody says there's a problem getting wave functions to obey Schrödinger's equation. It's collapse, that's the problem. But understanding why the wave function of a subsystem does indeed evolve according to Schrödinger's equation when the subsystem is suitably decoupled from its environment is a bit tricky. Nonetheless it is true, though we shall not go into any details here, see Chap. 2.

The main point we wish to have conveyed in this section is that for a Bohmian universe the wave function of a subsystem of that universe, defined in terms of the wave function of the universe and additional resources available to Bohmian mechanics and absent in orthodox quantum theory—namely the actual configuration of the environment of the subsystem—behaves exactly the way wave functions in orthodox quantum theory are supposed to behave.

## 12.3 The Wave Function as Nomological

The main thing we want to discuss here is the status of the wave function: what kind of thing it is. And what we want to suggest one should think about is the possibility that it's nomological, nomic—that it's really more in the nature of a law than a concrete physical reality.

Thoughts in this direction might arise when one considers the unusual way in which Bohmian mechanics is formulated, and the unusual sort of behavior that the wave function undergoes in Bohmian mechanics. The wave function of course affects the behavior of the configuration, i.e., of the particles. This is expressed by the guiding Eq. (2.10), which in more compact form can be written

$$dQ/dt = v^\psi(Q). \tag{12.2}$$

But in Bohmian mechanics there's no back action, no effect in the other direction, of the configuration upon the wave function, which evolves autonomously via Schrödinger's Eq. (2.1), in which the actual configuration $Q$ does not appear. Indeed the actual configuration could not appear in Schrödinger's equation since this equation is an equation also of orthodox quantum theory and in orthodox quantum theory there is no actual position or configuration. That's one point.

A second point is that for a multi-particle system the wave function $\psi(q) = \psi(\mathbf{q}_1, \ldots, \mathbf{q}_N)$ is not a weird field on physical space, its a weird field on configuration space, the set of all hypothetical configurations of the system. For a system of more than one particle that space is not physical space. What kind of thing is this field on that space?[1]

The fact that Bohmian mechanics requires that one take such an unfamiliar sort of entity seriously bothers a lot of people. It doesn't in fact bother us all that much, but it does seem like a significant piece of information nonetheless. And what it suggests to us is that you should think of the wave function as describing a law and not as

---

[1] The sort of physical reality to which the wave function corresponds is even more abstract than we've conveyed so far. That's because the wave function, in both orthodox quantum theory and Bohmian mechanics, is merely a convenient representative of the more physical "quantum state." Two wave functions such that one is a (nonzero) scalar multiple of the other represent the same quantum state and are regarded as physically equivalent. Thus the quantum state is not even a field at all, but an equivalence class of fields. It is worth noting that equivalent wave functions define the same velocity (2.10). They also define, with suitable normalization, the same $|\psi|^2$-probabilities.

Moreover, for the treatment of identical particles such as electrons in Bohmian mechanics, it is best to regard them as unlabelled, so that the configuration space of $N$ such particles is not a high dimensional version of a familiar space, like $\mathbb{R}^{3N}$, but is instead the unfamiliar high-dimensional space $^N\mathbb{R}^3$ of $N$-point subsets of $\mathbb{R}^3$. This space has a nontrivial topology, which naturally leads to the possibilities of bosons and fermions—and in two dimensions anyons as well [2]. As a fundamental space it is odd, but not as a configuration space.

12.3 The Wave Function as Nomological

some sort of concrete physical reality. After all (12.2) is an equation of motion, a law of motion, and the whole point of the wave function here is to provide us with the law, i.e., with the right hand side of this equation.

Now we've said that rather cavalierly. There are lots of problems with saying it at this point. But before going into the problems let us make a comparison with a familiar situation where nobody seems to have much of a problem at all, namely classical Hamiltonian dynamics.

## 12.3.1 Comparison of $\psi$ with the Classical Hamiltonian H

The wave function is strange because it lives on configuration space, for an $N$-particle system a space of dimension $3N$. Well, there's a space in the classical mechanics of an $N$-particle system that has twice that dimension, its phase space, of dimension $6N$. On that space there's a function, the Hamiltonian $H = H(q, p) = H(\mathscr{X})$ of the system, and to define the equations of motion of classical mechanics you put $H$ on the right hand side of the equations of motion after suitably taking derivatives. We've never heard anyone complaining about classical mechanics because it invokes a weird field on phase space, and asking what kind of thing is that? Nobody has any problem with that. Everybody knows that the Hamiltonian is just a convenient device in terms of which the equations of motion can be nicely expressed.

We're suggesting that you should regard the wave function in exactly the same way. And if you want to have a sharper analogy you can think not of $\psi$ itself but of something like $\log \psi(q)$ as corresponding to the Hamiltonian $H(\mathscr{X})$. The reason we suggest this is because the velocity in Bohmian mechanics is proportional to the imaginary part of $\nabla \psi / \psi$ for a scalar wave function, a sort gradient of the log of $\psi$, some sort of derivative, der, of $\log \psi(q)$, so that (2.10) can be regarded as of the form

$$dQ/dt = \text{der}(\log \psi).$$

Similarly in classical mechanics we have an evolution equation of the form

$$d\mathscr{X}/dt = \text{der } H$$

where der $H$ is a suitable derivative of the Hamiltonian. (This is a compact way of writing the familiar Hamiltonian equations $d\mathbf{q}_k/dt = \partial H/\partial \mathbf{p}_k$, $d\mathbf{p}_k/dt = -\partial H/\partial \mathbf{q}_k$.)

It is also true that both $\log \psi$ and $H$ are normally regarded as defined only up to an additive constant: When you add a constant to $H$ it doesn't change the equations of motion. If you multiply the wave function by a scalar—which amounts to adding a constant to its log—the new wave function is generally regarded as physically equivalent to the original one. And in Bohmian mechanics the new wave function defines the same velocity for the configuration, the same equations of motion, as the original one.

Moreover, with suitably "normalized" choices $\psi(q)$ and $H(\mathcal{X})$, corresponding to appropriate choices of the constants, one associates rather similar probability formulas: In classical statistical mechanics there are the Boltzmann-Gibbs probabilities, given by $e^{-H/kT}$ when $H$ has been suitably normalized, where $k$ is Boltzmann's constant and $T$ is the temperature. One thus has that

$$\log Prob \propto -H.$$

And in quantum mechanics or Bohmian mechanics, with $|\psi|^2$–probabilities, one has that
$$\log Prob \propto \log |\psi|.$$

(You probably shouldn't take this last point about analogous probabilities too seriously. It's presumably just an accident that the analogy seems to extend this far.)

### 12.3.2 $\psi$ versus $\Psi$

There are, however, problems with regarding the wave function as nomological. Laws aren't supposed to be dynamical objects, they aren't supposed to change with time, but the wave function of a system typically does. And laws are not supposed to be things that we can control—we're not God. But the wave function is often an initial condition for a quantum system. We often, in fact, prepare a system in a certain quantum state, that is, with a certain wave function. We can in this sense control the wave function of a system. But we don't control a law of nature. This makes it a bit difficult to regard the wave function as nomological.

But with regard to this difficulty it's important to recognize that there's only one wave function we should be worrying about, the fundamental one, the wave function $\Psi$ of the universe. In Bohmian mechanics, the wave function $\psi$ of a subsystem of the universe is defined in terms of the universal wave function $\Psi$. Thus, to the extent that we can grasp the nature of the universal wave function, we should understand as well, by direct analysis, also the nature of the objects that are defined in terms of it, and in particular we should have no further fundamental question about the nature of the wave function of a subsystem of the universe. So let's focus on the former.

### 12.3.3 The Universal Level

When we consider, instead of the wave function of a typical subsystem, the wave function $\Psi$ of the universe itself, the situation is rather dramatically transformed. $\Psi$ is not controllable. It is what it is! And it may well not be dynamical either. There may well be no "$t$" in $\Psi$.

The fundamental equation for the wave function of the universe in canonical quantum cosmology is the Wheeler-DeWitt equation [3],

## 12.3 The Wave Function as Nomological

$$\mathcal{H}\Psi = 0,$$

for a wave function $\Psi(q)$ of the universe, where $q$ refers to 3-geometries and to whatever other stuff is involved, all of which correspond to structures on a 3-dimensional space. In this equation $\mathcal{H}$ is a sort of generalized Laplacian, a cosmological version of a Schrödinger Hamiltonian $H$. And like a typical $H$, it involves nothing like an explicit time-dependence. But unlike Schrödinger's equation, the Wheeler-DeWitt equation has on one side, instead of a time derivative of $\Psi$, simply 0. Its natural solutions are thus time-independent, and these are the solutions of the Wheeler-DeWitt equation that are relevant in quantum cosmology.

That this is so is in fact the *problem of time* in quantum cosmology. We live in a world where things change. But if the basic object in the world is a timeless wave function how does change come about? Much has been written about this problem of time. A great many answers have been proposed. But what we want to emphasize here is that from a Bohmian perspective the timelessness of $\Psi$ is not a problem. Rather it is just what the doctor ordered.

The fundamental role of the wave function in Bohmian mechanics is to govern the motion of something else. Change fundamentally occurs in Bohmian mechanics not so much because the wave function changes but because the thing $Q$ it's governing does, according to a law

$$dQ/dt = v^{\Psi}(Q) \tag{12.3}$$

determined by the wave function. The problem of time vanishes entirely from a Bohmian point of view. And it's just what the doctor ordered because laws are not supposed to change with time, so we don't want the fundamental wave function to change with time. It's good that it doesn't change with time.

There may be another good thing about the wave function of the universe: it may be unique. It is of course the case that, together with being uncontrollable, a timeless wave function of our actual universe would be the one wave function that it is. But we mean more than that: While the Wheeler-DeWitt equation presumably has a great many solutions $\Psi$, when supplemented with additional natural conditions, for example the Hartle-Hawking boundary condition [4], the solution may become unique. And such uniqueness fits nicely with the conception of the wave function as law.

But there is a difficulty that arises in connection with the stationarity of the universal wave function. In Sect. 2.14 we argued that irreversibility and the arrow of time in our universe—with the law of entropy increase and its other manifestations—originates from a low entropy initial condition for the universal wave function and not from a low entropy initial configuration of the universe, which we claimed was in quantum equilibrium—in equilibrium relative to the wave function. But if the initial wave function is stationary, it can't be responsible for the irreversible behavior of our universe. This would seem to suggest that the latter must then originate in quantum nonequilibrium and this may well be so. At the same time our remark in the last paragraph of Sect. 11.4 about the meaning of quantum equilibrium in the absence of a physical external time parameter, as well as the recent work of Carroll and Chen [5], in which it is argued that it may be possible to explain the observed irreversibility of our world without invoking the past hypothesis requiring a low entropy universal initial condition, suggest that in fact a pure equilibrium analysis (such as that in Chap. 2) may suffice. These matters are difficult and much remains to be done, however not here.

### 12.3.4 Schrödinger's Equation as Phenomenological (Emergent)

Now we can well imagine someone saying, OK fine, in this Bohmian theory for the universe stuff changes—particles move, the gravitational field changes, the gravitational metric evolves, whatever. But we know that the most important equation in quantum mechanics, and one of the most important equations in our quantum world, is the time-dependent Schrödinger equation, describing wave functions that themselves change with time. Where does that come from in a theory in which the only fundamental wave function that you have is the timeless wave function $\Psi$?

But that question has already been answered here, in the last paragraph of Sect. 12.2. If you have a wave function of the universe obeying Schrödinger's equation, then in suitable situations, those in which a subsystem is suitably decoupled from its environment (and the Hamiltonian $\mathscr{H}$ is of the appropriate form), the wave function $\psi_t(x) = \Psi(x, Y_t)$ of the subsystem will evolve according to Schrödinger's equation for that subsystem. For $\Psi$ not depending upon time, the wave function of the subsystem inherits its time-dependence from that of the configuration $Y_t$ of the environment. And the crucial point here is that a solution $\Psi_t$ of Schrödinger's equation can be time-independent. These are the solutions $\Psi_t$ that are the same wave function $\Psi$ for all t, the solutions for which $\partial \Psi_t/\partial t$ is 0 for all $t$, corresponding precisely to solutions of the Wheeler-DeWitt equation.

But in this situation the time-dependent Schrödinger evolution (2.1) is not fundamental. Rather it is emergent and phenomenological, arising—as part of a good approximation for the behavior of suitable subsystems—from a Bohmian dynamics (12.3) for the universe given in terms of a suitable wave function of the universe, one which obeys the Wheeler-DeWitt equation.

And even this time-independent equation might not be fundamental—that it appears to be might be an illusion. What we have in mind is this: We've got a law of motion involving a vector field $v^\Psi$ (the right-hand-side of a first-order equation of motion), a vector field that can be expressed in terms of $\Psi$. If $\Psi$ is a nice sort of wave function it might obey all sorts of nice equations, for example the Wheeler-DeWitt equation or something similar. From a fundamental point of view, it might be a complete accident that $\Psi$ obeys such an equation. It might just happen to do so. The fact that the equation is satisfied might have nothing to do with why the fundamental dynamics is of the form (12.3). But as long as $\Psi$ does satisfy the equation, by accident or not, all the consequences of satisfying it follow.

## 12.3 The Wave Function as Nomological

So it could turn out, at the end of the day, that what we take to be the fundamental equation of quantum theory, Schrödinger's equation, is not at all fundamental for quantum theory, but rather is an emergent and accidental equation.

### 12.3.5 Two Transitions

We'd like to focus a bit on the change of perspective that occurs when we make the transition from orthodox quantum theory, which seems to involve only the wave function $\psi$, to (conventional) Bohmian mechanics, which is usually regarded as involving two types of physical entities, wave functions $\psi$ and the positions of particles, forming a configuration $Q$, to universal Bohmian mechanics, where the wave function $\Psi$ is taken out of the category of concrete physical reality and into that of law, so you've got just $Q$ as describing elements of physical reality:

$$\begin{array}{ccccc} \text{OQT} & \to & \text{BM} & \to & \text{UBM} \\ \psi & & (\psi, Q) & & Q \end{array}$$

You start with just $\psi$, you end with just $Q$.

And our original question, about the wave function—What kind of thing is that?—is rather dramatically transformed when we make this transition to the universal level, since we're then asking about a very different object, not about a wave function of a subsystem of the universe but about the universal wave function,

$$\begin{array}{ccc} ? & & ? \\ ? \ \psi \ ? & \to & ? \ \Psi \ ? \\ ? & & ? \end{array}$$

which is actually, so we are supposing, just a way of representing the law of motion. So, now we may ask, does any kind of question about $\Psi$ remain?

Here's one question: Why should the motion be of the form (12.3), involving $\Psi$ in the way that it does? Why should the law of motion governing the behavior of the constituents of the universe be of such a form that there is a wave function in terms of which the motion can be compactly expressed? We think that's a good question. And of course we have no definitive answer to it. But an answer to this question would provide us with a deep understanding of why our world is quantum mechanical.

The view that the wave function is nomological has another implication worth considering. This is connected with the question of how we ever come to know what the wave function of a system is. There must be some algorithm that we use. We don't directly see wave functions. What we see (more directly) are particles, at least from a Bohmian perspective. We should read off from the state of the primitive ontology, whatever it may be, what the relevant wave function is. There should be some algorithm connecting the state of the primitive ontology, for Bohmian mechanics the relevant configuration $Q(t)$ over, say, some suitable time interval, with the relevant wave function.

Now let's go to the universal level. You might think there should be some algorithm that we can use to read off what the universal wave function is from the state of the primitive ontology of the universe, whatever that may be. But in fact we kind of doubt there is any such algorithm. So far as we know, nobody has proposed any such algorithm. And from the point of view of the wave function $\Psi$ being nomological, you wouldn't expect there to be any such algorithm. That's because if the wave is nomological, specifying the wave function amounts to specifying the theory. You wouldn't expect there to be an algorithm for theory formation.

### 12.3.6 Nomological versus Nonnomological

Now, we can imagine—and in fact we're quite sure—that many physicists would respond to the question about whether or not the wave function is fundamentally nomological with a big "Who cares? What difference does it make?"

Well, we think it does matter. Being nomological has important implications. Laws should be simple. If we believe that the wave function of the universe is nomological, this belief should affect our expectations for the development of physics. We should expect somehow to arrive at physics in which the universal wave function involved in that physics is in some sense simple—while presumably having a variety of other nice features as well.

Now simplicity itself is sort of complicated. There are a number of varieties of simplicity. For example, the universal wave function could be simple in the sense that it has a simple functional form—that it's a simple function of its arguments. That's one possibility. Another, quite different, is that it could be a more or less unique solution to a simple equation. Or, a similar kind of thing, it could more or less uniquely satisfy some compelling principle, maybe a symmetry principle. See Chap. 10.

### 12.3.7 Relativistic Bohmian Theory

We want to say a bit about Lorentz invariance and a problem that arises in connection with it. It's widely said—and it's natural to think—that you can't have a Lorentz invariant Bohmian theory. That's basically because of the crucial role played in such a theory by the configuration of the system: the positions of its particles—or the detailed description of the primitive ontology of the theory, whatever that may be—*at a given time*.

Now you can also consider configurations determined, not by a $t = $ constant hypersurface but by a general space-like hypersurface. For example, for a particle ontology, the configuration corresponding to such a surface would be given by the space-time points on the surface at which the world-lines of the particles cross the surface, for $N$ particles, $N$ points. So if in fact you had somehow at your disposal

## 12.3 The Wave Function as Nomological

a Lorentz invariant foliation of space-time into space-like hypersurfaces, you could play a Bohmian game and define a Bohm-type dynamics for the evolution of configurations defined in terms of that foliation. In this way one could obtain a Lorentz invariant Bohmian theory, see Chap. 9.

To actually have such a thing the best possibility is perhaps the following: You have a Lorentz invariant rule for defining in terms of the universal wave function a foliation of space-time, a covariant map fol from wave functions to foliations:

$$\text{(Lorentz) covariant map } \Psi \xrightarrow{\text{fol}} \mathscr{F} = \mathscr{F}(\Psi).$$

(For the map to be *covariant* means that the diagram

$$\begin{array}{ccc} \Psi & \xrightarrow{\text{fol}} & \mathscr{F} \\ \Lambda_g \downarrow & & \downarrow g \\ \Psi_g & \xrightarrow{\text{fol}} & \mathscr{F}_g \end{array}$$

is commutative. Here g is any Lorentz transformation, on the right acting naturally on the foliation by moving the points on any leaf of the foliation, and hence the leaves themselves and the foliation itself, around according to g, while $\Lambda_g$ is the action of g on wave functions, given by a representation $\Lambda$ of the Lorentz group.)

Lorentz invariant Bohmian theories formed in this way, by utilizing such a covariant foliation map, have the virtue of being seriously Lorentz invariant. The point here is that any theory can be made Lorentz invariant in a trivial nonserious way by introducing suitable additional space-time structure beyond the Lorentz metric. The question then arises as to what kinds of structure are unproblematic. James Anderson has addressed this question by distinguishing between absolute and dynamical structures, and identifying the serious Lorentz invariance of a theory with the nonexistence in the theory of any additional absolute structures [6].

What exactly these are is not terribly relevant for our purposes here. That's because, for the sort of theory proposed here, what seems to be additional space-time structure, namely the foliation, is not an additional structure at all, beyond the wave function. To the extent that the wave function is a legitimate structure for a Lorentz invariant theory—and this is generally assumed to be the case—so are covariant objects defined solely in terms of the wave function.

Here are some examples of possibilities for covariant foliations. You could form a typical quantum expectation in the Heisenberg picture, involving the universal wave function and some sort of operator-valued Fermi field $\psi(x)$. The simplest such object is perhaps

$$j_\mu(x) = \langle \Psi | \bar{\psi}(x) \gamma_\mu \psi(x) | \Psi \rangle,$$

involving the Dirac matrices $\gamma_\mu$, defining a time-like vector field on space-time. You could also put suitable products in the middle to form tensors of various ranks. Ward Struyve [7] has suggested using the stress energy tensor

$$t_{\mu\nu}(x) = \langle \Psi | T_{\mu\nu}(x) | \Psi \rangle$$

and integrating that over space-like hypersurfaces to obtain a time-like vector (that in fact does not depend upon the choice of surface). There are a variety of such proposals for extracting from the wave function a vector field on space-time in a covariant manner. And a vector field on space-time is just the sort of thing that could define a foliation, namely into hypersurfaces orthogonal to that vector field, so that we have the following scheme for a map fol:

$$\Psi \rightarrow j_\mu \rightsquigarrow \mathscr{F}$$

Now Struyve's proposal works as is, but for other proposals you would have to do a lot of massaging to get the scheme to work: In order to define a foliation the vector field would have to be what is called "in involution." That can be achieved, but in so doing you would like the resulting vector field to remain time-like (so that the corresponding foliation would be into space-like hypersurfaces), and that is certainly not automatic.

The bottom line is that there is lots of structure in the universal wave function, enough structure certainly to typically permit the specification of a covariant rule for a foliation.

### 12.3.8 Wave Function as Nomological and Symmetry

But there is a problem: there is a conflict between the wave function being nomological and symmetry demands. The problem arises from the difference between having an action of the Lorentz group $G$ (or whatever other symmetry group we have in mind) on the Hilbert space $\mathscr{H}$ of wave functions (or on a suitable subset of $\mathscr{H}$, the domain of the foliation map)—which is more or less all that is usually required for a Lorentz invariant theory—and having the trivial action, always carrying $\Psi$ to itself: the difference between an action of $G$ on $\mathscr{H}$ and the $G$-invariance of $\Psi$. If the universal wave function represents the law, then that wave function itself, like an invariant law, should be $G$-invariant. (Actually, any change of the wave function that leaves the associated velocity vector field alone would be fine, for example multiplication by a constant scalar, but we shall for simplicity ignore this possibility here.)

It's not hard to see that that's incompatible with the covariance of the foliation map. No foliation can be Lorentz invariant, since there is always some Lorentz transformation that will tilt some of its leaves, at least somewhere. But if $\Psi$ is g-invariant so must be any foliation associated with $\Psi$ in a covariant manner. Thus a Lorentz invariant wave function $\Psi$ can't be covariantly associated with a foliation.

This is in sharp contrast with the situation for a generic wave function of the universe. Such a wave function will not be symmetric, and there is no obstacle to its being in the domain of a covariant foliation map. But if the universal wave function is nomological, it is not generic, and it must be too symmetric to permit the existence of a covariant foliation map.

## 12.3.9 Possible Resolutions

There seems to be a conflict between (i) having a Bohmian quantum theory, (ii) the universal wave function for that theory being nomological, and (iii) fundamental Lorentz invariance. Something, it would seem, has to give. And from a Bohmian point view the thing that gives wouldn't be the Bohmian part.

Here are some possible resolutions that would allow us to continue to regard the wave function as nomological. You could abandon fundamental Lorentz invariance, as many people have suggested. Another possibility is to make use of a Lorentz invariant foliation, but not one determined by the wave function but rather defined in terms of additional dynamical structure beyond the wave function, for example some suitable time-like vector field definable from the primitive ontology or perhaps transcending the primitive ontology, or something like a "time function" in general relativity, defined in terms of the gravitational metric and stuff in space-time. Or, the most likely possibility: something nobody has thought of yet.

And, of course, there is the possibility that we will have to abandon our attempt to regard the wave function as nomological. Many, for example Travis Norsen, would then insist that the wave function be eliminated in favor of something like exclusively local beables [8]. That's not how we feel. If it should turn out that the wave function can't be regarded as nomological—because it's too complicated or whatever—our reaction would probaby be: OK, that's just the way it is. It's not nomological but something different.

In fact, we think in fact that if someone gave us a Bohmian kind of theory, involving a complicated collection of exclusively local beables, and then someone else pointed out to us that the complicated local beables can be repackaged into a simple mathematical object of a nonlocal character—like a wave function on configuration space—our reaction would likely be that we would prefer to regard the wave function in that simpler though more unfamiliar way, just because of its mathematical simplicity.

## 12.3.10 $\psi$ as Quasi-Nomological

Suppose we accept that the universal wave function $\Psi$ is nomological. What then about the status of the wave function $\psi$ of subsystems of the universe—the wave functions with which we're normally concerned in applications of quantum theory? To this question we have several responses.

Our first response is this: You can decide for yourself. We are assuming the status of $\Psi$ is clear. The status of the primitive ontology is certainly clearer still. Therefore, since $\psi$ is defined in terms of $\Psi$ and the primitive ontology (specifically, the configuration $Y$ of the environment), the status of $\psi$ must follow from an analysis of its definition.

We don't insist that everyone would agree on the conclusion of such an analysis. It may well be that different philosophical prejudices will tend to lead to different conclusions here. Our point is rather that once the status of the wave function of the universe has been settled, the question about the status of $\psi$ is rather secondary—something about which one might well feel no need to worry.

Be that as it may, we would like to regard $\psi$ as quasi-nomological. We mean by this that while there are serious obstacles to regarding the wave function of a subsystem as fully nomological, $\psi$ does have a nomological aspect in that it seems more like an entity that is relevant to the behavior of concrete physical reality (the primitive ontology) and not so much like a concrete physical reality itself.

But we can say more. The law governing the behavior of the primitive ontology of the universe naturally implies a relationship between the behavior of a subsystem and the configuration of its environment. It follows from its definition (2.39),

$$\psi(x) = \Psi(x, Y),$$

that the wave function of the subsystem captures that aspect of the environment that expresses this relationship—that component of the universal law that is relevant to the situation at hand, corresponding to the configuration of the environment.

## 12.4 The Status of the Wave Function in Quantum Theory

Let's return to the possibilities for the wave function, mentioned in Sect. 12.1. It could be nothing. While not exactly nothing, it could be merely subjective or epistemic, representing our information about a system. Or it could be something objective. If it is objective, it could be material or quasi-material, or it could be nomological, or at least quasi-nomological.

Rather than deciding in absolute terms which of these possibilities is correct or most plausible—concerning which our opinion should be quite clear—we conclude by stressing that one's answer to this question should depend upon one's preferred version of quantum theory. Here are some examples:

- In orthodox quantum theory the wave function is quasi-nomological. It governs the results of quantum "measurements"—it provides statistical relationships between certain macroscopic variables.
- In Everett the wave function is quasi-material. After all, there it's all that there is. In Everett there's only the wave function. (We say "quasi-material" here instead of plain material since in Everett the connection between the wave function and our familiar material reality is not at all straightforward. In fact for some Everettians part of the appeal of their approach is the extensive conceptual functional analysis that it requires, see, e.g., [9].)
- In Bohmian mechanics as we understand it, as well as in decoherent or consistent histories and in causal set theory, the wave function is either nomological or quasi-nomological: In these theories the wave function governs the behavior of something else, something more concretely physical.

- However, in David Albert's version of Bohmian mechanics [10], in which what we call configuration space is in fact a very high-dimensional *physical* space, on which the wave function lives as a physical field, the wave function is material or quasi-material.
- In GRW theory [11] or CSL [12]—the theory of continuous spontaneous localization—the wave function is quasi-material, since there it either is everything or at least determines everything.
- And in the quantum information approach to quantum theory, the wave function is quasi-subjective—"quasi" because quantum information theorists differ as to how subjective it is.

This means that if you want to grasp the status of the wave function in quantum theory, you need to know exactly what it is that quantum theory says. If you're not clear about quantum theory, you shouldn't be worrying about its wave function.

We have suggested seriously considering the possibility that the wave function is nomological. One psychological obstacle to doing so is this: It seems to be an important feature of wave functions that they are variable and that this variability—from system to system and not just over time—leads to the varieties of different behaviors that are to be explained by quantum theory. But the behavior of the primitive ontology of a Bohmian theory, and all of the empirical consequences of the theory, depend on the universal wave function only via the one such wave function that exists in our world and not on the various other universal wave functions that there might have been. The variability we see in wave functions is that of wave functions of subsystems of the universe. This variability originates in that of the environment $Y$ of the subsystem as well as that of the choice of subsystem itself. So this variability does not conflict with regarding the wave function as fundamentally nomological, but rather is explained by it.

## References

1. H. Everett. Relative State Formulation of Quantum Mechanics. *Reviews of Modern Physics*, 29:454–462, 1957.
2. D. Dürr, S. Goldstein, J. Taylor, R. Tumulka, and N. Zanghì. Topological Factors Derived from Bohmian Mechanics. *Annales Henri Poincaré*, 7:791–807, 2006.
3. B. S. DeWitt. Quantum Theory of Gravity. I. The Canonical Theory. *Physical Review*, 160:1113–1148, 1967.
4. S. Hawking J. Hartle. Wave Function of the Universe. *Physical Review D*, 28:2960–2975, 1983.
5. S. M. Carroll and J. Chen. Does Inflation Provide Natural Initial Conditions for the Universe? *Gen. Rel. Grav.*, 14:2335–2340, 2005.
6. J. L. Anderson. *Principles of Relativity Physics*. Academic Press Inc., New York, 1967.
7. D. Dürr, S. Goldstein, T. Norsen, W. Struyve, and N. Zanghì. Can Bohmian Mechanics Be Made Relativistic? In preparation, 2012.
8. T. Norsen. The Theory of (Exclusively) Local Beables. *Foundations of Physics*, 40:1858–1884, 2010.
9. D. Wallace. Everett and Structure. *Studies in History and Philosophy of Modern Physics*, 34:87–105, 2003.

10. D. Z. Albert. Elementary quantum metaphysics. In J. Cushing, A. Fine, and S. Goldstein, editors, *Bohmian Mechanics and Quantum Theory: an Appraisal*. Kluwer, Dordrecht, 1996.
11. G. C. Ghirardi, A. Rimini, and T. Weber. Unified Dynamics for Microscopic and Macroscopic Systems. *Physical Review D*, 34:470–491, 1986.
12. G. C. Ghirardi, P. Pearle, and A. Rimini. Markov Processes in Hilbert Space and Continous Spontaneous Localization of System of Identical Particles. *Physical Review A*, 42:78–89, 1990.

# Index

**A**

Absolute uncertainty, 15, 25, 37, 49, 57, 59, 60, 63, 67, 68, 70, 82
Aharonov, Yakir, 119, 142, 196
Aharonov-Bohm effect, 206, 216, 217
Albert, David, xii, 150, 277
Albeverio, Sergio, xii
Allori, Valia, xi, xii
Anandan, Jeeva, 119, 142
Anderson, James, 273
Anyons, 213, 266
Arnowitt, Richard, 248
Arnowitt-Deser-Misner formalism, 248
Ashtekar, Abhay, 248, 250
Aspect, Alain, 156, 166
Asymptotic velocity, 128, 138

**B**

Bacciagaluppi, Guido, xii
Banks, Tom, 259
Barrett, Jonathan, 12
Bauer, Gernot, 13
Beables, local, 2–6, 275
Bell's inequality, 6, 154, 156
Bell, John Stewart, vi, ix, x, 2, 4–6, 8–10, 13, 16, 18, 24, 29, 34, 37, 42, 44, 46, 63, 65, 80, 82, 87, 92, 126, 136, 146, 151, 152, 154–157, 165–167, 171, 202, 226, 237, 239
Bell-type quantum field theory, 239, 240, 242–244
Beltrametti, Enrico, xii
Bohm, David, vi, vii, x, 6–10, 24, 28, 67, 87, 128, 151, 154, 165, 167, 223, 225, 243
Bohm-Dirac model (BD), 224, 237
Bohmian mechanics
 equations of motion, 3, 31, 83, 173, 195, 207, 242, 254
 for particles with spin, 83, 89, 92, 124
 for vector-valued wave functions, 214
 on manifolds, 32, 206, 207, 212, 214, 215
 relativistic,
  Relativistic Bohmian theory
  Bell-type quantum field theory
  Hypersurface Bohm-Dirac model
  Bohmian quantum gravity, 223
Bohmian quantum gravity, 253
Bohr, Niels, 2, 7, 33, 54, 66, 67, 126, 151, 152, 158, 163, 164, 202
Boltzmann, Ludwig, 28, 61, 268
Born
 interpretation, 183
 rule, 80, 184
 statistical law, 24, 45, 46
Born, Max, 6–8, 28, 165, 166
Bosons, 206, 212, 213, 226, 241, 257, 266
Braid group, 212, 213
Bricmont, Jean, xii
Bux, Kai-Uwe, xii

**C**

Callender, Craig, 254
Caprile, Bruno, xii
Carroll, Sean, 270
Casher, Aharon, 216
Chen, Jennifer, 270
Classical limit, 17, 171, 173–182
Collapse
 of the wave function, 11, 26, 41, 66, 70, 88, 114, 194, 197, 265
 postulate/rule, 4, 11, 253
 random, 4, 144
Combes, J. M., 186

Complementarity, 2, 163, 164, 247
Configuration space
 for Bell-type quantum field theories, 240
 multiply-connected, 18, 205, 206, 209
 of identical particles, 212
Consciousness, v, ix
Consistent histories, vi, 113, 144, 276
Contextuality, 148–153
Continuity equation, 10, 32, 33, 84, 136, 183, 189, 195, 201, 207, 225, 231–235, 254
Covering space, universal, 206, 208, 209, 213, 217, 219
Creation and annihilation of particles, 18, 240, 241, 245
Crossing
 measure, 225
 probability, *see* Probability, 225
Current positivity condition (CPC), 187
Cushing, James, v

**D**
Daumer, Martin, xi, xii
Davies, Edward Brian, 81
De Broglie
 relation, 10
  local, 182
 wave length, 174–176
De Broglie, Louis, vi, vii, 6–8, 13, 27, 167
De Broglie-Bohm theory, *see* Bohmian mechanics, 3
De Gosson, Maurice, xii
Decoherence, 73, 86–88, 180
Decoherent histories, 113, 144, 276
Dell'Antonio, Gianfausto, xii
Density matrix, 15, 42, 59, 85, 132–137
Deser, Stanley, 248
Determinism, 2–5, 8, 9, 23, 146, 148, 155, 166
Dewdney, Chris, 13
DeWitt, Bryce, 250, 251, 254, 257, 260
DeWitt, Cécile, 206
Diffeomorphism invariance, 250–252
 problem of observables, 256
Dirac
 algebra, 259, 260
 constraint quantization, 250
 current, 224, 228–233, 236, 237
 electron, 131
 equation, 223, 224, 227, 231, 236
 matrices, 273
 particles, 18, 223, 226, 227, 236, 237
 spinor, 224
 wave function, 237, 243

Dirac, P.A.M., 101
DLR equations, 61
Dobrushin, Roland, 61
Dollard, John, 184
Dorato, Mauro, xii
Dowker, Fay, xii

**E**
Ehrenfest theorem, 171, 175
Einstein equations, 249, 250, 260
Einstein, Albert, 2, 5, 6, 23, 28, 60, 66, 67, 82, 154, 155, 157, 164, 264
Einstein-Podolsky-Rosen argument, 6, 154–156
Empirical distribution, 25, 36, 47, 49, 50, 55, 56, 58, 69, 80
Empirical distributions, 242
EPR, *see* Einstein-Podolsky-Rosen argument, 156
EPRB experiment, 108, 154
Equilibrium
 quantum, *see* Quamtum equilibrium, 24
 thermodinamic, *see* Thermodynamic equilibrium, 24
Equivariance, 32–35, 45, 49, 62, 83, 84, 175, 207, 211, 231, 234, 236, 242, 243, 256
Ergodicity, 34, 45, 62, 63
Escape time, *see* Measurement of time, 12
Everett's Many-Worlds, 42, 276
Everett, Hugh, 42, 276
Experiments
 associated with operators, 98, 99, 102, 104
 Bohmian, 83, 115, 133
  general framework, 97
 discrete, 92
 equal-time, 55
 formal, 116
 multi-time, 50
 real-world, 81, 99, 114, 124, 145
 scattering, 146, 184–191
 sequence of, 56
 Stern-Gerlach, 89–108
 strong, 134
Eying, Gregory, xii

**F**
Faris, William, xii
Fermions, 206, 212, 217, 237, 241, 257, 266
Feynman diagrams, 240
Feynman, Richard, vi–viii, 12, 13, 128
Figari, Rodolfo, xii

# Index

Foliation of space-time, 18, 226, 227, 229–232, 235–238, 250, 251, 255, 258, 259, 273–275
Fröhlich, Jürg, xii
Fundamental group, 18, 205, 206, 210, 212, 217, 219

## G

Galilean invariance, 10, 30–32
Gallavotti, Giovanni, xii
Garrido, Pedro, xii
Gell-Mann and Hartle (GMH), 72–74
Gell-Mann, Murray, 72, 73, 87, 112
Ghirardi, GianCarlo, xii, 27
Gibbs
  distribution, 35
  postulate, 61, 62
Gibbs, Josiah Willard, 28, 35, 61, 62, 137, 268
Gleason, Andrew, 146
God, 68, 243, 268
Goldin, Gerald, 206
Goldstein, Rebecca, xii
Greenberger, D. M., 147
Griffiths, Robert, vi, 73, 87, 112
Ground state, 50, 127
GRW, 277
GRW1986, 27
Guiding
  equation, viii, 3, 6, 7, 10, 83, 84, 154, 206, 223, 224, 266
  field, 6, 28

## H

Hall, Ned, xii
Hamilton-Jacobi
  equation, 6, 10, 174, 259
  function, 6
  theory, 6, 10
Hamiltonian
  classical, 61
  Dirac, 243
  Pauli, 90, 125
  Schrödinger, 3, 83, 105, 136, 243, 269
Hardy, Lucien, 148
Hartle, James, 72, 73, 87, 112, 269
Hawking, Stephen, 269
Heisenberg
  equations, 128, 129
  uncertainty principle, 11, 15, 66, 167
Heisenberg, Werner, x, 1, 6, 7, 11, 15, 25, 28, 60, 64, 67, 97, 128, 163
Hemmick, Doug, xii
Hibbs, Albert, 128
Hidden variables, 8, 12, 24, 144–156, 263

Hiley, Basil, 13, 151
Holevo, Alexander, 81
Holland, Peter, 254
Holonomy-twisted representation, 18, 205, 206, 209, 213, 215–217, 219
Horne, M.A., 147
Hypersurface Bohm-Dirac model (HBD), 226, 227, 229–231, 235, 237

## I

Identical particles, 12, 14, 17, 18, 83, 208, 212, 216, 219, 241, 245, 266
Indistinguishable particles , see Identical particles, 83
Instrument, 135
Isham, Christopher, 252

## J

Joos, Erich, 87

## K

Kiessling, Michael, xii
Kochen, Simon, 8, 16, 146, 147, 165
Kochen-Specker theorem, 8, 16, 146, 147
Kowalski-Glikman, Jerzy, 254
Kraus, K., 81
Kruskal, Martin, xii
Krylov, Nikolay Sergeyevich, 61
Kuchar, Karel, 252, 259
Kupiainen, Antti, xii

## L

Lüders, Gerhart, 113
Lahee, Angela, xii
Laidlaw, Michael, 206
Landau, Lev, 1, 132
Lanford, Oscar, 61
Lang, Reinhard, xii
Laudisa, Federico, xii
Lazarovici, Dustin, xii
Lebowitz, Joel, xii
Leggett, Anthony, 26, 87
Lifshitz, Evgeny, 1
Local plane wave structure, see Plane wave, local, 171
Loewer, Barry, xii
Loose, Franck, xii
Lorentz invariance, 15, 18, 27, 226, 237, 239, 272–275
Lyle, Stephen, xii

## M

Münch-Berndl, Karin, xi, xii
Maes, Christian, xii

Many-Worlds interpretation, *see* Everett's Many-Worlds, 42
Markov
   process, 18, 53, 244
   property, 53
Maudlin, Tim, xii, 226
Maudlin, Vishnya, xii
Maxwellian velocities, *see* Maxwellian velocity distribution, 80
Maxwellian velocity distribution, 70, 80
Measurability
   necessary condition for, 138
   of Bohmian trajectories, 14
   of velocity, 139
   of wave function, 139, 140
Measurement
   of an operator, 102, 104, 189
   of position, 143
   of position operator, 142, 143
   of time, 12, 114, 187
   function of, 108
   genuine, 17, 81, 99, 107, 112, 130, 131, 137–144, 149, 152, 153, 195, 196, 200, 202
   ideal, 96, 101, 105, 106, 109, 112, 113, 137, 141
   normal, 102
   of a commuting family of operators, 106
   of operators with continuous spectrum, 110
   problem, 1, 3, 4, 26, 81, 87, 157, 158, 252
   sequential, 110
   standard, 102
   strong, 104, 109, 110, 118, 194, 196
   von Neumann, 81, 105, 197
   weak, 13, 17, 119, 193–198, 200, 202
   weak formal, 104, 110
   weak of velocity, 196
Mermin, David, vi, viii, 147, 152, 153
Misner, Charles, 248
Mixing, 34, 62
Moser, Tilo, xii

# N
Nelson, Edward, 7, 206
Newton, Roger, 186
No-go theorems, 8
Nonlocality, 6, 8, 18, 27, 65, 148, 154, 156, 166
Norsen, Travis, xii, 275
Number operator, *see* Operators, 245

# O
Observables,
   *see* oerators as observables
   measurement of an operator, 11

Olivieri, Giuseppe, xii
Omnes, Roland, vi, 73, 112
Ontology, 71–74, 252, 254
   field, 25, 257
   particle, 25, 253, 257, 272
   primitive, 29, 253, 257, 259, 271, 272, 275–277
Operators
   measurements of, *see* measurement of an operator, 104
   position, 242
   as observables, 12, 16, 79–82, 92, 96, 114, 202, 253
   associated with experiments, 98, 99, 102, 104
   momentum, 128
   naive realism about, 16
   naive realism about, 157–159
   number, 18, 245
   position, 114, 127, 128, 142, 144
   quantum field, 19, 244
   spin, 129
Oppenheimer, Robert, 9

# P
Page, Don, 112
Particle creation/annihilation, *see* creation and annihilation of particles, 240
Pauli
   equation, 216
   Hamiltonian, *see* Hamiltonian, Pauli, 125
   matrices, 89, 108
Pauli, Wolfgang, 7, 10, 28, 102
Penrose, Roger, 26, 35, 68, 70
Peres, Asher, 147
Philippidis, C., 13
Photon, 6, 27, 131, 158, 240–242
Pilot-wave theory, *see* Bohmian mechanics, 3
Plane wave, local, 171, 176–179
Podolsky, Boris, 6, 154, 155
Positive-operator-valued measure (POVM), 81, 114–127, 130–132, 138, 242
Primitive ontology, *see* Ontology, 253
Probability
   meaning in terms of typicality, 69
   and relative frequencies, *see* Empirical distributions, 242
   crossing, 225, 231, 236
   subjective, 45
Projection postulate, *see* Collapse, 114
Projection-valued measure (PVM), 98, 100, 101, 107–109, 114, 116–118, 120, 122–125, 127, 130–132, 146, 244, 245

# Index

and self-adjoint operator, *see* spectral theorem, 98
Pusey, Matthew, 12

## Q

Quantum equilibrium, 14, 16, 25, 34, 49, 62, 64, 67, 68, 70, 71, 74, 79, 80, 84, 106, 225, 231, 237, 269
 analysis, 70, 256
 distribution, 34, 44, 45, 49, 68, 69, 84, 122, 146, 225, 227
 hypothesis, 34–36, 44, 46, 57, 59, 61, 62, 74, 80, 84, 93, 265
 vs. quantum nonequilibrium, 64, 74, 269
 vs. thermodynamic equilibrium, 14, 15, 60–64, 74, 80
Quantum gravity, 248–252
 Bohmian, *see* Bohmian quantum gravity, 253
Quantum potential, 6, 10, 174
Quantum probability current, 10, 17, 33, 183–191, 195, 254

## R

Random systems, 51–55, 71, 72, 256
Randomness
 appearance of, 15, 25, 80
 intrisic, 11, 71, 253
 irreducible, 140, 164
 quantum, 8, 15, 23, 26, 27, 68, 71, 84, 85, 132, 167
Reduced density matrix, 15, 42, 132–134
Relativistic Bohmian theory, 272
Rimini, Alberto, 27
Rosen, Nathan, 6, 154, 155
Rovelli, Carlo, 247
Rudolph, Terry, 12
Ruelle, David, 61

## S

Scattering theory, 184–191
Schamel, Folker, xii
Schrödinger's
 equation, viii, 3, 4, 9, 11, 15, 23, 26–33, 41, 65, 66, 83–88, 144, 171, 173, 189, 195, 207, 208, 210, 242, 250–252, 254, 256–260, 265, 266, 269, 271
 cat, vi, ix, 1, 3, 26, 252
 evolution, 4, 26, 122, 139, 177–181, 214, 223, 258
Schrödinger, Erwin, 2, 26, 157, 163, 164
Schulman, Lawrence, 206
Schwartz, Jacob, 44
Shimony, Abner, 147

Shtokhamer, R., 186
Smith, Penny, xii
Soffer, Avy, xii
Solvay Congress, 7, 10
Specker, Ernst, 8, 16, 146, 147, 165
Spectral theorem, 98, 101, 107, 108, 114, 116
Speer, Eugene, xii
Spin, *see* Bohmian mechanics for particles with spin, 83
Spohn, Herbert, xii
Spontaneous localization, 144, 277
Squires, Euan, 226, 254
Stapp, Henry, 26
Statistical mechanics, 28, 35, 61, 64, 164, 181, 190, 268
Steinberg, Aephraim, 13
Stochastic mechanics, 7, 71, 144, 191, 206
Struyve, Ward, xii, 273
Sussman, Hector, xii

## T

Tausk, Daniel, xii
Taylor, James, xi, xii
Teta, Alessandro, xii
Teufel, Stefan, xii, 259
Thermodynamic equilibrium, 14, 25, 60–62, 80
Thermodynamics, 28, 60–64
Time, problem of
 in quantum gravity, 19, 251, 259
 in quantum mechanics, 187
 physical time vs. external time, 15, 256, 270
Topological factors, 18
Topological factors, 205, 206, 209, 213, 216, 219
Tumulka, Roderich, xi, xii, 237
Two-slit experiment, vi, vii, 11–13, 74
Typicality, 25, 36, 49, 68, 69, 241

## U

Uncertainty principle, *see* Heisenberg uncertainty principle, 167

## V

Vaidman, Lev, 119, 142, 193, 196
Valentini, Antony, 74
Vector bundle, 206, 208, 209
Velocity vector field, 30, 31, 206, 207, 209, 219, 274
Vink, J, C., 254
Von Neumann, John, ix, x, 8, 23, 24, 66, 81, 113, 132, 146, 147, 164–166

**W**

Wave function
  as nomological, 257, 259, 266, 268, 271, 272, 274–277
  conditional, 15, 16, 41–45, 48, 53, 55, 56, 59, 67, 68, 72, 85, 86, 88, 111, 119, 136, 181, 199, 256, 258, 259
  conditional/effective with spin, 42
  effective, 25, 37, 38, 40–44, 53, 55, 56, 58, 59, 66–68, 72, 86–88, 94, 136
  universal, 36–40, 44, 64, 67–69, 71, 73, 256, 257, 259, 264, 268, 269, 271–274, 277
Weak measurement, *see* Measurement, weak, 140
Weber, Tullio, 27
Weinberg, Steven, 26, 239
Weingard, Robert, 254
Wheeler, John Archibald, 250, 251, 254, 257, 260
Wheeler-DeWitt equation, 250, 251, 254, 268–270
Wien, Wilhelm, 2, 164
Wigner
  distribution, 11
  formula, 66, 72, 73, 112, 113
  paradox of Wigner's friend, 74
Wigner, Eugene, 26, 165, 166
Wiseman, Howard, 17, 193–195

**Z**

Zeh, Heinz-Dieter, 87
Zeilinger, Anton, 147
Zurek, Wojciech, 87

Quantum Physics Without Quantum Philosophy

Printed by Printforce, the Netherlands